COSPAR-IAU-IAG/IUGG-IUTAM

Dynamics of Satellites (1969)

Proceedings of a Symposium
held in Prague, May 20-24, 1969

Edited by Bruno Morando

Springer-Verlag
Berlin · Heidelberg · New York 1970

Bruno Morando, Docteur ès Sciences
Astronome au Bureau des Longitudes, Paris

This symposium is a continuation of a similar symposium organized in Paris in 1965 ("Trajectories of Artificial Celestial Bodies", Editor: J. Kovalevsky).

ISBN-13: 978-3-642-99968-0 e-ISBN-13: 978-3-642-99966-6
DOI: 10.1007/978-3-642-99966-6

With 89 Figures

 Library of Congress Catalog Card Number 75-111 895. Title No. 1664
Softcover reprint of the hardcover 1st edition 1970

Preface

This book gathers the proceedings of a symposium on Dynamics of satellites which took place in Prague in May 1969 during the twelfth COSPAR meeting.

This symposium was sponsored by the International Astronomical Union, the International Association of Geodesy, the International Union of Theoretical and Applied Mechanics and COSPAR (Committee on Space Research).

The organizing committee was composed of Dr. Kovalevsky chairman, Dr. Yu. V. Batrakov representing IAU, Dr. A. H. Cook for IAG, Dr. D. King-Hele for COSPAR, Prof. M. Roy for IUTAM and Dr. Rosenberg.

I wish to take advantage of the opportunity to thank, on behalf of all the participants, the organizing committee members, Prof. Buchar, Dr. Rajski and Dr. Sehnal, for the kindness and efficiency of their welcome. The interpreters who translated with virtuosity during the whole symposium also deserve our gratitude.

I am grateful also for the care and skill with which Springer-Verlag has printed this volume.

Paris, April 1970

Bruno Morando

Table of Contents

Theory of the Motion of an Artificial Satellite

By

B. Morando

Bureau des Longitudes, Paris, France

This symposium being mainly concerned with artificial satellites of the Earth that move in the near vicinity of this planet, we shall confine ourselves to studying the motion of these bodies; space probes, lunar satellites or other objects being excluded.

Moreover the motion around their center of mass of these satellites will not be considered and it will be assumed that they are points with negligible masses.

The forces to which such bodies are submitted are of two kinds: gravitational forces due to the newtonian attraction of the Earth, the Moon and the Sun and non-gravitational forces such as air-drag and radiation pressure. As it happens most of these forces are small in comparison with the attraction of the Earth and, by choosing proper approximations of this attraction, it is possible to adapt the classical methods of Celestial Mechanics to the problem involved. The idea is to find a proper preliminary orbit by integration of a simplified problem and, then, to correct this orbit by adding to the constants that define it, some functions of time which satisfy the actual differential equations of the motion. In practice these functions of time are developed in powers of some small parameters related to the problem, the coefficient being periodic functions themselves developed in Fourier Series of one or many linear functions of time. Sometimes terms that are proportional to time or powers of the time appear in the solution; they are called secular terms.

Gravitational forces

Development of the potential. The gravitational forces derive from a potential. The potential of the Earth, which is by far the most important, may be developed in spherical harmonics in the following form

$$U = \frac{\mu}{r}\left[1 - \sum_{n=2}^{\infty} \frac{J_n a_e^n}{r^n} P_n(\sin\varphi) + \sum_{n=2}^{\infty}\sum_{k=1}^{n} \frac{J_{n,k} a_e^k}{r^n} P_{n,k}(\sin\varphi)\cos k(\lambda - \lambda_{n,k})\right]. \tag{1}$$

In this formula r, φ, λ are the spherical coordinates of a satellite, r being the radius vector from the center of the Earth to the satellite, φ the latitude above the plane of the equator and λ the longitude of the satellite. $P_n(\sin\varphi)$ is Legendre's polynomial of order n of $\sin\varphi$, $P_{n,k}(\sin\varphi)$ Legendre's associated function; a_e is the equatorial radius of the Earth and μ the product of the constant of gravitation by the mass of the Earth. The coefficients J_n, $J_{n,k}$ and $\lambda_{n,k}$ are constant coefficients that define the potential of the Earth; those terms in U that have the J_n as factors are known as the zonal harmonics, those that are proportional to the $J_{n,k}$ are tesseral harmonics. The highest coefficient is J_2 which is of the order of 10^{-3}, the others being at least of the order of 10^{-6}. It is to be noted, in Eq. (1), that there are no terms corresponding to $n = 1$; it may be shown that these terms vanish if the origine of coordinates is taken at the center of masses of the Earth.

It has been suggested by Duboshin (1966) to develop U in Lamé functions instead of spherical harmonics. A paper on this by Walter will be found in this volume. On the other hand Dufour in Paris is trying to replace the above development by the sum of all the potentials created by a finite number of given masses. Anyway as no theories of the motion of a satellite using these methods of development have yet been published we shall confine ourselves to the expression given by Eq. (1).

Vinti type of preliminary orbit

If U, as given by Eq. (1), is restricted to the zonal harmonics for which n are even numbers, the potential of the Earth is very close to the potential created by two fixed points with equal masses; the distance between the masses is purely imaginary. This peculiar circumstance was pointed out by Vinti (1959) and its interest lies in the fact that the problem of the motion of a particle attracted by two fixed masses may be solved completely using elliptic functions. The solution obtained includes a certain number of constants and may be taken as what we defined earlier as a preliminary orbit. The difference between the actual potential of the Earth and the potential of two fixed points alters slightly the preliminary orbit, the constants of integration becoming functions of the time. The advantage of this method is to introduce right at the beginning the effects on the motion of the satellite due to the main zonal harmonics. Work on these lines have been carried out in the U.S.S.R. [Aksionov, Grebenikov, Diomin] and in France [Marchal 1966].

Keplerian preliminary orbits

If only μ/r is retained in the expression of the potential, the problem of the motion is very simple to solve. As it is well known the orbit is a conic section. We shall assume that it is an ellipse which is then characterized by six constants: the semi-major axis a, the eccentricity e, the inclination i of the plane of the orbit with respect to the equator, the right ascension of the ascending node Ω, the angular distance between the ascending node and the perigee ω and the mean anomaly at a given time M_0. The motion on this ellipse, called a Keplerian ellipse is given by Kepler's equation

$$E - e \sin E = M = n t + M_0$$

where E is the eccentric anomaly on the ellipse and n is given by the following relation

$$n^2 a^3 = \mu .$$

In order to know the actual motion of the satellite it is necessary to take into account the remaining part of the potential, that is

$$R = U - \frac{\mu}{r} .$$

R is called the disturbing function. Two methods have been favoured to study the perturbations of the Keplerian notion.

1. Lagrange's equations methods

These well known equations relate the derivatives of the six Keplerian elements a, e, i etc. to the partial derivatives of R (see BROUWER and CLEMENCE 1961). For example we have

$$\frac{d e}{d t} = \frac{-\sqrt{1 - e^2}}{n a^2 e} \frac{\partial R}{\partial \omega} + \frac{1 - e^2}{n a^2 e} \frac{\partial R}{\partial M}$$

and five other such relations.

It is obvious that R is a complicated function of the Keplerian elements and of time through Greenwich Sideral time T if the tesseral harmonics are taken into account. The disturbing function R has to be developed in Fourier series of linear combinations of the angular parameters (M, Ω, ω, T) the coefficient being known functions of a and i but also of the eccentricity e in such a way that developments in powers of e of these functions have to be used. The computations of the development of the disturbing function and the integration of Lagrange's equations have been carried out by many authors [KOZAI (1959), KAULA, IZSAK, CHALLE and LACLAVERIE (1969)]. First the Keplerian elements are taken as constants in the right-hand members of Lagrange's equations which are integrated once. The first order solution is then obtained, it is introduced into the right hand members of the equations, a new integration is performed and so on. If only the first order solution with respect to J_2 is taken into account it is shown that each element a, e, etc. appears as a sum of periodic terms, the angular elements including also secular terms which means that the node and the perigee revolve around the axis of the Earth. The periodic terms are called short period terms if their period is of the same order of magnitude that the period of the satellite on its Keplerian orbit, long period terms if their period is comparable with the periods of revolution of the node and the perigee around the axis of the Earth. The long period terms of the first order in J_2 only appear after the second approximation.

The introduction in R of other harmonics than the first does not usually lead to new difficulties. The terms that come from the tesseral harmonics introduce mean period terms, the period of which being of the same order of magnitude that the period of rotation of the Earth around its axis.

When the initial eccentricity is very small, difficulties arise. They have been overcome by BATRAKOV (1966).

2. Von Zeipel's method

Instead of the Keplerian elements a, e, i, etc. let us take the following variables

$$L = \sqrt{\mu a}, \qquad G = \sqrt{\mu a (1 - e^2)}, \qquad H = \sqrt{\mu a (1 - e^2)} \cos i,$$
$$l = M, \qquad g = \omega, \qquad h = \Omega.$$

These variables are called Delaunay's variables and they satisfy a system of differential equations which may be obtained from Lagrange's equations

$$\left.\begin{aligned} \frac{dL}{dt} &= \frac{\partial F}{\partial l}, & \frac{dG}{dt} &= \frac{\partial F}{\partial g}, & \frac{dH}{dt} &= \frac{\partial F}{\partial h}, \\ \frac{dl}{dt} &= -\frac{\partial F}{\partial L}, & \frac{dg}{dt} &= -\frac{\partial F}{\partial G}, & \frac{dh}{dt} &= -\frac{\partial F}{\partial H}. \end{aligned}\right\} \tag{2}$$

These equations are canonical and the Hamiltonian F is

$$F = \frac{\mu^2}{2L^2} + R.$$

Eq. (2), called Delaunay's equations, have been solved by BROUWER (1959). The idea is to make a canonical change of variables on Eqs. (2) in such a way that the new Hamiltonian of the new equations includes one angular variable less than the Hamiltonian of system (2). This process is carried out again on the new system until a canonical system of equations is obtained in which the Hamiltonian depends only on the new metric variables L', G', H'. It is then very easy to solve such a system and to go back to the first variables.

Resonance cases. The methods described to solve the problem of the motion are usually quite straight forward, although the algebra involved may be quite heavy. However some difficulties may be met when the initial conditions have special values.

First if the initial value of the inclination i is very close to i_0, where

$$1 - 5\cos^2 i_0 = 0 \qquad (i_0 = 63°\,26')$$

the motion of the perigee is completely altered. Instead of revolving around the Earth axis the perigee librates backwards and forwards around a mean position and the methods used to solve the equations of motion have to be changed. Hori (1960) studied this case, called the critical inclination case, using Bohlin's method. Elliptic functions are introduced and the motion of the perigee is fairly similar to the motion of a pendulum.

If, now, the period of the satellite is close to 24 h, which happens when the semi-major axis is roughly 40000 km, a resonance occurs between this period as it appears in the mean anomaly and the period of the rotation of the Earth which appears in the tesseral harmonics. The coefficient of time in some of the terms of the disturbing function being very small, if not zero, it is impossible to use the ordinary methods of integration. This problem has been solved by many authors: Musen (1962), Morando (1963). Bohlin's method may be used, the satellite itself librates around some meridians.

If the period of the satellite on its Keplerian orbit is commensurable with 24 h analogues resonance effects take place also.

Perturbations of the satellit by the Sun and the Moon. This is in fact quite a classical problem of Celestial Mechanics. The disturbing function introduced is similar to the disturbing function in the motion of the Moon around the Earth when disturbed by the Sun. The problem here is simplified as, for close Earth Satellites, the disturbance is small and only a few terms of the disturbing function have to be taken into account. These questions have been studied by Cook (1964), Kozai (1963) etc.

Non gravitational forces

Air drag

On fairly low satellites (under 600 km) the friction of the atmosphere has a very important effect. Of course the density of the atmosphere at a given point has to be known and models for such atmospheres have to be used. The disturbing force due to air drag does not derive from a potential and its three components on the tangent to the Keplerian orbit, the normal and the perpendicular to the plane of the orbit at a given point are used. The right hand members of Lagrange's equations are modified accordingly and a process of integration is carried out (see King-Hele 1964). Brouwer showed how Delaunay's canonical equations may be used.

The main effects of the air drag are the decreasing of the eccentricity which tends towards zero and the appearance of a secular term in the semi-major axis which decreases causing the satellite to be eventually consumed in the low layers of the atmosphere.

Radiation pressure

The impact of the photons carried by the light of the Sun produces a disturbing force which is not negligible when the mass of the satellite is low and its surface of section large. This force may be considered as constant and parallel to the direction Sun–satellite but it vanishes when the satellite is in the shadow of the Earth. This problem has been studied by Ferraz-Mello (1963). New and interesting results have recently been found by Sehnal and Lala and are published in this volume.

References

Batrakov, Y. V. – 1966: Trajectories of artificial celestial bodies as determined from observation, p. 174. Proc. of Symp. Paris, Apr. 1965, Berlin/Heidelberg/New York: Springer.

Brouwer – 1959: Solution of the problem of artificial satellite theory without drag. Astron. J. 64, p. 378.

Brouwer, Clemence – 1961: Methods of Celestial Mechanics. Academic Press.

Challe, Laclaverie – 1969: Astronomy and Astrophysics, a European Journal, Vol. 3, No. 1, p. 15, Berlin/Heidelberg/New York: Springer.

Cook, G. E., Allan, R. R. – 1964: Proc. R. Soc. A. 280, 97.

Duboshin, G. B. – 1966: Sur le développement du potential de la Terre par les fonctions de Lamé, p. 68. Trajectories of Artificial celestial bodies. Proc. of Symp. Paris, Apr., 1965, Berlin/Heidelberg/New York: Springer.

Hori, G. – 1960: The motion of an artificial satellite in the vicinity of the critical inclination. Astron. J. 65, p. 291.

King-Hele, D. – 1964: Theory of satellite orbits in an atmosphere. London: Butterworths.

Kozai, Y. – 1959: The motion of a close Earth satellite. Astron. J. 64, p. 867.

Kozai, Y. – 1963: Smithsonian Contrib. Astrophys. Vol. 6, p. 47.

Marchal, Ch. – 1966: Bull. Astron. 3ᵉ série. Tome 1, fasc. 3, p. 189.

Morando, B. – 1963: Orbites de résonance des satellites de 24 h. Bull. Astron. XXIV, fasc. 1, p. 47.

Musen, P., Bailie, A. 1963 –: On the motion of a 24 h satellite, p. 62. Proceedings the first international symposium on the use of artificial satellites for Geodesy. Amsterdam: North-Holland Publishing Company.

Vinti, J. P. – 1959: J. Res. Natl. Bur. Std., 63B, p. 105.

Advances in Rapid Orbit Prediction and Applications to Satellite Visibility Studies

By

F. Morrison

Geodetic Research and Development Laboratory – Coast and Geodetic Survey, Rockville, MD, USA

Abstract. Methods are given for including the effects of the square of the oblateness and of radiation pressure in an orbit prediction program. Optimum orbit predictions are obtained by fitting a rapid orbit prediction model to the results of a DOI (Smithsonian Differential Orbit Improvement Program) orbit. The advantage of this procedure is that the effect of biases due to the use of different orbital models for short and long term analyses are minimized and the average rates of change of angular variables are precisely matched in the two models.

The immediate application of the improved long range orbit predictions is for planning the logistics of an observing program. Other uses being developed are the determination of tesseral harmonic coefficients producing long beat periods and the measurement of high altitude (above 1000 km) atmospheric densities by means of the orbits of the satellites PAGEOS, Echo I and Echo II.

1. Introduction

The potential of the earth (or any other body) may be written in the form

$$V = \sum_{n=0}^{\infty} \sum_{m=0}^{n} V_n^m$$

$$V_n^m = \frac{\mu}{r}\left(\frac{R}{r}\right)^n P_n^m(\sin\beta)\left[C_n^m \cos m\lambda + S_n^m \sin m\lambda\right].$$

Each term of the series may be expressed in orbital elements as

$$V_n^m = \frac{\mu R^n}{a^{n+1}} \sum_{p=0}^{n} F_{nmp}(I) \sum_{q=-\infty}^{+\infty} G_{npq}(e)\, \Phi_{nmpq}(\omega, M, \Omega, \theta).$$

Important properties of this expansion are

a) $$G_{npq}(e) = O(e^{|q|}),$$

b) $$\Phi_{nmpq} = C_n^m \cos[(n-2p)\omega + (n-2p+q)M + m(\Omega-\theta)] + S_n^m \sin[(n-2p)\omega + (n-2p+q)M + m(\Omega-\theta)]$$

if $(n-m)$ is even, but $-S_n^m$ replaces C_n^m and C_n^m, S_n^m if $(n-m)$ is odd.

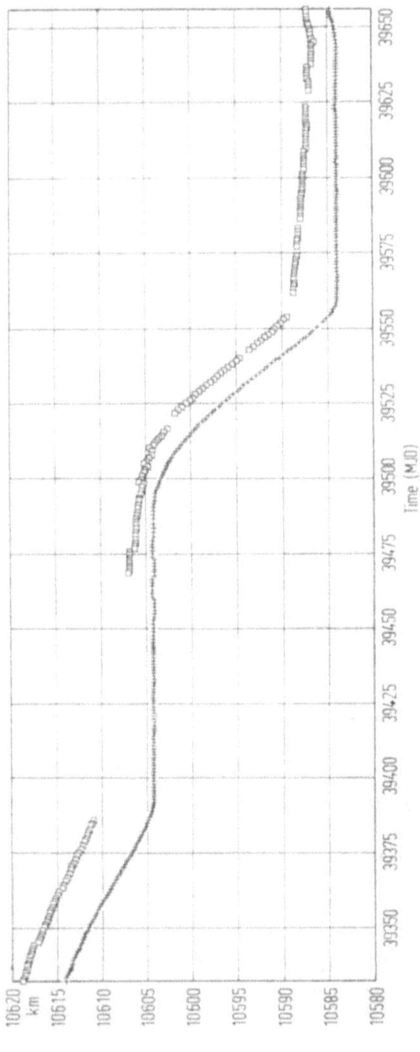

Fig. 1. Semi-major axis.

c) $F_{mnp}(I)$ is a trigonometric polynomial of order n. These conditions are sufficient to insure that we can obtain good approximations to V_n^m by truncating the expansion.

This development of V in orbital elements is needed for expressing the equations of motion of an artificial satellite by *Lagrange's Planetary Equations*. A parallel development in Delaunay's elements has been done by GARFINKEL [6]. Analytic integrations of the resulting equations have been given by KAULA [9] and GARFINKEL [7]. Kaula's formulas have been utilized by the Smithsonian Astrophysical Observatory [10] to determine tesseral harmonic coefficients. Numerical integration of the exact equations in this form would be more complex than a direct application of Cowell's method, and the time step used could be no shorter. What we can obtain readily from these expansions is a set of averaged equations of motion. All we do is use selected terms of V_n^m, and only the V_n^m that are very important. All short period terms are readily determined and thereby dropped. Long period, secular, and resonant terms are retained – their integration will give the qualitative properties of the orbit.

2. The Rapid Orbit Prediction Program (ROPP)

A program utilizing these developments with an Adams-Moulton integration with automatic time step variation has been written for NASA by TRW, Inc. [12]. Luni-solar perturbations are included by

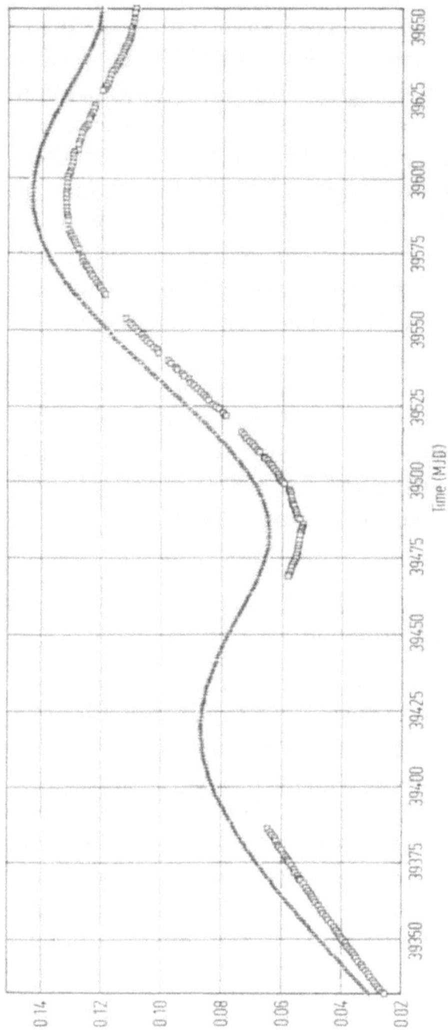

Fig. 2. Eccentricity.

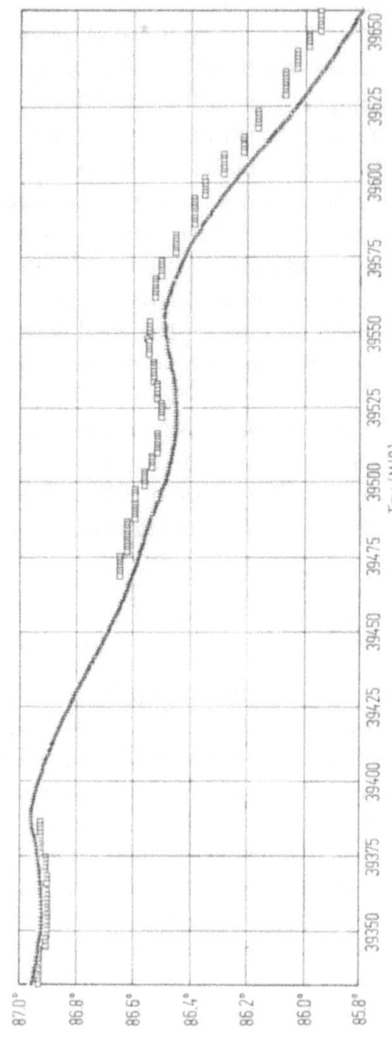

Fig. 3. Inclination.

analogous developments for that disturbing potential. An elaborate technique for computing drag perturbations by an averaging technique was included, but this has been discarded due to the presence of a defect in the model used. Mr. Bernard Chovitz, of the Geodetic Research and Development Lab, at the suggestion of Mr. BRUCE DOUGLAS, of Wolf Research and Development Corp, has programmed a Gauss-quadrature solution of the drag equations of STERNE [11]. The results compare favorably with those from the original formulation but contain no singularities for any orbits. Since we at the Geodetic Research and Development Lab have been working primarily with balloon satellites, we must take radiation pressure into account. Averaged equations of motion have been developed by BRYANT [3]. However, we have used a formulation devised by KAULA [8] which is convenient in that it can be incorporated in Lagrange's equations. The places where the satellite enters and leaves the earth's shadow is determined by explicitly solving a quartic equation devised by R. P. YEREMENKO [14].

Another addition to the program has been for terms proportional to J_2^2 – these cause secular and long period effects as pronounced as J_4. The necessary equations were obtained by taking Brouwer's [1] transformed Hamiltonian F_2^* and computing $\partial F_2^*/\partial a$, $\partial F_2^*/\partial e$, $\partial F_2^*/\partial I$, and $\partial F_2^*/\partial \omega$ and inserting these derivatives in Lagrange's equations.

3. Orbit Determination

The immediate application we have had for the ROPP program has been long range orbit predictions. Long term plans to occupy observing sites are needed as well as anticipated satellite lifetimes. Before launching one can experiment with the nominal elements and vary them to get some idea about the stability of the satellite lifetime. Once the satellite has been launched we have more information – its present orbital elements and past history. Until now no long range orbital prediction program has made use of the accumulated data on the satellite's orbit. To get something out of this we have developed an orbit determination procedure for the ROPP program. Results of a ROPP integration are compared with mean elements obtained from DOI (Differential Orbit Improvement program, see [5] and [10]) runs, and the initial conditions for ROPP, $a(T)$, $e(T)$, $I(T)$, $\Omega(T)$, $\omega(T)$, and $M(T)$ are adjusted to obtain a least squares fit. The state-transition matrix is obtained not by integrating variational equations, but by integrating a total of seven orbits – the original orbit and six others with one initial condition varied in each.

The change in any element due to a change in initial conditions is therefore known for any time. For example, when we change e_0 by Δe, we observe changes Δa, Δe, ΔI, $\Delta\Omega$, $\Delta\omega$, ΔM, and compute approximations $\Delta a/\Delta e_0$, $\Delta e/\Delta e_0$, $\Delta I/\Delta e_0$, ... etc., to $\partial a/\partial e_0$, $\partial e/\partial e_0$, $\partial I/\partial e_0$, ..., etc. The ROPP-generated elements are taken as computed quantities and the

DOI elements for "observed". For any given time

$$\Delta \bar{p} = A \, \Delta \bar{p}_0$$

where $\Delta \bar{p}$ = observed − computed, A = state transition matrix at time t, $\Delta \bar{p}_0$ = changes in initial conditions. Then we multiply through by A^T to obtain

$$A^T \Delta \bar{p} = A^T A \, \Delta \bar{p}_0 .$$

This is summed over all time points considered to obtain normal equations

$$\sum_t (A^T \Delta \bar{p}) = \Delta \bar{p}_0 \sum_t (A^T A)$$

which may be solved for $\Delta \bar{p}_0$ by inverting the matrix $\Sigma(A^T A)$ and multiplying through the above. No method has been tried to optimize the choice of Δa_0, Δe_0, ΔI_0, $\Delta \Omega_0$, $\Delta \omega_0$, ΔM_0 to speed convergence. If we could guess the corrections right the first time, only one iteration would be needed; obviously a clever selection of the variations could speed the rate of convergence.

The advantage of this method is that the size and complexity of the program is increased very little and that the technique could be tested quickly, without waiting for a long, extended programming effort. No problems with convergence have been encountered so far. The initial variations chosen are in Table 1.

Table 1. *Initial Conditions Variations*

Element		Variation
a	semi-major axis	10^{-5} earth radii
e	eccentricity	0.001
I	inclination	0.1 degrees
Ω	R.A. of ascending node	1.0 degree
ω	argument of perigee	1.0 degree
M	mean anomaly	1.0 degree

4. Applications of Results

After the orbit determination ceases to improve the fit, the numerical integration is extrapolated into the future. Our initial tests have consisted of fitting to a 30-day period and extrapolating for a year. Since we already have elements for PAGEOS, Echo I, and Echo II for a two-year period (MJD 39332–40028) it is easy to test the prediction technique. At our present state of development the mean anomaly drifts off by 90° in about one year – for PAGEOS. The other elements remain quite close to the predicted values.

Orbital elements, once obtained, can be used in a series of programs that evaluate look angles and count the number of useable appearances of the satellite at the given observing stations. The methods of analysis were devised by W. M. KAULA [9] and revised by B. CHOVITZ. The simplification used is that all elements but the mean anomaly are assumed constant for any given day. The possible appearances of the satellite are sought by determining the maxima of Z_s – the 3-coordinate of the satellite in the *local altazimuth system*. An iteration technique is needed to find these maxima, since the earth's rotation enters into the solution. Of course the whole procedure breaks down for 24-hour satellites and those at still higher altitudes.

Geophysical quantities may be obtained using these techniques. CARL WAGNER has investigated the possibility of determining resonance producing geopotential coefficients using the ROPP program [13]. Observe the graph of a versus time. The value of a is not only biased by about $4^1/_2$ km, but the computed values of $-da/dt$ are too small. Most of this effect is certainly due to drag. We have made some preliminary estimates of the atmospheric density at high altitudes (about 2500 km); final values await a more elaborate analysis [4].

5. Further Improvements

The graphs of a and e reveal biases. Most of the corrections in a and e are used up in making the values of M, ω, and Ω agree with observed values. All the deficiencies of the model are absorbed into a, e, and probably I to a lesser extent. Little of the fitting capability is left to match the observed values of a and e, since $\partial a/\partial a_0 \approx 1$, $\partial e/\partial e_0 \approx 1$, but $\partial M/\partial e_0 \approx t\,(\partial n/\partial a) = -1.5\,t\,(n/a)$ and $\partial M/\partial e_0 \approx t\,(\partial n/\partial e) = t\,O(J_2)$. To solve this problem we may make the model very elaborate, but this will only cut down our efficiency – the limiting case of model improvement being Cowell's method. What we can do is introduce new adjustable parameters to absorb the effects of model deficiency. Auxiliary "dynamical" values of a, e, and I can be used in Lagrange's equations, together with the geometric values of a, e, and I, which correspond to those given by the DOI, for a total of 9 parameters. Using *two* dynamical values of a would probably be profitable – one for the part of dM/dt that is $\sqrt{\mu/a^3}$, the other for the rest of Lagrange's equations.

The residuals in M show a definite bias. A large part of this is due to drag – d^2M/dt^2 being larger than it should be from radiation pressure and J_2. Variable drag parameters will be needed, undoubtedly. The model will be satisfactory when the residuals in M are truly random.

With the fitting being done to Keplerian elements, undue weight is given to ω and M, particularly if e is small. While the integrations would continue to be done in Keplerian elements, we may fit to an alternative set of elements. A good non-singular one is [2]:

$$\begin{array}{ll} L & M + \omega + \Omega \\ \sqrt{2(L-G)}\cos(\omega+\Omega) & -\sqrt{2(L-G)}\sin(\omega+\Omega) \\ \sqrt{2(G-H)}\cos\Omega & -\sqrt{2G-H)}\sin\Omega \end{array}$$

with

$$L = \sqrt{\mu a},\ G = L\sqrt{1-e^2},\ H = G\cos I.$$

This set is canonical and has the advantage that there is only one fast variable. Alternative methods are available. One might obtain a weight matrix from the DOI. Going to direct fitting to observations would be too complex. Just using the non-singular elements recommends itself due to simplicity. Using a weight matrix from DOI runs would scarcely improve the quality of long range predictions, but might be justified in attempts to determine geophysical parameters.

Provisions need to be made for precession and nutation, since these runs may be for extended periods – six months up to five years.

6. Conclusions

Graphs 4 and 5 show the improvement made in the prediction of the mean anomaly by the fitting process. These results prove the usefulness of the procedure even in its present unrefined form.

The ROPP program indicates the great potential of utilizing the numerical integration of mean elements to analyze orbits over long time spans (thousands of revolutions). Very long period effects that cannot be examined effectively by Cowell-generated orbits or DOI-type semi-empirical orbits may be treated by this method. For deep resonance and multiple resonance cases and long term radiation pressure effects this method is the only one applicable at all. Better long term predictions may be made by adding the orbit determination procedure to the integration of mean elements – these often have been needed but were hitherto unavailable. Care must be taken when utilizing mean elements techniques

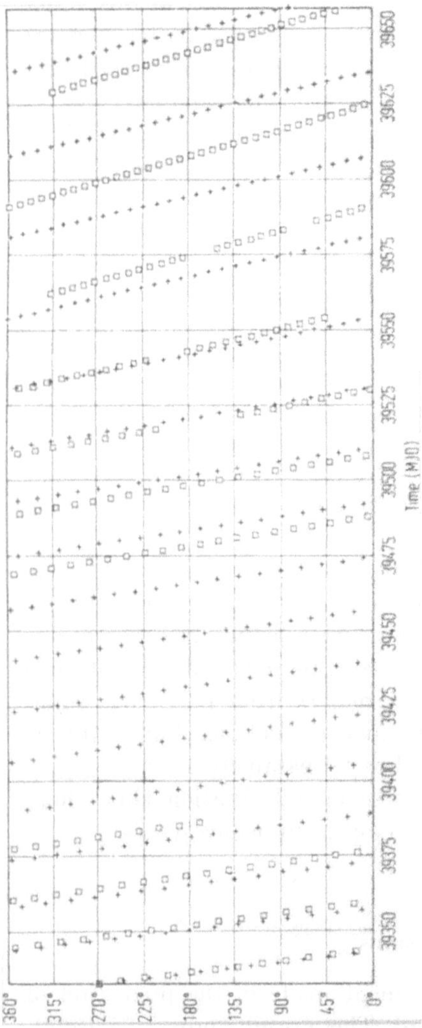

Fig. 4. Mean anomaly no fitting procedure.

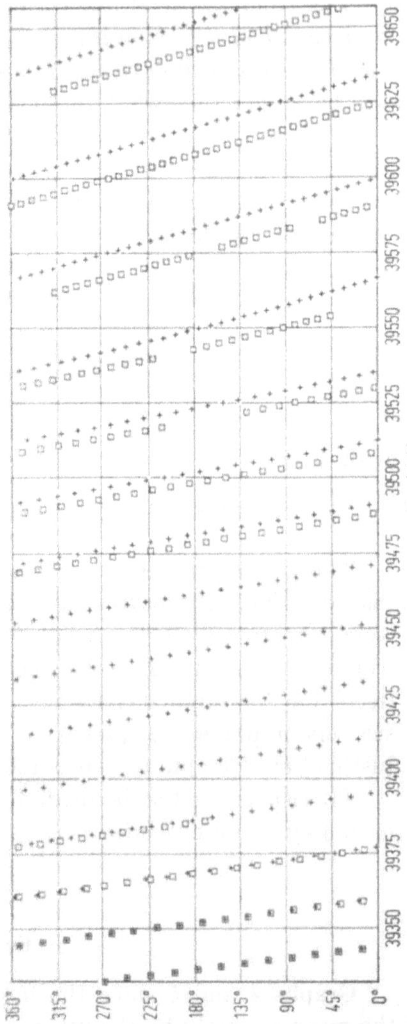

Fig. 5. Mean anomaly fitting employed.

to determine geophysical parameters, but one must avoid the temptation to throw every conceivable effect into the integration. Numerical experimentation is the quickest and most reliable way to see what is important and what may be neglected.

7. Acknowledgements

Considerable technical advice was given by my colleague, BERNARD CHOVITZ, and by CARL A. WAGNER of NASA. Much of the programming and program modification was done by Mrs. DOROTHY DUBESTER; the graphing program was supplied by MYRON W. LAWRENCE.

Appendix I

Symbols Undefined in the Text

An orbital element with a zero-subscript indicates epoch value, e.g. $a_0 = a(T)$.

a semi-major axis
C_n^m geopotential coefficient
e eccentricity
F_2^* second order part of Hamiltonian in Brouwer's satellite theory [1], a function of mean elements
I inclination
J_2 oblateness coefficient $= -C_2^0$
J_3 $= -C_3^0$
J_4 $= -C_4^0$
M mean anomaly
P_n^m associated Legendre function of degree n and order m
r radial coordinate
R dynamic earth radius
S_n^m geopotential coefficient
t time from epoch
T epoch
V potential
V_n^m component of geopotential
β geocentric latitude
θ Greenwich sidereal angle
λ longitude
μ (gravitational constant) $\times$ (mass of earth)
ω argument of perigee
Ω right ascension of ascending node

Appendix II

Results from a Trial Run of the Rapid Orbit Prediction Fitting Program

The mean elements obtained by fitting to three overlapping DOI arcs included perturbations due to J_2, J_3, J_4, $(J_2)^2$, C_8^8 and S_8^8, luni-solar gravitation as well as radiation pressure. Graphs 1, 2, and 3 show the evolution of a, e, and I, respectively. Graph 4 compares the mean anomalies *at midnight* U.T. for a run done *without* the fitting procedure; graph 5 shows the superior results obtained when fitting is utilized. All the graphs were drawn on a Calcomp plotter controlled by an IBM 1620 computer.

Table 2 is the residuals in mean anomaly for the final fitting of the orbit. Table 3 is all the data from the three arcs.

The key for the graphs 1–5 is: □ observed values, + computed values.

Table 2. *Residuals in Mean Anomaly*

Time	ΔM in degrees		
MJD	1st arc	2nd arc	3rd arc
39332	−0.36°	—	—
39333	−0.27	—	—
39334	−0.19	—	—
39335	−0.11	—	—
39336	−0.038	—	—
39337	0.020	—	—
39338	0.066	—	—
39339	0.097	—	—
39340	0.12	—	—
39341	0.11	—	—
39342	0.086	—	—
39343	0.043	—	—
39344	−0.022	—	—
39345	−0.11	0.26°	—
39346	—	0.29	—
39347	—	0.31	—
39348	—	0.31	—
39349	—	0.30	—
39350	—	0.28	—
39351	—	0.23	—
39352	—	0.17	0.00036°
39353	—	0.088	−0.14
39354	—	−0.015	−0.29
39355	—	−0.14	−0.45
39356	—	—	−0.63

References

1. Brouwer, D.: Solution of the Problem of Artificial Satellite Theory without Drag. Astron. J. 64 (1959) pp. 378–397.
2. Brouwer, D.: Notes of the Summer Institute in Dynamical Astronomy. 1959, pp. 173–187.

Table 3. *The Data for the Fit*

Time MJD	a [m]	e	I	Ω	ω	M [rev.]	
39332.0	10618867.9	.02563300	86.94680°	334.09960°	223.11000°	.752578000	↑
39333.0	10618768.6	.02642700	86.94680	334.01173	223.05200	.687690080	
39334.0	10618669.4	.02722100	86.94680	333.92386	222.99400	.622913520	
39335.0	10618570.1	.02801500	86.94680	333.83599	222.93600	.558248320	
39336.0	10618470.9	.02880900	86.94680	333.74812	222.87800	.493694480	
39337.0	10618371.6	.02960300	86.94680	333.66025	222.82000	.429252000	
39338.0	10618272.4	.03039700	86.94680	333.57238	222.76200	.364920880	
39339.0	10618173.1	.03119100	86.94680	333.48451	222.70400	.300701120	1st arc
39340.0	10618073.9	.03198500	86.94680	333.39664	222.64600	.236592720	
39341.0	10617974.7	.03277900	86.94680	333.30877	222.58800	.172595680	
39342.0	10617875.4	.03357300	86.94680	333.22090	222.53000	.108710000	
39343.0	10617776.2	.03436700	86.94680	333.13303	222.47200	.044935680	
39344.0	10617677.0	.03516100	86.94680	333.04516	222.41400	.981272720	
39345.0	10617577.8	.03595500	86.94680	332.95729	222.35600	.917721120	↓
39345.0	10617230.8	.03574589	86.92647	332.93733	222.05450	.918759566	↑
39346.0	10617092.7	.03650110	86.92647	332.84988	221.88100	.855730073	
39347.0	10616954.6	.03725631	86.92647	332.76244	221.70751	.792855562	
39348.0	10616816.6	.03801152	86.92647	332.67499	221.53401	.730136031	
39349.0	10616678.5	.03876673	86.92647	332.58754	221.36052	.667571481	
39350.0	10616540.4	.03952193	86.92647	332.50009	221.18702	.605161912	2nd arc
39351.0	10616402.3	.04027714	86.92647	332.41264	221.01353	.542907324	
39352.0	10616264.3	.04103235	86.92647	332.32519	220.84003	.480807716	
39353.0	10616126.2	.04178756	86.92647	332.23774	220.66653	.418863090	
39354.0	10615988.2	.04254276	86.92647	332.15030	220.49304	.357073444	
39355.0	10615850.1	.04329797	86.92647	332.06285	220.31954	.295438780	↓
39352.0	10616408.0	.04117428	86.92597	332.31793	221.00133	.480335324	↑
39353.0	10616253.0	.04191421	86.92597	332.22826	220.86837	.418239022	
39354.0	10616098.0	.04265414	86.92597	332.13860	220.73541	.356316692	3rd arc
39355.0	10615943.0	.04339408	86.92597	332.04893	220.60245	.294568334	
39356.0	10615788.1	.04413401	86.92597	331.95926	220.46949	.232993948	↓

3. Bryant, R. W.: The Effect of Solar Radiation Pressure on the Motion of an Artificial Satellite. Astron. J. 66 (1961) pp. 430–432.
4. Chovitz, B., Morrison, F. F.: The Orbit of PAGEOS – Preliminary Analysis. [Abstract in Trans. Amer. Geophys. Union, 50, No. 4, p. 122 (1969)].
5. Gaposchkin, E. M.: Differential Orbit Improvement. SAO Special Report No. 161 (1964).
6. Garfinkel, B.: The Disturbing Function for an Artificial Satellite. Astron. J. 70 (1965) pp. 699–704.
7. Garfinkel, B.: Tesseral Harmonic Perturbations of an Artificial Satellite. Astron. J. 70 (1965) pp. 784–786.
8. Kaula, W. M.: Development of the Lunar and Solar Disturbing Functions for a Close Satellite. Astron. J. 67 (1962) pp. 300–303.
9. Kaula, W. M.: Theory of Satellite Geodesy. Blaisdell, 1966.
10. Lundquist, C. A., Veis, G.: Geodetic Parameters for a 1966 Smithsonian Standard Earth. SAO Special Report No. 200 (1966).
11. Sterne, T. E.: Am. Rocket Soc. J. 29 (1959) pp. 777–789.
12. TRW Systems, Inc.: Rapid Orbit Prediction Program (Program Description).
13. Wagner, C. A.: Preliminary Results from Mean Element Analysis of 12-Hour Resonant Orbits. [Abstract in Trans. Amer. Geophys. Union, 50, No. 4, p. 122 (1969)].
14. Yeremenko, R. P.: Precise solution of the shadow equation (Abstract). Bulletin Instituta Teoreticheskoi Astronomii Vol. 10, No 6 (1965) pp. 446–449.

Determination of the Most Probable Elements of a Satellite Orbit from Incomplete Systems of Elements

By

Y. V. Batrakov

Institute for Theoretical Astronomy, Leningrad, USSR

Abstract. Normal equations are ill-conditioned if the observations embrace a small arc of the satellite orbit. In order to avoid the great scattering of the results in this case some elements are to be fixed. Then the non-fixed elements, determined with small random errors, form an incomplete system of elements which is not sufficient to calculate satellite coordinates definitely. When computing the orbit a number of such incomplete systems refered to different time intervals may be obtained. In this case there arises a problem of determining the most probable set of elements which is sufficient for the satellite coordinates to be found. This problem has been solved by the method of maximum likelihood in general form which allows to use the formulae for determining both elements and disturbing parameters such as the coefficients of the earth potential, air drag coefficients and so on.

As it is known the observing stations are distributed mainly in the northern hemisphere of the Earth and only a small part of the earth satellite orbit near the earth terminator can be observed from them. Therefore the normal equations for the corrections of elements obtained when calculating the orbit are often ill-conditioned and the elements can be determined with considerable errors. In order to reduce scattering of the results the usual procedure is to fix some elements. Then the remaining elements can be determined with small errors.

In the case of a small arc of the orbit observed from many stations (Sochilina [3]) there are two sets of mutually connected corrections, one including the corrections of the inclination and the longitude of node, and that including the corrections of the eccentricity and the argument of perigee. In order to improve the conditioning of the normal equations, when only a short arc has been observed, one element in every set mentioned above must be fixed, for example, the inclination and the eccentricity.

We shall call a system of elements in which some elements are fixed an incomplete set of elements. The cause for this designation is that the nonfixed elements are not sufficient for determining the positions of the satellite. If all the elements are nonfixed and they are sufficient for determining the satellite positions we shall designate such systems of elements as complete.

We shall consider here the mathematically strict theory of the use of the incomplete systems of the elements for determining the most probable values of the elements and the perturbing parameters based on the principle of the maximum likelihood. For this purpose we shall give first of all the generalization of the method of least squares for the case of correlated observations.

Let the system of linear conditional equations be

$$\sum_{i=1}^{n} a_{ij} \varepsilon_i = l_j \quad (j = 1, \ldots, N;\ N > n), \tag{1}$$

where a_{ij} are known numerical coefficients, ε_i are the unknowns, l_j are the components of the N-dimensional vector $\boldsymbol{l}$ of observed quantities which are concrete realizations of the normal random values with the averages $\bar{l}_j$ and the correlation matrix

$$\sigma^2 B = \sigma^2 \begin{Vmatrix} b_{11}, \ldots, b_{1N} \\ b_{N1}, \ldots, b_{NN} \end{Vmatrix} \quad (b_{ij} = b_{ji}), \tag{2}$$

where σ^2 is unknown and all the l_{ij} are known.

Let Eq. (1) be satisfied by some system of values $\bar{\varepsilon}_i$, when $l_j = \bar{l}_j$. The solution of (1) is such that the likelihood function is maximum [2].

$$L = -\frac{N}{2} \ln 2\pi - \frac{1}{2} \ln \det B - \frac{N}{2} \ln \sigma^2 - \frac{1}{2\sigma^2} \Delta^T B^{-1} \Delta, \tag{3}$$

where Δ is a column matrix with the components $l_j - \bar{l}_j$, T means the transposition, and B^{-1} is the inverse matrix. After differentiating L with respect to ε_i and equating the partial derivatives to zero we obtain the matrix equation

$$A^T B^{-1} A \varepsilon = A^T B^{-1} l, \tag{4}$$

where A is the $n \times N$ matrix of coefficients a_{ij}, ε, l – are the column matrices of unknowns ε_i and the right-hand parts l_j. Eq. (4) is the generalized normal equation for the case of correlated observations. The solution of (4) which we denote by $\tilde{\varepsilon}$ gives the effective estimate for the vector $\boldsymbol{\varepsilon}$, because it corresponds to the maximum of the likelihood function. This estimate will also be unbiased.

Now let L be differentiated with respect to σ^2 and the derivative be equated to zero. From the equation obtained the following biased estimate for σ^2 can be deduced

$$\sigma^2 = \frac{\Delta^T B^{-1} \Delta}{N}. \tag{5}$$

The corresponding unbiased estimate will be

$$\sigma_0^2 = \frac{\Delta^T B^{-1} \Delta}{N - n}, \tag{6}$$

where $\Delta = l - \tilde{l}$, $\tilde{l}_j = \sum_i a_{ij} \tilde{\varepsilon}_i$ must be taken instead of $\Delta = l - \bar{l}$ because the $\bar{l}_j$ are not known. It is clear that σ_0 is the generalization of the concept of the unit weight error.

The correlation matrix $\sigma_0^2 B_\varepsilon$ of vector $\tilde{\varepsilon}$, as can be easily shown, is

$$\sigma_0^2 B_\varepsilon = \sigma_0^2 (A^T B^{-1} A)^{-1}. \tag{7}$$

Eqs. (4), (6), (7) give the complete solution of the problem of determining the most probable values of the unknowns in (1). If the right-hand sides of (1) are uncorrelated, random values Eqs. (4), (6), (7) turn into the corresponding formulae of the usual method of least squares. An analogous method was developed by A. V. Brykov [1].

We shall consider now the problem of fixing the badly determined unknowns. Let one of the diagonal coefficients of the normal system of equations be equal or near zero. It is clear then, from the very construction of the normal system, that all the coefficients in the corresponding column or line are also equal or near zero. The corresponding unknown either cannot be determined at all or can be determined with a considerable error overlapping the interval of admissible values for the unknown. It is then advisable not to determine this unknown but to fix it giving to it the approximate value used when constructing the equations of conditions. When fixing this unknown some unknown error is introduced but its influence on the accuracy of the remaining unknowns is negligible because this error enters the normal equations with equal to or near zero coefficients.

In general a zero diagonal coefficient appears only after the several first unknowns have been eliminated from the normal equations when solving them by the method of successive elimination of unknowns. Let us suppose that during the process of solving the normal Eq. (4) $k - 1$ first unknowns have been eliminated until a zero or near zero diagonal coefficient appears. Then we have two systems of equations

$$\left.\begin{aligned} a_{11}\varepsilon_1 + a_{12}\varepsilon_2 + \cdots + a_{1k}\varepsilon_k + \cdots + a_{1n}\varepsilon_n &= l_1, \\ a_{222}\varepsilon_2 + \cdots + a_{2k2}\varepsilon_k + \cdots + a_{2n2}\varepsilon_n &= l_{22}, \\ \cdots \\ a_{k-1,k-1,k-1}\varepsilon_{k-1} + a_{k-1,k,k-1}\varepsilon_k + \cdots + a_{k-1,n,k-1}\varepsilon_n &= l_{k-1,k-1}, \end{aligned}\right\} \tag{8}$$

$$\left.\begin{aligned} a_{k,k,k-1}\varepsilon_k + \cdots + a_{k,n,k-1}\varepsilon_n &= l_{k,k-1}, \\ a_{n,k,k-1}\varepsilon_k + \cdots + a_{n,n,k-1}\varepsilon_n &= l_{n,k-1}, \end{aligned}\right\} \tag{9}$$

where the last indices of coefficients a_{ijm}, l_{im} show that the corresponding coefficients have been obtained as the result of elimination of the first m unknowns.

In system (8) all the coefficients a_{mmm} are different from zero and it is possible to determine $\varepsilon_1, \ldots, \varepsilon_{k-1}$ as linear functions of $\varepsilon_k, \ldots, \varepsilon_n$. System (9) is the normal system of equations in which at least one of the diagonal coefficients is equal to or near zero. According to what was said above we must fix the corresponding unknowns and eliminate the corresponding equations from (9). This gives a system of normal equations the order of which is less than that in (9); the diagonal coefficients of the new system are all different from zero and it is possible to begin the elimination of the unknowns again.

It must be noted, however, that the fixation of the unknowns is to be done only when solving (9). In (8) the corresponding unknowns must be left in literal form because the fixation errors of these unknowns may influence considerably the accuracy of $\varepsilon_1, \ldots, \varepsilon_{k-1}$.

This process being repeated gives in the end, a system of the type of (8) the order of which is equal to the number of the non-fixed unknowns. The solution of this system gives the non-fixed ε_i as linear functions

$$\varepsilon_i = \sum_j c_{ij}\varepsilon_j + d_i\,, \tag{10}$$

where ε_j are the unknowns fixed when solving the equations of the type of (9), and d_i are the components of vector $\boldsymbol{d}$ with the known correlation matrix, the order of which being equal to the number of the non-fixed unknowns.

In particular when the determinant of (4) is essentially different from zero the order of vector $\boldsymbol{d}$ is equal to that of system (4). In this case the numerical values of all the unknowns are determined unambiguously and the complete system of ε_i is obtained.

Now, for the sake of definition, we shall consider ε_i in (1) as the corrections to the mean elements of the earth satellite orbit. We shall suppose also that the mean elements are calculated with the unique system of the initial elements and of the perturbing parameters such as zonal coefficients of the earth potential, air drag coefficients and so on. Then, neglecting the higher order terms, the ε_i may be expressed as linear combinations of the corrections of the initial elements and the perturbing parameters. After these expressions have been put into (10) and all the unknown corrections have been gathered at the left-hand sides of (10), we shall have equations of condition of type (1) for the corrections mentioned. The right-hand sides of these equations will be the numerical components of vector $\boldsymbol{d}$ with the known correlation matrix. When having a sufficiently great number of such equations of conditions the generalized method of least squares described above is to be used in order to find the most probable values of the elements.

Thus the problem of the use of incomplete system of elements may be considered to have been solved from a theoretical point of view. As the complete systems are particular cases of the incomplete ones they can be used together with the latter within the framework of the described method. Therefore this method is more general than the known methods of determining elements from complete systems [4].

According to the developed method the complete and incomplete systems of elements can be considered as some kind of normal positions replacing the observations. The use of them gives an opportunity of reducing the computation work noticeably.

It is clearly seen from the discussion in this paper that the use of the incomplete systems for determining the elements and perturbing parameters is legitimate only in the case when the non-fixed elements are known as functions of the fixed ones, and the correlation matrix of the non-fixed elements is also known. So the corresponding data are highly desirable to be given in the publications of elements.

References

1. Brykov, A. V.: Iskysstv. sputn. Zemli, 16, 1963 (in Russian).
2. Linnik, Yu. V.: The method of least squares and principles of the theory of treatment of observations. 2nd edition (in Russian), Moscow 1962.
3. Sochilina, A. S.: Bull. Instit. Teoret. Astron. 11 (1967) NI (124) (in Russian).
4. Blackman, R. B.: Bell. System Techn. J. 43 (1964) 3.

On the Influence of Distribution of Observations on Accuracy of Orbital Elements of the Earth's Artificial Satellites

By

A. S. Sochilina

Institute for Theoretical Astronomy, Leningrad, USSR

Abstract. The formulae of weight coefficients for the elements determined by differential correction procedure are given. The analysis of these formulae can be useful for the discussion of the accuracy of the elements obtained.

The accuracy of the orbital elements of artificial satellites determined by the differential correction procedure depends upon not only the accuracy of observations themselves but also upon the distribution of the observations.

The distribution of the observations influences mainly the conditioning of the normal equations for the corrections of the elements and consequently the values of the weight coefficients.

The analytical expressions for the weight coefficients have been derived under some assumptions to analyse the accuracy of the corrections obtained.

Let the observations of some satellite be made in an horizontal coordinate system A, z, ϱ, where A is the azimuth, z is the zenith distance, ϱ is the topocentric distance and let the orbital elements i, Ω, $M + \omega$, $a_x = e \cos\omega$, $a_y = e \sin\omega$, n at the epoch t_0 be improved, where i = inclination, Ω = ascension of node, M = mean anomaly, ω = argument of perigee, e = eccentricity, n = mean motion.

Suppose that

1. The total number of the observations is equal to N $(N = mk)$, where m is the number of stations and k is the number of points uniformly distributed along the orbit.

2. At any time t in the interval (t_1, t_2) the satellite is observed from all m stations for which the zenith distance of the satellite does not surpass some limit ζ.

This condition can be changed into a more simple one, namely, that the observer situated at the subsatellite point can see the satellite in m points uniformly distributed in the sky.

3. All k points are distributed on a short arc of orbit Δu, where $\Delta u = u_2 - u_1$, and u is the argument of latitude.

4. The epoch t_0 of the elements to be improved is selected in the middle of the interval (t_1, t_2) and $t_2 - t_1 = 2T$.

5. The eccentricity in the equation of condition is considered to be equal to 0.

6. Twofold summation when the normal system is formed, is substituted by the integrations first with respect to A and z and then with respect to u.

7. The standard error σ_0 of the least-squares solution is

$$\sigma_0 = \sqrt{\frac{\sum_{i=1}^{N} (\varrho \sin z \, \mathrm{d}A)_i^2 + (\varrho \, \mathrm{d}z)_i^2}{2N - \nu}}$$

where ν is the number of unknowns.

Then the general expressions of the weight coefficients under condition that $\Delta u \ll 1$ become as presented in Table 1, where $U = u_1 + u_2$ is the angle reckoned from the node and characterizing the position of the arc Δu in the orbit, a is the semi-major axis, parameters β and γ characterize the kind of observation: 1. with three observed quantities (A, z, ϱ); 2. with two observed quantities (A, z); 3. with one observed quantity (A or z or ϱ).

These expressions have been given in some papers by the same author [1967, 1968]. The formulae of the weight coefficients show that all the elements but n are determined with a poor accuracy not only in the case of a small Δu but also in the case of nearly zenith observations, e.g. when $\cos \zeta = 1$. The weight coefficients do not depend upon ζ only when there are three observed quantities (A, z, ϱ).

As has been shown earlier [1967] the correlation between different groups of elements occurs when the observations are distributed on a short arc of orbit. In this case the elimination of one of the elements from improving permits only to derive the others with better accuracy.

The accuracy of the elements can be considerably increased when the observations occur to be distributed on two short arcs (u_1, u_2 and u_3, u_4) removed from each other at an arbitrary angle ΔW. In this case the expressions for

$$q(i), \, q(\Omega), \, q(M + \omega), \, q(a_x), \, q(a_y)$$

Table 1.

			β	γ
$q(i)$	$\frac{6(1+\cos U)}{\beta \Delta u^2}$	A, z, ϱ	$N a^2$	1
$q(\Omega)$	$\frac{6(1-\cos U)}{\beta \sin^2 i\, \Delta u^2}$	A, z	$\frac{N a^2 (4+\cos\zeta+\cos^2\zeta)}{6}$	$\frac{4-2\cos\zeta-2\cos^2\zeta}{4+\cos\zeta+\cos^2\zeta}$
$q(a_x)$	$\frac{6\left(\gamma+\frac{\Delta u^2}{3}\right)(1-\cos U)}{\beta \Delta u^2\left(\gamma^2+\frac{1}{45}\Delta u^4\right)}$	z	$\frac{N a^2 (1+\cos\zeta+\cos^2\zeta)}{6}$	$\frac{4-2\cos\zeta-2\cos^2\zeta}{1+\cos\zeta+\cos^2\zeta}$
$q(a_y)$	$\frac{6\left(\gamma+\frac{\Delta u^2}{3}\right)(1+\cos U)}{\beta \Delta u^2\left(\gamma^2+\frac{1}{45}\Delta u^4\right)}$	A	$\frac{N a^2}{2}$	0
$q(M+\omega)$	$\frac{48\left[\gamma \sin^2 i+\frac{\gamma^2}{8}\cos^2 i(1-\cos U)+\frac{\Delta u^2}{6}(1+\sin^2 i)\right]}{\beta \Delta u^2 \sin^2 i\left(\gamma^2+\frac{1}{45}\Delta u^2\right)}$	ϱ	$\frac{N a^2 (2-\cos\zeta-\cos^2\zeta)}{6}$	$\frac{2+2\cos\zeta+2\cos^2\zeta}{2-\cos\zeta-\cos^2\zeta}$
$q(n)$	$\frac{3}{\beta T^2}$			

are

$$q(i) = \frac{2(1 + \cos \Delta W \cos W)}{\beta \sin^2 \Delta W},$$

$$q(\Omega) = \frac{2(1 - \cos \Delta W \cos W)}{\beta \sin^2 \Delta W},$$

$$q(a_x) = \frac{2\left\{\left(1 + \frac{\gamma}{2}\right) - \cos \Delta W - \left[1 - \left(1 - \frac{\gamma}{2}\right) \cos \Delta W\right] \cos W\right\}}{\beta(1 - \cos \Delta W)\left[\left(2 + \frac{\gamma}{2}\right) - \left(2 - \frac{\gamma}{2}\right) \cos \Delta W\right]},$$

$$q(a_y) = \frac{2\left\{\left(1 + \frac{\gamma}{2}\right) - \cos \Delta W + \left[1 - \left(1 - \frac{\gamma}{2}\right) \cos \Delta W\right] \cos W\right\}}{\beta(1 - \cos \Delta W)\left[\left(2 + \frac{\gamma}{2}\right) - \left(2 - \frac{\gamma}{2}\right) \cos \Delta W\right]},$$

$$\begin{aligned} q(M + \omega) = {} & \frac{2}{\beta \gamma \sin^2 i \sin^2 \Delta W (1 - \cos \Delta W)} [(1 + \cos \Delta W) \\ & \times (\sin^2 i - 2 \sin^2 i \cos^2 \Delta W + 2 \cos^2 i \cos \Delta W \cos W)] \\ & + \frac{\gamma}{2} (1 - \cos \Delta W) [1 + \cos^2 i - \cos^2 \Delta W \sin^2 i \\ & - 2 \cos^2 i \cos \Delta W \cos W], \end{aligned}$$

where

$$\Delta W = \frac{u_3 + u_4}{2} - \frac{u_1 + u_2}{2},$$

$$W = \frac{u_1 + u_2}{2} + \frac{u_3 + u_4}{2},$$

the parameters β and γ are the same as those employed in Table 1. These formulae are valid only when $\sin \Delta W \neq 0$. The analysis of the weight coefficients can be useful for the discussion of the accuracy of the elements obtained and also for recommendation connected with observations of the satellites and distribution of the station positions.

References

Sochilina, A. S. – 1967: On the accuracy of orbital elements of ASE. Bull. ITA II, N 1 (124) (Russ.).

Sochilina, A. S. – 1968: The investigation of the accuracy of orbital elements. (The report, Prague, April 1968) (Russ.).

The Use of Ellipsoidal Harmonics for the Representation of the Geopotential

By

H. G. Walter

European Space Operations Centre Darmstadt, FRG

Abstract. The potential of the Earth is developed in ellipsoidal harmonics, and the mathematical tools required, which are the generation of Lamé's functions and the relationship between rectangular and ellipsoidal coordinates, are compiled. Brief reference is made to a procedure utilized for the determination of the gravity coefficients in the expansion ot the geopotential in ellipsoidal harmonics when precise satellite tracking data is available.

With the aim of carrying out a numerical integration of the Lagrangian equations of planetary motion, the functional dependence of the disturbing earth potential on the orbital elements for elliptic motion is given. In particular, formulae for the partial derivatives of the disturbing potential with respect to the orbital elements are derived, thus making possible the numerical calculation of these partial derivatives from orbital elements.

1. Introduction

In recent years the knowledge of the geopotential has been considerably widened by the fact that it is possible to determine the gravity coefficients in the expansion of the potential in spherical harmonics from the apparent irregularities in the motion of artificial satellites. Insight has not only been gained into the zonal coefficients [Kozai 1967, King-Hele et al. 1963, 1968], but also into the tesseral and sectorial coefficients [Kaula 1961, Iszak 1964]. A generally adopted representation of the potential is its expansion in a series of Legendre associated polynomials. For near earth satellites this series does not converge rapidly, and therefore for refined work in orbital analysis a great number of gravity coefficients must be known.

It is conceivable that the speed of convergence can be increased if the potential is expanded in a series of ellipsoidal harmonics which are more adaptable to the shape of the geoid [Doubochine 1966, Hotine 1967]. Aligned with the ellipsoidal harmonics are the ellipsoidal coordinates. They are the independent variables of Lamé's functions, which form the constituents of the ellipsoidal harmonics.

The subsequent sections deal with the preparation of the means needed for the numerical integration of the differential equations of satellite motion, i.e. the expansion of the potential in ellipsoidal harmonics, the ellipsoidal coordinates and the mechanism for the generation of the ellipsoidal harmonics and their derivatives. This is followed by the outlining of a least squares process for the determinations of the ellipsoidal gravity coefficients from the residuals resulting from the observed and computed positions of the satellites.

2. Expansion of the Geopotential in Ellipsoidal Harmonics

The expansion of the geopotential in a series of surface harmonics is well known as a generally adopted representation [Hagihara, 1962]. It is denoted here as

$$U(P) = \frac{GM}{r} \sum_{n=0}^{\infty} \left(\frac{R}{r}\right)^n S_n(\theta, \psi) \tag{1}$$

with

$$S_n(\theta, \psi) = \sum_{m=0}^{n} P_n^m(\cos\theta)\,[C_{nm}\cos m\psi + S_{nm}\sin m\psi]\,, \tag{2}$$

where

P is a point exterior to the Earth within the Earth's gravitational field

GM is the product of the gravitational constant G and the mass M of the Earth

R is the Earth's equatorial radius

r, θ, ψ are the spherical coordinates of P in an Earth-centred coordinate system and are related to the geocentric rectangular coordinates by

$$\left.\begin{aligned} x &= r\sin\theta\cos\psi \\ y &= r\sin\theta\sin\psi \\ z &= r\cos\theta\,, \end{aligned}\right\} \tag{3}$$

i.e. ψ represents the geographical longitude and θ is the co-latitude: $\theta = \pi/2 - \varphi$, where φ is the geographical latitude.

P_n^m is the associated Legendre polynomial of degree n and order m

C_{nm} and S_{nm} are spherical gravity coefficients typical for the shape and structure of the Earth.

The expansion of the potential in ellipsoidal harmonics is represented by [Doubochine, 1966].

$$U(P) = GM \sum_{n=0}^{\infty} \frac{V_n(\lambda, \mu, \nu)}{r^{2n+1}} \tag{4}$$

with

$$V_n(\lambda, \mu, \nu) = \sum_{s=0}^{2n} G_{ns} E_n^s(\lambda) E_n^s(\mu) E_n^s(\nu) \tag{5}$$

and

$$r = (\lambda + \mu + \nu + a^2 + b^2 + c^2)^{1/2}, \tag{6}$$

where

λ, μ, ν are ellipsoidal coordinates
a, b, c are the semi-axes of the reference ellipsoid
E stands for Lamé's functions
G_{ns} are the ellipsoidal gravity coefficients, typical for the shape and structure of the Earth.

3. Definition of Ellipsoidal Coordinates

For given rectangular coordinates x, y, z the cubic equation in ω

$$\frac{x^2}{a^2+\omega} + \frac{y^2}{b^2+\omega} + \frac{z^2}{c^2+\omega} = 1, \tag{7}$$

where a, b, c are the semi-axes of an ellipsoid ($a \geq b > c$), allows for three real roots λ, μ, ν with

$$-a^2 \leq \nu \leq -b^2 \leq \mu \leq -c^2 \leq \lambda < \infty. \tag{8}$$

In (7) a, b, c are the semi-axes of the reference ellipsoid

$$\frac{x^2}{a^2} + \frac{y^2}{b^2} + \frac{z^2}{c^2} = 1. \tag{9}$$

The semi-axes can be chosen so as to make (9) approximate the surface of the Earth as closely as possible.

If the roots λ, μ, ν are substituted in (7), three confocal surfaces are defined in the rectangular coordinates x, y, z: an ellipsoid, a hyperboloid of one sheet and a hyperboloid of two sheets forming a threefold orthogonal system of surfaces.

Introducing λ, μ, ν into (7) and solving for x, y, z yields

$$\left.\begin{aligned} x^2 &= \frac{(a^2+\lambda)(a^2+\mu)(a^2+\nu)}{(a^2-b^2)(a^2-c^2)} \\ y^2 &= \frac{(b^2+\lambda)(b^2+\mu)(b^2+\nu)}{(a^2-b^2)(c^2-b^2)} \\ z^2 &= \frac{(c^2+\lambda)(c^2+\mu)(c^2+\nu)}{(a^2-c^2)(b^2-c^2)}. \end{aligned}\right\} \tag{10}$$

If one compares (7) with the definition of ellipsoidal coordinates $\xi_1(=\varrho)$, $\xi_2(=\mu)$, $\xi_3(=\nu)$ given by Hobson [1955], the following association is found:

$$\left.\begin{aligned} \lambda &= \xi_1^2 - a^2 \\ \mu &= \xi_2^2 - a^2 \\ \nu &= \xi_3^2 - a^2. \end{aligned}\right\} \tag{11}$$

4. The Generation of Lamé's Functions

Several definitions exist for Lamé's functions in so far as they refer to different representations of Lamé's differential equation. The line of thought pursued here has been adopted from HEINE [1961] and defines Lamé's functions in ellipsoidal coordinates ξ_1, ξ_2, ξ_3.

The symbols E_n^s, $s = 0, 1, \ldots, 2n$; introduced in (5) stand for Lamé's functions of the arguments λ, μ or ν. By the appropriate substitutions according to the relations (11), these functions can be calculated from Lamé's functions of the arguments ξ_i, $i = 1, 2, 3$. They divide into four classes which for a given n comprise $2n + 1$ linear independent functions. Let

$$\sigma = \begin{Bmatrix} n/2 & \text{for } n \text{ even} \\ (n-1)/2 & \text{for } n \text{ odd}. \end{Bmatrix} \tag{12}$$

Then the classes are represented by

$$\begin{aligned} &\sigma + 1 \text{ functions } K_n^s(\xi_i), \quad s \equiv 0 \bmod 4 \\ &n - \sigma \text{ functions } L_n^s(\xi_i), \quad s \equiv 1 \bmod 4 \\ &n - \sigma \text{ functions } M_n^s(\xi_i), \quad s \equiv 2 \bmod 4 \\ &\sigma \text{ functions } N_n^s(\xi_i), \quad s \equiv 3 \bmod 4, \end{aligned}$$

and K, L, M and N are abbreviations for the expressions defining Lamé's functions of first, second, third and fourth class:

$$K_n^s(\xi_i) = a_{os}\xi_i^n + a_{1s}\xi_i^{n-2} + \cdots + \begin{cases} a_{\sigma s} & \text{for } n \text{ even} \\ a_{\sigma s}\xi_i & \text{for } n \text{ odd}, \end{cases} \tag{13}$$

$$L_n^s(\xi_i) = \sqrt{\xi_i^2 - h^2}\left[b_{os}\xi_i^{n-1} + b_{1s}\xi_i^{n-3} + \cdots + \begin{cases} b_{n-\sigma-1,s}\xi_i & \text{for } n \text{ even} \\ b_{n-\sigma-1,s} & \text{for } n \text{ odd} \end{cases}\right], \tag{14}$$

$$M_n^s(\xi_i) = \sqrt{\xi_i^2 - k^2}\left[c_{os}\xi_i^{n-1} + c_{1s}\xi_i^{n-3} + \cdots + \begin{cases} c_{n-\sigma-1,s}\xi_i & \text{for } n \text{ even} \\ c_{n-\sigma-1,s} & \text{for } n \text{ odd} \end{cases}\right], \tag{15}$$

$$N_n^s(\xi_i) = \sqrt{\xi_i^2 - h^2}\sqrt{\xi_i^2 - k^2} \times \left[d_{os}\xi_i^{n-2} + d_{1s}\xi_i^{n-4} + \cdots + \begin{cases} d_{\sigma-1,s} & \text{for } n \text{ even} \\ d_{\sigma-1,s}\xi_i & \text{for } n \text{ odd} \end{cases}\right]. \tag{16}$$

The coefficients a_{ks}, b_{ks}, c_{ks} and d_{ks} in (13) to (16) can be considered as known, apart from an arbitrary constant factor which is common to each of them. They are the solution of equations of conditions [HOBSON,

1955], the setting up of which falls outside the scope of this paper. A technique for solving these equations, which transforms the task into an algebraic eigenvalue problem, as well as a computer program for the generation of the four classes of Lamé's functions, have been developed and will be described elsewhere.

5. The Equations of Motion

It is assumed that the only perturbing forces acting on an artificial satellite are caused by the disturbing Earth potential. The differential equations for the perturbed motion of a satellite can be given in the form of Lagrange's planetary equations [BROUWER and CLEMENCE, 1961]. Let R represent the disturbing function appearing in the equations of motion. It can be inferred from (4) and (5) to be

$$R = R(\lambda, \mu, \nu) = U(P) - \frac{GM}{r}. \tag{17}$$

Since essentially

$$E_0^0(\lambda) = E_0^0(\mu) = E_0^0(\nu) = 1 \tag{18}$$

and the origin of the coordinate system coincides with the centre of mass of the geoid, the coefficient G_{00} is equal to 1 by necessity. Thus one obtains as a disturbing function

$$R = GM \sum_{n=1}^{\infty} \frac{1}{r^{2n+1}} \sum_{s=0}^{2n} G_{ns} E_n^s(\lambda) E_n^s(\mu) E_n^s(\nu), \tag{19}$$

where λ, μ, ν are functions of the orbital elements and time.

6. Partial Derivatives of the Disturbing Function

For the evaluation of Lagrange's planetary equations one requires the partial derivatives of the disturbing function with respect to the orbital elements. The disturbing function (19) is implicitly dependent on the orbital elements and time via three stages: firstly, the ellipsoidal coordinates λ, μ, ν are functions of the Earth-rotating rectangular coordinates x, y, z according to (10). Secondly, x, y, z are functions of the inertial geocentric rectangular coordinates and the sidereal angle. Thirdly, these quantities are expressible by the orbital elements and time.

The partial derivatives of R with respect to the orbital elements can be found by means of the chain rule. It is only in connection with the formation of the partial derivatives

$$\frac{\partial\sigma}{\partial x}, \frac{\partial\sigma}{\partial y}, \frac{\partial\sigma}{\partial z}, \qquad \sigma = \lambda, \mu, \nu, \tag{20}$$

that special effort is required, because closed expressions for λ, μ and ν as functions of x, y and z cannot easily be established. The problem is overcome by resorting to properties of threefold orthogonal systems of surfaces, to which the system of ellipsoidal coordinates adheres. Accordingly, for the partial derivatives in question holds [LENSE, 1950]

$$\left.\begin{aligned} \frac{\partial \sigma}{\partial x} &= H_\sigma^2 \frac{\partial x}{\partial \sigma} \\ \frac{\partial \sigma}{\partial y} &= H_\sigma^2 \frac{\partial y}{\partial \sigma} \\ \frac{\partial \sigma}{\partial z} &= H_\sigma^2 \frac{\partial z}{\partial \sigma} \end{aligned}\right\} \quad \sigma = \lambda, \mu, \nu. \tag{21}$$

In (21) the factors H_σ, $\sigma = \lambda, \mu, \nu$ are in turn [HOBSON, 1955)]

$$\left.\begin{aligned} H_\lambda^2 &= 4\,\frac{(a^2+\lambda)(b^2+\lambda)(c^2+\lambda)}{(\lambda-\mu)(\lambda-\nu)} \\ H_\mu^2 &= 4\,\frac{(a^2+\mu)(b^2+\mu)(c^2+\mu)}{(\mu-\nu)(\mu-\lambda)} \\ H_\nu^2 &= 4\,\frac{(a^2+\nu)(b^2+\nu)(c^2+\nu)}{(\nu-\lambda)(\nu-\mu)} \end{aligned}\right\} \tag{22}$$

There still remains the determination of the partial derivatives on the right-hand side of the equation (21). They are found by formal differentiation of (10) with respect to λ, μ and ν.

7. Determination of Ellipsoidal Gravity Coefficients

With the establishment of the partial derivatives, the equations of motion are prepared for numerical integration provided that the gravity coefficients G_{ns} of the geopotential expanded in ellipsoidal harmonics are known. At this stage it can be assumed that a formal association of these ellipsoidal gravity coefficients with the gravity coefficients C_{nm} and S_{nm} relating to the expansion of the potential in spherical harmonics has been achieved [WALTER, in preparation]. Thus the numerical integration of the equations of motion can be executed in the sense of calculation of special perturbations. Furthermore, a comparison of the resulting orbital elements with those obtained by numerical integration of the equations of motion with a disturbing potential expanded in spherical harmonics enables one to check the correctness of the formal association of the ellipsoidal to the spherical gravity coefficients.

A different interpretation of the equations of motion can lead to a direct determination of the gravity coefficients G_{ns}. The numerical approach is similar to the one of ANDERLE [1966] and its essence is the determination of gravity coefficients or their improvements by comparing the observed and computed satellite positions and subsequent minimizing of the residuals by making adequate alterations to the initially assumed values of the ellipsoidal gravity coefficients. In principle, the equations of motion must be integrated numerically on the basis of unknown or only approximately known gravity coefficients in the disturbing function. The decision as to which of the G_{ns} should be improved depends on the orbit types being studied, and these orbit types will enable one to specify which harmonics make the main contribution to the gravitational perturbations of the individual orbits. As a first approximation for the ellipsoidal gravity coefficients, those values found by the formal association referred to above will be used.

The positions thus obtained as a function of time are transformed into the coordinate system in which the individual satellite observations $\theta^{(i)}$ are taken, yielding the residuals

$$v^{(i)} = \theta_o^{(i)} - \theta_c^{(i)}, \qquad i = 1, 2, \ldots, N, \tag{23}$$

where θ refers to any type of satellite tracking data and the subscripts mean "observed" and "computed" positions respectively; N is the number of observations to be processed. Finally, the residuals are minimized in the least squares sense as follows.

For the purpose of a differential correction of the gravity coefficients, the term $\theta_c^{(i)}$ must be treated as a function of the coefficients G_{ns}, i.e.

$$\theta_c^{(i)} = L^{(i)}(G_{ns}), \qquad n = 1, 2, \ldots; \; s = 1, 2, \ldots, n; \; s \leq n. \tag{24}$$

Throughout the integration process $\theta_c^{(i)}$ is available as a numerical quantity. The least squares fit of the G_{ns} to the observations gives rise to corrections ΔG_{ns} and results in the computed positions

$$\tilde{\theta}_c^{(i)} = L^{(i)}(G_{ns} + \Delta G_{ns}) = L^{(i)}(G_{ns}) + \sum_{n,s} \frac{\partial L^{(i)}}{\partial G_{ns}} \Delta G_{ns} + \text{higher order terms} \tag{25}$$

with

$$\sum_{i=1}^{N} (\tilde{v}^{(i)})^2 = \sum_{i=1}^{N} \left(\theta_o^{(i)} - \tilde{\theta}_c^{(i)}\right)^2 = \min. \tag{26}$$

If one suppresses the higher order terms in (25), the standard least squares technique leads to the normal equations which are solved for the desired corrections ΔG_{ns}. Subsequent iterations take into account nonlinearity effects. An important constituent of the normal equations is the partial derivatives in (25). Refined methods of numerical differentiation [Rutishauser, 1963] must be applied for their calculation in order to avoid instabilities in the solution of the normal equations.

8. Final Remarks

At first glance the numerical implications and the amount of computer programming work required for the determination of the gravity coefficients in the expansion of the potential in ellipsoidal harmonics appear to be enormous. Nevertheless, the efforts involved lie wholly within the capabilities of the computerized methods practised nowadays.

Apart from a possible increase in the speed of convergence of the series of ellipsoidal harmonics, into which the geopotential is expanded, the comparison between the empirically determined gravity coefficients and those formally inferred from the transformation of the spherical gravity coefficients into the ellipsoidal ones may lead to useful conclusions for future theoretical investigations. Also, the expansion of the potential into ellipsoidal harmonics will show its superiority in conjunction with the representation of the potential of celestial bodies which are shaped even more like an ellipsoid than the Earth.

References

ANDERLE, R. J.: Computational methods employed with Doppler observations and derivation of geodetic results, in IUTAM Symposium Trajectories of Artificial Celestial Bodies (pp. 178–193). Paris, Apr. 1965. Berlin/Heidelberg/New York: Springer 1966.

BROUWER, D., CLEMENCE, G. M.: Methods of celestial mechanics, New York: Academic Press 1961.

DOUBOCHINE, G. N.: Sur le développement du potentiel de la terre par les fonctions de Lamé, in IUTAM Symposium Trajectories of Artificial Celestial Bodies, Paris Apr. 1965, Berlin/Heidelberg/New York: Springer 1966.

HAGIHARA, Y.: Recommendations on notation of the earth potential. Astron. J. 67 (1962) 108.

HEINE, E.: Handbuch der Kugelfunctionen, Würzburg: Physica Verlag 1961.

HOBSON, E. W.: The Theory of Spherical and Ellipsoidal Harmonics, New York: Chelsea Publishing Company 1955.

HOTINE, M.: Downward Continuation of the Gravitational Potential (paper presented at the International Association of Geodesy, Lucerne, Switzerland, 1967).

ISZAK, I. G.: Tesseral Harmonics of the Geopotential and Corrections to Station Coordinates. J. Geophys. Res. 69 (1964) 2621–2630.

KAULA, W. M.: Analysis of Gravitational and Geometric Aspects of Geodetic Utilization of Satellites. Geophys. J. 5 (1961) 104–133.

KING-HELE, D. G., COOK, G. E., REES, J. M.: Determination of the Even Harmonics in the Earth's Gravitational Potential. Geophys. J. 8 (1963) 119–145.

KING-HELE, D. G., COOK, G. E., SCOTT, D. W.: Evaluation of Odd Zonal Harmonics in the Geopotential, of Degree less than 33, from the Analysis of 22 Satellite Orbits. Royal Aircraft Establishment, Tech. Rep. No. 68202, 1968.

KOZAI, Y.: Improved Values for Coefficients of Zonal Spherical Harmonics in the Geopotential, in Geodetic Satellite Results during 1967, edited by C. A. Lundquist. Smith. Astrophys. Obs., Special Report No. 264 (1967) 43–56.

LENSE, J.: Kugelfunktionen, p. 2, Leipzig: Akademische Verlagsgesellschaft 1955.

RUTISHAUER, H.: Ausdehnung des Rombergschen Prinzips. Numerische Mathematik 5 (1963) 48–54.

WALTER, H. G.: The Association of Gravity Coefficients in the Expansion of the Geopotential in Spherical and Ellipsoidal Harmonics (prov. title), publication in preparation.

Computer Derivation of Short-Period Lunar Perturbations*

By

J. R. Cherniack and **E. M. Gaposchkin**

Smithsonian Institution, Astrophysical Observatory, Cambridge, MA, USA

Abstract. The CDC 6400 computer was used to verify the literal algebra of Kozai's short-period lunar perturbations. Most of Kozai's results were verified; some errors were found and corrected. The output of the literal algebra program (SPASM) was Fortran arithmetic statements, which were compiled to produce a program that would evaluate the perturbations. SPASM operates directly on symbolic expressions. Some advantages of this approach over table-driven algebraic processors are observed.

1. Introduction

The use of the general-perturbation theory for satellite-trajectory analysis requires a great deal of tedious algebra to produce the formulas that are evaluated. These formulas must be transcribed for computer interpretation, and the resulting program is eventually incorporated into the general computer analysis. Needless to say, these steps are error prone.

The Smithsonian Astrophysical Observatory (SAO) has traditionally done analytical work of some complexity. An example is the perturbations of a close-earth satellite by the moon. These perturbations were originally developed by Izsak [1] and contained the dominant 14-day period. With the intention of achieving greater accuracy, Kozai [2] developed new formulas incorporating the motion of the moon's inclination with respect to the equator. These formulas never provided orbit accuracy greater than that achieved with the Izsak expressions.

Our attempt to improve the Kozai formulas afforded an opportunity to verify SPASM with a reasonably large problem. The essential properties of this problem are the same as those of all trajectory analyses. The fundamental basis lies in the Lagrange planetary equations. We have the opportunity to specify the problem in the simplest terms, i.e., the number of spherical harmonic terms in which we express $1/\varrho$. Adopting

* This work was supported in part by grant NGR 09-015-002 from the National Aeronautics and Space Administration.

the approach of KOZAI, we generate the disturbing function in coordinates, then elements, substitute it into Lagrange's equation, and finally perform the integration.

2. The Problem

The key analytical concept is the variation of constants applied to the Lagrangian form of the equations of motion. Oversimplifying, the resulting equations are of the form

$$\frac{d(\delta e_i)}{dt} = g_i(\bar{e}) \frac{\partial R(\bar{e})}{\partial \alpha_i} + h_i(\bar{e}) \frac{\partial R(\bar{e})}{\partial \beta_i}, \qquad i = 1, \ldots, 6, \tag{1}$$

where $\bar{e} = (e_1, \ldots, e_6)$ are satellite elements, and δe_i are the desired perturbations; α_i, $\beta_i \; \varepsilon \; \{e_1, e_2, \ldots, e_6\}$; and $g_i(\bar{e})$, $h_i(\bar{e})$, and $R(\bar{e})$ are such that the equations are immediately integrable.

The R derived by IZSAK afforded positional accuracy to 10 m. The R derived by KOZAI (R_K) was theoretically accurate to 0.5 m but in practice was not demonstrably accurate to 10 m. We did not know whether the positional inaccuracy obtained from R_K was due to an error in

1) the derivation of R_K,
2) the algebra required by developing and integrating Eq. (1), or
3) the programming of the results of (2).

The computations are exceedingly tedious though algebraically simple. The algebraic programming system discussed by GAPOSCHKIN [3] came to fruition early this year. The first large problem attempted was steps (2) and (3) above. The first of 6 pages of R_K is

$$\begin{aligned}
R = n'^2 m' r^2 \Big\{ & \Big(1 - \frac{3}{2} \sin^2 i\Big) [0.189 - 0.024 \cos N \\
& + 0.031 \cos M' - 0.002 \cos (M' + N) - 0.002 \cos (M' - N) \\
& + 0.003 \cos 2M' + 0.059 \cos 2\lambda' + 0.026 \cos (2\lambda' - N) \\
& + 0.003 \cos 2(\lambda' - N) + 0.011 \cos (2\lambda' + M') \\
& + 0.005 \cos (2\lambda' + M' - N)] \\
& + \sin 2i [0.136 \cos \Omega + 0.027 \cos (\Omega - N) - 0.004 \cos (\Omega + N) \\
& + 0.011 \cos (\Omega + M') + 0.011 \cos (\Omega - M') \\
& + 0.002 \cos (\Omega + M' - N) + 0.002 \cos (\Omega - M' - N) \\
& - 0.141 \cos (\Omega - 2\lambda') - 0.030 \cos (\Omega + N - 2\lambda') \\
& - 0.003 \cos (\Omega - N - 2\lambda') - 0.027 \cos (\Omega - 2\lambda' - M') \\
& - 0.006 \cos (\Omega + N - 2\lambda' - M') + 0.004 \cos (\Omega - 2\lambda' + M') \\
& - 0.004 \cos (\Omega - 2\lambda' - 2M') + 0.006 \cos (\Omega + 2\lambda') \\
& + 0.004 \cos (\Omega + 2\lambda' - N)]
\end{aligned}$$

$$+ \sin^2 i\,[0.029 \cos 2\,\Omega + 0.013 \cos (2\,\Omega - N)$$
$$+ 0.002 \cos (2\,\Omega + M') + 0.002 \cos (2\,\Omega - M')$$
$$+ 0.341 \cos 2\,(\Omega - \lambda') - 0.006 \cos (2\,\Omega - 2\,\lambda' - N)\}\,.$$

A SPASM program was written that read R and produced FORTRAN statements for the perturbations to the elements. The set of all FORTRAN statements produced by the algebraic processor was combined in a subroutine, albeit inefficient, and compiled. The compiled program was then used to evaluate the perturbations. Positional accuracy was improved. We need an improved gravity field to tell if R_K achieves the 0.5-m accuracy.

The computer-produced perturbations were compared to the Kozai-produced perturbations. We found one large error, some missing small terms, and some rounding errors in the coefficients in the Kozai perturbations.

3. Algebra on the Computer

The Smithsonian Package for Algebraic and Symbolic Manipulation (SPASM) [4] is a collection of FORTRAN-accessible programs for the CDC 6400 that symbolically manipulate expressions that often arise in celestial mechanics. Every operation described by Davis in his recent survey [5] is available in SPASM.

An admissible expression looks much like the right-hand side of the FORTRAN arithmetic statement. We have no restrictions on the number of variables or on the nesting of parenthesized expressions; given this generality, we try very hard to minimize the storage required. The usual operations of addition, subtraction, multiplication, division, differentiation, substitution, etc. are allowed. We can also rearrange arithmetic expressions.

Of more interest is that the output can be in the form of FORTRAN statements. Thus, the user can output programs. This is especially useful when the result is a 6-page expression for, say, a perturbation.

Of special interest to celestial mechanicians are provisions for manipulating linear trigonometric series (Poisson series). Such series can have arbitrary expressions as coefficients and arguments. Identities such as $\cos(-x) = \cos x$ and $\cos x \cos y = {}^1/_2\,[\cos(x+y) + \cos(x-y)]$ are applied automatically to ensure closure of Poisson series under the elementary operations.

Unique to the system, as far as we know, is a provision for efficient conversion of trigonometric polynomials to Poisson series [6]. Straightforward application of the above identities does not suffice.

SPASM is a symbol manipulator as well as an algebraic processor. We have already seen that the output of SPASM can be FORTRAN statements. When our FORTRAN compiler was first given the 150-card FORTRAN statement for δi, it responded that statements could be only 20 cards long. Because breaking up long FORTRAN statements is an error-prone procedure, we used SPASM to read in long FORTRAN statements, to isolate all references to trigonometric terms of the same argument, and to write out short FORTRAN statements.

When trying to produce programs by computer, we must interact with compilers that were written for people. The designer of our compiler never imagined that anyone would write a 150-card statement. To achieve this interaction, we must manipulate symbols as well as algebraic expressions.

We emphasize that SPASM does literal algebra rather than table-driven algebra. This is important because SPASM makes no assumptions about the number of variables. Any expression can be substituted for any variable.

Also, the freedom of representation allowed by literal algebra permitted us to transcribe R_K directly from Kozai's paper to punched cards without any rearrangement and to compare the computer output directly with Kozai's. Had the form of the output been less flexible, it would have been considerably more difficult to observe and correct the differences between the computer and the Kozai results.

The price of such generality is computer time and memory. We have not found the time excessive, but we need more memory on our 65K machine. We are extending SPASM to allow long Poisson series to reside on our 75-million-character disk.

With this extension, SPASM compares quite favorably with special-purpose table-driven programs, although it may be somewhat slower. This factor may be balanced by the more rapid obsolescence of special-purpose programs.

References

1. Gaposchkin, E. M.: Orbit Determination. Smithsonian Astrophys. Obs. Spec. Rep. No. 200, vol. 1 (1966) 185.
2. Kozai, Y.: Lunisolar Perturbations with Short Periods. Smithsonian Astrophys. Obs. Spec. Rep. No. 235 (1966).
3. Gaposchkin, E. M.: Space Research VIII, p. 76, Amsterdam: North-Holland 1968.
4. Hall, N. M., Cherniack, J. R.: SPASM User's Manual. Smithsonian Astrophys. Obs. Spec. Rep. No. 291 (1969).
5. Davis, M. S.: Astron. J. 73 (1968) 195.
6. Cherniack, J. R.: Converting Trigonometric Polynomials to Linear Trigonometric Series (to be published).

Perturbations du Second Ordre d'un Satellite Artificiel de la Terre

By

B. Morando, J. Chapront, P. Bretagnon

Bureau des Longitudes, Paris, France

Résumé. Un programme de traitement des séries trigonométriques littérales sur un calculateur a été écrit au Bureau des Longitudes. Ce programme utilisé principalement pour la théorie du mouvement de la Lune et des planètes est adaptable à la théorie du mouvement d'un satellite et, en substituants la théorie analytique du premier ordre en J_2 dans les seconds membres des équations différentielles donnent la solution du second ordre en J_2^2.

Ce travail sera publié dans «Astronomy and Astrophysics, a European Journal», édité par Springer-Verlag.

A Short Study on the Motion of Artificial Satellites of Mercury and Venus

By

E. Buchar

Technical University, Prague, ČSSR

Abstract. The author studies the motion of possible artificial satellites of the planets Mercury and Venus under the perturbing action of the Sun. All bodies are considered dimensionless, the orbital eccentricities of both planets are neglected. Approximative secular perturbations of node and pericentre are deduced, periodic perturbations are not investigated. The semi-major axis of critical orbits are determinated on the base of closed Hill's limiting curve. Variation, evection and eccentricity are taken into account here.

The critical semi-major axis of the Mercury satellite orbits are, respectively, 126000 km for the direct motion and 142000 km for the retrograde motion. The corresponding orbital periods are about 22 days and 26 days, respectively. The critical semi-major axis of the direct and retrograde Venus satellite orbits are 581 000 km or 655000 km, respectively. The corresponding orbital periods are about 57 days and 68 days. These values diminish with increasing eccentricity. It is convenient to note the greater stability of retrograde satellites.

The effect of solar radiation pressure is obviously more intensive than in the case of satellites of the Earth. The resonance condition for the secular perturbation in eccentricity occurs in the instability area.

This paper will be published in Bulletin of the Astronomical Institutes of Czechoslovakia.

Results on the Mass and the Gravitational Field of the Moon as Determined from Dynamics of Lunar Satellites

By

W. H. Michael, Jr., W. Th. Blackshear, and J. P. Gapcynski

NASA Langley Research Center, Langley Station, Hampton, VA, USA

Abstract. This paper presents a review of procedures and a summary of recent results on the mass and gravitational field of the moon. Gravitational field solutions as represented by spherical harmonic expansions of the potential through various degrees and order are available, and typical solutions obtained by the different procedures are presented. The relative and absolute precisions of the solutions in fitting the data are illustrated by residual plots and other considerations. Lunar surface elevation plots, based on the gravitational coefficients (assuming homogeneity), illustrate the distribution and magnitude of mass anomalies over the lunar sphere. For a 13th-order field, the contour plots show considerable correlation with published mascon results in regions where such results are available. The recently obtained gravitational field results are applied to determination of the moments of inertia of the moon, which correspond closely to overall homogeneous density distribution in the moon, and to determination of other properties related to mass distribution.

1. Introduction

The use of artificial satellites for lunar geodetic research is still a relatively new field of endeavor with a history, at least with respect to actual analyses of satellite data, of only about 3 years. Although data from the United States Ranger lunar impact probes provided independent dynamical estimates of the mass of the moon, lunar gravitational field analyses required the establishment of lunar satellites. The first artificial lunar satellite was the Soviet Luna 10, placed in lunar orbit in April 1966. This was soon followed by the United States Lunar Orbiter I, placed in orbit in August 1966, the first in a series of five successful lunar satellites established at 3-month intervals during 1966 and 1967. Orbital characteristics of these and other lunar satellites are shown in Table 1.

A specific objective of the analyses of lunar satellite dynamics is to determine a sufficient number of coefficients to define the lunar gravitational potential function as expressed in terms of spherical harmonics:

$$U = \frac{\mu}{r}\left[1 + \sum_{n=2}^{\infty}\sum_{m=0}^{n}\left(\frac{a}{r}\right)^{n} P_{n,m}(\sin\Phi)\,(C_{n,m}\cos m\lambda + S_{n,m}\sin m\lambda)\right] \quad (1)$$

Table 1. *Lunar Satellite Summary*

Configuration	Semi-major axis [km]	Eccentr.	Incl. [deg]	Period [min]	Lunar orbit injection date
Lunar Orbiter I	2670	0.327	12	206	8/14/66
Lunar Orbiter II	2702	.341	18	210	11/10/66
Lunar Orbiter III a	2688	.332	21	208	2/08/67
Lunar Orbiter III b	1968	.062	21	130	8/30/67
Lunar Orbiter IV a	6150	.654	85	721	5/8/67
Lunat Orbiter IV b	3751	.516	84	344	6/5/67
Lunar Orbiter V	2832	.317	85	225	8/5/67
Explorer XXXV	5984	.576	169	660	7/22/67
Apollo 8	1850	.002	168	119	12/24/68
Luna 10	2420	.137	72	178	4/3/66
Luna 11	2498	.241	10	187	8/28/66
Luna 12	2658	.309		205	10/25/66
Luna 14	2253	.158	42	160	4/10/68

where μ is the gravitational constant times the mass of the moon, a is the mean radius of the moon, r is radial distance from the lunar center of mass, $P_{n,m}$ are the associated Legendre polynomials, Φ is latitude (measured with respect to the lunar equator), and λ is longitude (measured positive eastward with zero longitude at the mean moon-earth line). Although this is not the only representation which could be used for the definition of the lunar gravitational field, it is the form recommended by the International Astronomical Union and is that which is being used by most investigators.

While considerable progress has been made toward the specific objective noted above, several complications exist in arriving at a precise definition of the lunar potential function. A major complication is the nature of the moon itself, in that it exhibits numerous localized mass anomalies near its surface and perhaps in its interior, and these anomalies are difficult to represent in a finite-series approximation. Another complication is that the variety of orbital parameters covered by the satellites established thus far is relatively limited. In addition, there is little direct tracking data coverage on the back side of the moon, which complicates the dynamical analyses.

Irrespective of the complications, the progress has been considerable, and some of the results given later in this paper are quite interesting. However, the results are not yet considered to be definitive and it is clear that additional research effort will be required for the precise definition of the lunar gravitational field.

2. Procedures for Lunar Gravitational Field Analyses

In common with most procedures involving analysis of tracking data, analyses of the components of the lunar gravitational field generally involve the differential correction technique, in which initial estimates of the parameters are iterated until the changes become acceptably small. The observational data used, the prodedures of trajectory integration, the calculation of the partial derivatives, and other procedures differ somewhat among various approaches.

Two general approaches to the determination of gravitational field constants arise from a distinction in the type of observational data used. In the "direct" approach the basic observables, usually Doppler or range tracking data, are used in a direct manner in that the partial derivatives of the observables with respect to each parameter to be determined are used in the formulation and in the differential correction process. In another approach, the "long-period and secular" approach, the "direct" procedure is used for relatively short data arcs to derive a set of mean orbital elements, and the long-period and secular variations in the mean elements are then used in a differential correction to derive the gravitational coefficients. In the determination of the gravitational field of the earth, the zonal harmonics are best determined by analysis of the long-period and secular effects, while the tesseral harmonics are best determined by use of short-period effects. For the moon, because of its relatively slow rotation about its axis, tesseral harmonics as well as zonal harmonics can be determined from analysis of long-period effects. Both types of coefficients can likewise be determined by the direct procedure, so an option is available. The "direct" procedure has some advantages in that it constitutes both a short-period and long-period approach to the problem; short-period because many observations are made during a single orbit of the spacecraft around the moon, and long-period because the procedure can be applied to data arcs of as long as a month or perhaps more. On the other hand, the "long-period" approach has advantages in that data arcs of one or several months are much more easily handled from a computational standpoint.

As pointed out by Michael and Tolson [1], both the direct and long-period and secular procedures have been formulated for use at the NASA Langley Research Center, although the more recent analyses have been conducted using the direct approach. The direct approach uses a 12th-order numerical technique for trajectory integration, numerical integration of perturbation equations for obtaining the partial derivatives, and the techniques of weighted least squares for the solutions for the parameters. Lorell [2] uses a long-period approach based on analysis of orbital elements averaged over about one-day arcs, and in which short-period effects are neglected in the trajectory integration, allowing rapid computations. He has analyzed arcs as long as 4 months with this procedure. KAULA [3] has suggested a quasi-analytic procedure which is essentially a long-period approach, in which short-period effects are averaged out analytically, allowing numerical integration of a nonlinear intermediate orbit with relatively long time steps. In his procedure, short-period perturbations required in forming the partial derivatives at the observation times are obtained analytically. Although details are lacking on his procedure, AKIM [4] has used a long-period technique covering a 2-month tracking period, apparently with analytical procedures for orbit integration.

For somewhat different purposes, MULLER and SJOGREN [5] have used a more direct approach than the "direct" procedure described above. They determined the Doppler residuals for half-orbit fits to actual tracking data using a simplified potential model, and then used a polynomial fit to the residuals to derive the line of sight component of the spacecraft accelerations. Very interesting correlations between the derived accelerations and lunar surface features have been obtained. This procedure has provided results which have been applied to obtain estimates of local mass anomalies on the near side of the moon. If the mass anomalies are expanded in terms of spherical harmonics, an estimate of the gravitational potential can be obtained. Although there are some limitations inherent in this procedure, it may contribute to the overall determination of the lunar potential. The latest results from this analysis are presented by MULLER and SJOGREN [6] at this COSPAR meeting.

3. Results on the Gravitational Field of the Moon

The results which have been obtained for the coefficients in the lunar gravitational potential to date are considerably affected both by the number of coefficients which are included in the approximations and by the extent of global coverage of the moon inherent in the data used for the analyses. Various authors have used different degrees of approximations to the infinite series expansion of the potential and no two solutions published so far have numerical values for the same set of coefficients. Solutions based on data from a single lunar satellite, or on data from satellites with similar orbital parameters, generally exhibit high correlations between various coefficients. These factors indicate that term by term comparison of various solutions is not particularly meaningful, and that any comparisons made should be based on the complete sets of coefficients. The degree of applicability of various potential models is best evaluated by testing the models on data arcs which have not been included in the solutions, particularly on arcs from different satellites, and by comparing long-period results with short-arc fits and vice versa.

3.1. Results from the Long-period Approach

Lunar gravitational field results by use of the long-period and secular approach have been obtained by AKIM [4], by BLACKSHEAR et al. [7], and by LORELL amd SJOGREN [8].

Akim's results, based on analysis of about 2 months of data from Luna 10 and consisting of a set of 11 coefficients, are shown in Table 2.

Table 2. *Long-period Results on the Lunar Gravitational Field*

		Luna 10 (AKIM)		Lunar Orbiter (LORELL and SJOGREN)	
n	m	$C_{n,m} \times 10^4$	$S_{n,m} \times 10^4$	$C_{n,m} \times 10^4$	$S_{n,m} \times 10^4$
2	0	−2.06	0	−2.026	0
	1	−0.157	−0.036	−0.088	0.015
	2	0.140	−0.014	0.129	0.131
3	0	−0.363	0	−0.222	0
	1	0.568	0.178	0.364	0.074
	2	0.118	−0.007	−0.026	−0.020
	3	—	—	−0.027	−0.050
4	0	0.333	0	0.094	0
	1	—	—	−0.124	0.056
	2	—	—	0.036	0.005
	3	—	—	0.164	−0.028
	4	—	—	0.009	0.008
5	0	—	—	−0.161	0
6	0	—	—	−0.109	0
7	0	—	—	0.173	0
8	0	—	—	−0.201	0

These results have been tested with data from Lunar Orbiter I, by using Akim's coefficients to predict variations in orbital elements and then comparing the predictions with the actual orbital elements derived from the Lunar Orbiter I tracking data. The comparisons using Akim's coefficients as presented were poor. Subsequent independent calculations of the lunar profiles presented by AKIM suggested that his zero reference for longitude differs by 180° from that given in Eq. (1), indicating that a change in sign should probably be made for those coefficients with odd values of m. The values for Akim's coefficients listed in Table 2 have the signs of coefficients with odd m changed from his published values. After making such changes, the comparisons of predicted and calculated variations of the elements for Lunar Orbiter I were considerably improved, but were still not particularly satisfactory, as may now be expected in consideration of the differences in orbital parameters of the satellites involved in the comparison. An additional example of these phenomena has been noted by comparing the results of BLACKSHEAR et al. [7], consisting of a fourth degree and order field (21 coefficients) derived from Lunar Orbiter I data, with subsequent Lunar Orbiter data, in which the comparisons are also not particularly satisfactory. These comparisons illustrate the point that solutions based on single lunar satellites do not represent a general solution to the problem.

LORELL and SJOGREN [8] have published results of a 25-coefficient solution based on a combination of data from Lunar Orbiters I through IV, also shown in Table 2. They note that their solution provides a resonably good fit to the variations in the orbital elements of these satellites over long data arcs, including one data arc of approximately 4 months. This is probably the best long-period model available to date. However, comparisons of their results for orbit predictions and orbit fits for relatively short arcs of a few orbits or a few days' duration, on data arcs not included in the solution, are again not wholly satisfactory, indicating that the solution is probably not of general applicability.

In an effort to derive a simple three- or four-coefficient potential model for the special purpose of rapid calculations for real-time operational orbit determination and prediction, RISDAL [9] has presented an emperically derived four-coefficient model, taking essentially nominal values for $C_{2,0}$ and $C_{2,2}$ and adding the coefficients $C_{3,0}$ and $C_{3,1}$. In this model: $C_{2,0} = -2.071 \times 10^{-4}$, $C_{2,2} = 0.207 \times 10^{-4}$, $C_{3,0} = -0.21 \times 10^{-4}$, and $C_{3,1} = 0.34 \times 10^{-4}$. This model is not as generally applicable as the solution by LORELL and SJOGREN, but does provide some improvement for orbit determination and orbit prediction over the nominal two-coefficient model ($C_{2,0}$ and $C_{2,2}$), available prior to lunar satellite solutions.

In view of the now recognized local mass anomalies in the moon, it is probably not to be expected that long-period results should provide entirely satisfactory solutions applicable for short arc fits and predictions. However, the long-period solutions should contribute to determination of global effects, such as the zonal harmonics, and should provide some information pertaining to the far side of the moon where little or no direct tracking data are available.

3.2. Results from the Direct Approach

Analyses using the direct approach for lunar gravitational field determination have been underway at the NASA Langley Research Center since 1966, commencing with receipt of tracking data from the first Lunar Orbiter spacecraft. Results based on Lunar Orbiter I data, consisting of coefficients through fourth degree and order (21 coefficients) were obtained by MICHAEL, TOLSON, and GAPCYNSKI [10]. Results for coefficients through fifth degree and order (32 coefficients), with data from Lunar Orbiters I, II, and III were also obtained by these authors [11], followed by another fifth degree and order solution, using additional data and including data from Lunar Orbiter IV by TOLSON and GAPCYNSKI [12]. A seventh degree and order (60 coefficients) solution, using some data from all five Lunar Orbiters, was obtained by GAPCYNSKI, BLACKSHEAR, and COMPTON [13].

The most recent solution is that by BLACKSHEAR [14], which is a 13th-degree and order (192 coefficients) solution, based on data from Orbiters III, IV, and V. The coefficients in this solution by BLACKSHEAR are given in Table 3. The solution is based on 15 separate data arcs, each of 2 to 4 days' duration, with a total of 12600 coherent two-day Doppler observations covering about 45 days of satellite tracking. In common with other analyses using Lunar Orbiter data, low-altitude data for this analysis is limited to latitudes between about ± 21° because of specification of spacecraft pericenter locations to obtain equatorial zone photography, and little direct radar data are available on the far side of the moon because of occultation by the moon. This solution is being extended to include additional data and additional coefficients.

The series of solutions noted above has been characterized by continual improvement in the quality of the fits to the Doppler tracking data as the representations of the gravitational field have been extended to higher degrees and orders. But even with the 13th-degree field solution,

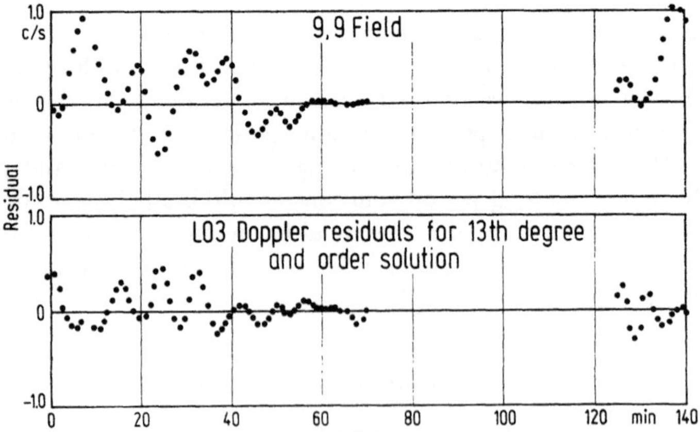

Fig. 1. LO3 Doppler residuals.

Table 3. *Coefficients of 13th Degree and Order Solution for Lunar Gravitational Field*

n	m	$C_{n,m}$	$S_{n,m}$
2	0	-2.0707×10^{-4}	—
	1	$-.4425 \times 10^{-6}$	-0.4573×10^{-5}
	2	$.2242 \times 10^{-4}$	$.2119 \times 10^{-6}$
3	0	$-.6303 \times 10^{-5}$	—
	1	$.2437 \times 10^{-4}$	$.2301 \times 10^{-5}$
	2	$.5016 \times 10^{-5}$	$.2031 \times 10^{-5}$
	3	$.1657 \times 10^{-5}$	$-.6798 \times 10^{-6}$
4	0	$.1938 \times 10^{-4}$	—
	1	$-.5732 \times 10^{-5}$	$.6802 \times 10^{-5}$
	2	$-.9628 \times 10^{-6}$	$-.1505 \times 10^{-5}$
	3	$.3332 \times 10^{-6}$	$-.5601 \times 10^{-6}$
	4	$-.1097 \times 10^{-6}$	$.1436 \times 10^{-6}$
5	0	$-.7459 \times 10^{-5}$	—
	1	$.1361 \times 10^{-5}$	$-.1211 \times 10^{-5}$
	2	$.1161 \times 10^{-5}$	$-.2239 \times 10^{-6}$
	3	$.3677 \times 10^{-7}$	$.3212 \times 10^{-6}$
	4	$-.4407 \times 10^{-7}$	$-.6055 \times 10^{-7}$
	5	$.1065 \times 10^{-7}$	$-.2249 \times 10^{-7}$
6	0	$-.1078 \times 10^{-5}$	—
	1	$-.3608 \times 10^{-6}$	$.2489 \times 10^{-5}$
	2	$-.8893 \times 10^{-6}$	$-.6553 \times 10^{-6}$
	3	$-.3013 \times 10^{-6}$	$-.1234 \times 10^{-6}$
	4	$-.9780 \times 10^{-8}$	$-.9199 \times 10^{-8}$
	5	$.1160 \times 10^{-7}$	$-.8338 \times 10^{-9}$
	6	$-.2027 \times 10^{-8}$	$.3899 \times 10^{-8}$
7	0	$.2408 \times 10^{-4}$	—
	1	$.7825 \times 10^{-5}$	$.8072 \times 10^{-5}$
	2	$-.2876 \times 10^{-7}$	$.4082 \times 10^{-6}$
	3	$.9552 \times 10^{-7}$	$.1925 \times 10^{-6}$
	4	$.3536 \times 10^{-7}$	$.2545 \times 10^{-7}$
	5	$.3427 \times 10^{-8}$	$.4139 \times 10^{-8}$
	6	$-.1046 \times 10^{-8}$	$-.3027 \times 10^{-9}$
	7	$.1202 \times 10^{-9}$	$-.3228 \times 10^{-9}$
8	0	$-.2655 \times 10^{-4}$	—
	1	$.3915 \times 10^{-5}$	$-.1929 \times 10^{-5}$
	2	$-.4236 \times 10^{-6}$	$.1021 \times 10^{-6}$
	3	$-.1278 \times 10^{-6}$	$-.8495 \times 10^{-7}$
	4	$.4497 \times 10^{-9}$	$-.2423 \times 10^{-7}$
	5	$-.3431 \times 10^{-8}$	$-.3146 \times 10^{-8}$
	6	$-.3267 \times 10^{-9}$	$-.8890 \times 10^{-9}$
	7	$.6019 \times 10^{-10}$	$.3593 \times 10^{-10}$
	8	$-.1700 \times 10^{-10}$	$.3003 \times 10^{-10}$
9	0	$-.1543 \times 10^{-5}$	—
	1	$.3113 \times 10^{-5}$	$-.5388 \times 10^{-5}$
	2	$-.3399 \times 10^{-7}$	$-.7719 \times 10^{-7}$
	3	$-.1032 \times 10^{-7}$	$.7082 \times 10^{-7}$
	4	$-.7594 \times 10^{-9}$	$.3100 \times 10^{-8}$
	5	$-.5896 \times 10^{-9}$	$-.6877 \times 10^{-10}$
	6	$.1733 \times 10^{-9}$	$.3254 \times 10^{-9}$
9	7	-0.1802×10^{-10}	0.1056×10^{-9}
	8	$-.2858 \times 10^{-11}$	$.7883 \times 10^{-13}$
	9	$.9618 \times 10^{-12}$	$-.1309 \times 10^{-11}$
10	0	$.5634 \times 10^{-4}$	—
	1	$.9866 \times 10^{-6}$	-0.4387×10^{-5}
	2	$.4507 \times 10^{-8}$	$.1562 \times 10^{-6}$
	3	$.5270 \times 10^{-7}$	$.6100 \times 10^{-8}$
	4	$-.1263 \times 10^{-8}$	$.3405 \times 10^{-9}$
	5	$-.3326 \times 10^{-9}$	$-.1906 \times 10^{-9}$
	6	$.9548 \times 10^{-10}$	$.6065 \times 10^{-10}$
	7	$-.1612 \times 10^{-10}$	$-.2332 \times 10^{-10}$
	8	$.1589 \times 10^{-11}$	$-.6560 \times 10^{-11}$
	9	$-.1970 \times 10^{-13}$	$-.2194 \times 10^{-12}$
	10	$-.3677 \times 10^{-13}$	$.4656 \times 10^{-13}$
11	0	$-.2460 \times 10^{-4}$	—
	1	$-.3018 \times 10^{-5}$	$-.1162 \times 10^{-4}$
	2	$-.4904 \times 10^{-6}$	$-.1440 \times 10^{-6}$
	3	$-.1665 \times 10^{-7}$	$-.5797 \times 10^{-7}$
	4	$-.4579 \times 10^{-8}$	$-.5297 \times 10^{-9}$
	5	$-.5080 \times 10^{-9}$	$-.4178 \times 10^{-9}$
	6	$.4348 \times 10^{-10}$	$-.1719 \times 10^{-10}$
	7	$-.3006 \times 10^{-11}$	$-.1071 \times 10^{-10}$
	8	$.1018 \times 10^{-11}$	$.8618 \times 10^{-12}$
	9	$-.1568 \times 10^{-12}$	$.3522 \times 10^{-12}$
	10	$-.8114 \times 10^{-14}$	$.1011 \times 10^{-13}$
	11	$.3365 \times 10^{-15}$	$-.1121 \times 10^{-14}$
12	0	$.3299 \times 10^{-4}$	—
	1	$-.7492 \times 10^{-5}$	$.1741 \times 10^{-5}$
	2	$.2196 \times 10^{-6}$	$.3018 \times 10^{-6}$
	3	$.1767 \times 10^{-7}$	$.1406 \times 10^{-7}$
	4	$-.3808 \times 10^{-9}$	$.3799 \times 10^{-8}$
	5	$.1613 \times 10^{-9}$	$.2592 \times 10^{-9}$
	6	$.4248 \times 10^{-10}$	$.3727 \times 10^{-10}$
	7	$-.1548 \times 10^{-11}$	$-.2311 \times 10^{-11}$
	8	$.1040 \times 10^{-12}$	$.4489 \times 10^{-12}$
	9	$-.7071 \times 10^{-13}$	$-.3553 \times 10^{-13}$
	10	$.4963 \times 10^{-14}$	$-.1136 \times 10^{-13}$
	11	$.7496 \times 10^{-16}$	$-.2479 \times 10^{-15}$
	12	$-.3627 \times 10^{-17}$	$-.1553 \times 10^{-17}$
13	0	$-.5772 \times 10^{-4}$	—
	1	$-.1212 \times 10^{-5}$	$-.5758 \times 10^{-5}$
	2	$-.2697 \times 10^{-6}$	$-.4811 \times 10^{-6}$
	3	$-.3097 \times 10^{-8}$	$-.6567 \times 10^{-7}$
	4	$-.6767 \times 10^{-9}$	$.2535 \times 10^{-10}$
	5	$-.1816 \times 10^{-9}$	$-.1769 \times 10^{-9}$
	6	$.4368 \times 10^{-11}$	$-.1534 \times 10^{-10}$
	7	$-.1489 \times 10^{-11}$	$-.2404 \times 10^{-11}$
	8	$-.2117 \times 10^{-13}$	$.5572 \times 10^{-13}$
	9	$.4683 \times 10^{-15}$	$-.1922 \times 10^{-13}$
	10	$.2138 \times 10^{-14}$	$.1472 \times 10^{-14}$
	11	$-.2018 \times 10^{-15}$	$.1719 \times 10^{-15}$
	12	$-.1876 \times 10^{-17}$	$-.8119 \times 10^{-17}$
	13	$.9554 \times 10^{-18}$	$-.1706 \times 10^{-18}$

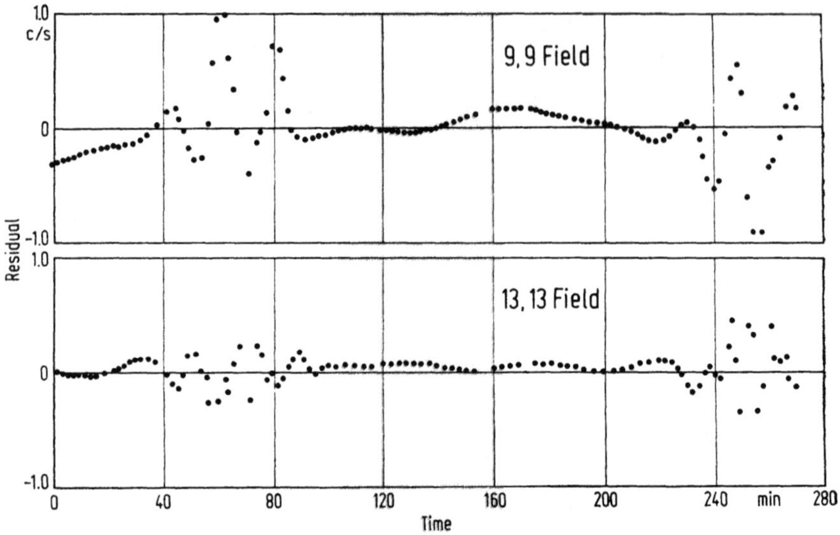

Fig. 2. LO5 Doppler residuals.

high-frequency oscillations in the residuals still exist, particularly near pericenter for high eccentricity orbits and generally throughout the orbit for low-altitude, nearly circular orbits. The residual plots shown in Figs. 1 and 2 illustrate the reduction in the magnitude of residuals between an intermediate 9th-degree field and the 13th-degree field of Table 3, and the characteristics of the residuals remaining after the 13th-degree field solution, for data from Lunar Orbiters III and V. The amplitude of the residuals at times where oscillations occur is well above the noise level of the data, indicating that additional information is contained in the data and that higher degree approximations are required for additional reductions in the residuals. Tests of the validity of this 13th-degree field for long-term predictions, using data arcs not included in the solution, show fair agreement with local one-day solutions for 1- or 2-week predictions, but show some secular deviations for longer predictions, indicating some lack of generality. The solution given in Table 3 represents considerable progress toward a general solution for the lunar gravitational field and it has some interesting applications, but it is clear that additional research is required for a more precise representation.

3.3. Some Comparisons of Results

Although term-by-term comparison of the higher degree coefficients is not particularly meaningful when the potential expansions extend to different degrees and orders, there is some justification for comparison

Table 4. *Comparison of Low-degree Gravitational Coefficients*

Coefficients ($\times 10^4$)	Akim	Lorell	Blackshear
$C_{2,0}$	−2.06	−2.026	−2.071
$C_{2,2}$	.140	.219	.224
$C_{2,1}$	−.157	−.088	−.004
$S_{2,1}$	−.036	.015	−.046
$S_{2,2}$	−.014	.131	.002
$C_{3,1}$	.568	.364	.244

of the low-degree coefficients, for which some previous information is available. A summary of second-degree coefficients and one third-degree coefficient is shown in Table 4. From studies of the physical librations of the moon, the expected value of $C_{2,0}$ is somewhat less than -2.0×10^{-4} and of $C_{2,2}$ is 0.205 to 0.231×10^{-4}, with approximately zero for the other second-degree coefficients. The values in Table 4 for $C_{2,0}$ are reasonable, and the more recent values for $C_{2,2}$ are within the expected range. The trend of the other second coefficients seems to be toward smaller values. The coefficient $C_{3,1}$ is one for which fairly large values have been obtained in most of the analyses, indicating some degree of consistency, and indicating that it should be considered in geophysical discussions.

4. Applications of Results

4.1. Moments of Inertia of the Moon

The second-degree coefficients in the gravitational potential function can be related to the moments and products of inertia of the moon, to

Table 5. *Relations between Gravitational Coefficients and Moments and Products of Inertia*

$A = I_{xx}$ about axis in eq. plane pointing toward Earth
$B = I_{yy}$ about $\perp$ axis in lunar eq. plane
$C = I_{zz}$ about lunar polar axis
$D = I_{yz}$ $\quad E = I_{xz}$ $\quad F = I_{xy}$

$$C_{2,1} = \frac{E}{Ma^2} \qquad S_{2,1} = \frac{D}{Ma^2} \qquad S_{2,2} = \frac{F}{2Ma^2}$$

$$C_{2,0} = \frac{1}{Ma^2}\left[\frac{A+B}{2} - C\right] \qquad C_{2,2} = \frac{1}{4Ma^2}(B - A)$$

give an indication of the overall mass distribution in the moon. Relations between coefficients and inertia parameters are shown in Table 5. With the five relations shown for the six parameters of interest, an additional

4*

relation is required. An additional relation is available from analyses of the physical librations of the moon, from which the quantity

$$\beta = \frac{C - A}{C}$$

is well determined. The value used here is that derived by KOZIEL [15] from heliometer observations, $\beta = 6.29 \times 10^{-4}$, which compares well with that derived by JEFFREYS [16] from the motion of the node and perigee of the orbit of the moon, $\beta = 6.279 \times 10^{-4}$.

Using the second-degree coefficients from Table 3, with $\beta = 6.29 \times 10^{-4}$, the following values are obtained for moments of inertia about the body axes of the moon defined in Table 5:

$$A = 0.400248\, Ma^2; \quad B = 0.400338\, Ma^2; \quad C = 0.400500\, Ma^2 .$$

The corresponding principal moments of inertia, as obtained from the eigenvalues of the inertia matrix, have the same values to the accuracy given. The principal axes are displaced less than two degrees from the body axes.

Results from previous analyses [12] and [17], have given values slightly less than 0.4, which is the moment of inertia for an homogeneous sphere. Therefore, the recent gravitational field analyses indicate that the moments of inertia of the moon correspond very closely to overall homogeneous density distribution in the moon.

4.2. Indications of Mass Anomalies

As shown by JEFFREYS [18], the spherical harmonic coefficients in the expansion for the gravitational potential can be simply related to the coefficients of surface harmonics, representing variations in the effective topography of the moon, if assumptions are made on the density distribution in the moon. With the assumption of uniform density distribution in the moon, the equation of the surface can be expressed as

$$a'(\Phi, \lambda) = a \left[1 + \sum_{n=2}^{\infty} \sum_{m=0}^{n} P_{n,m}(\sin \Phi) \frac{2n+1}{3} (C_{n,m} \cos m\lambda + S_{n,m} \sin m\lambda) \right].$$

The assumption of uniform density implies that positive contours represent regions of greater mass and negative contours regions of lesser mass than the average.

Early attempts were made to determine correlations between surface features and the topography plots, mostly based on the series of analyses at Langley Research Center. MICHAEL et al. [11] pointed out that there was a slight indication that the mare regions represent areas with mass exceeding the mean value and vice versa for the highland

Fig. 3. Lunar nearside surface contours derived from gravitational coefficients. Deviations in 0.5 km increments.

regions, based on a fifth-degree gravitational field, but these conclusions were very tentative. The first definite indication of local mass anomalies from contour plots was shown by GAPCYNSKI et al. [13] using preliminary results from an 11th-degree solution.

Plots of contours derived the 13th-degree field of Table 3 are shown in Figs. 3 and 4. Both positive and negative surface deviations are shown for the near side plot in Fig. 3, and a high degree of correlation between the positive contours and Mare Imbrium, Mare Serenitatis, and Mare Crisium is noted, in good agreement with acceleration results obtained by MULLER and SJOGREN [5] by a different procedure. Other positive contours occur in places where there are no maria. Preliminary estimates of the size and depth of point masses which produce effects comparable to the deviations shown in Fig. 3 are $40 \times 10^{-6} M$ at 180 km for Serenitatis, $33 \times 10^{-6} M$ at 183 km for Imbrium, and $14.5 \times 10^{-6} M$ at 128 km

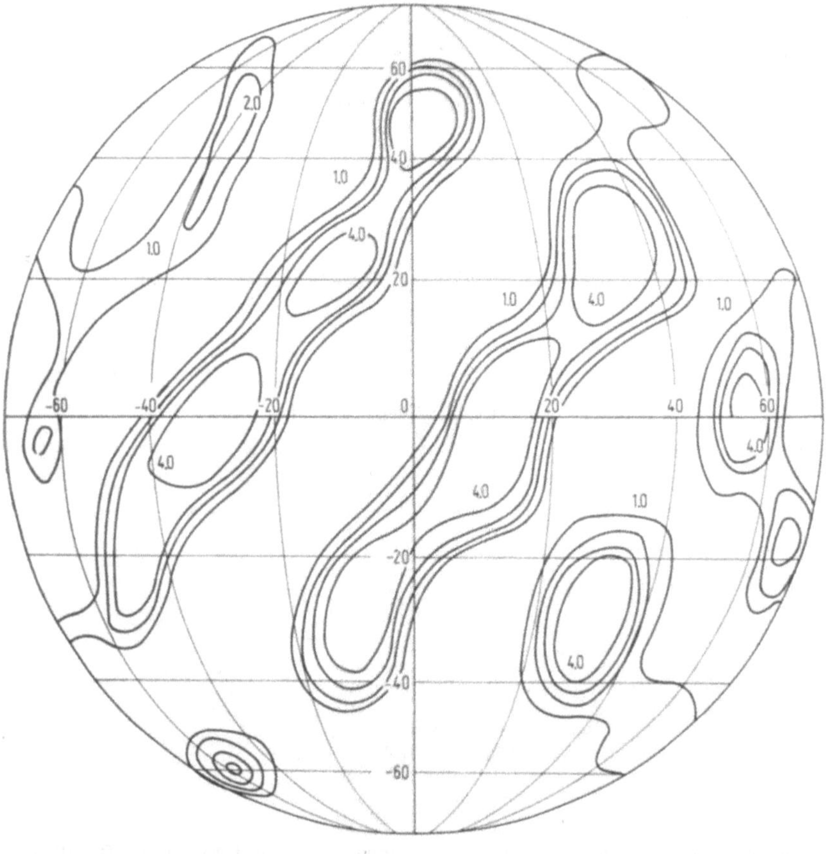

Fig. 4. Lunar farside surface contours derived from gravitational coefficients. Deviations in 1.0 km increments.

for Crisium. The positive contours on the far side of the moon, shown in Fig. 4, in conjunction with negative contours of comparable amplitude, which are not shown, indicate rather large undulations. Because of the lack of direct tracking data on the far side of the moon, the far side local mass anomalies are represented in the tracking data through their integrated effects, and are apparently not yet well determined. While the confidence in the results of Fig. 4 is rather low, it is expected that the addition of more data will improve the far side situation.

5. Results for the Lunar Gravitational Constant

One important utilization of the tracking data from the various space missions has been the attempt to define the astronomical constants to a higher degree of precision than previously obtainable. Of particular concern in this paper is the estimation of the lunar gravitational constant, GM (the product of the lunar mass and the universal gravitational constant). Estimates of GM may be obtained indirectly from planetary probe data by the determination of the lunar inequality and, in this respect, considerable effort has been devoted to the analysis of data from the Mariner and Pioneer missions [20]. For lunar missions, the estimates can be obtained directly, by observation of the effect of the central force term on the spacecraft motion. Results in this case have been obtained from analyses of the data from the Lunar Orbiter and Ranger series of spacecraft [12], [14], and [19].

Table 6. *Estimates of the Lunar Gravitational Constant*

Mission configuration	μ (km^3/sec^2)
Combined Rangers (VI, VII, VIII, IX)	4902.63 ± 0.07
Combined planetary probes (Mariner IV, V; Pioneer VII)	$4902.75 \pm .12$
Lunar Orbiters (I, II, IV) solution for 5, 5	$4902.64 \pm .11$
Lunar Orbiters (III, IV, V) solution for 13, 13	4902.87
Lunar Orbiters (III, V) combined solutions: 9, 9; 11, 11; 13, 13	$4902.73 \pm .14$

A comparison of the estimated values of GM from various space missions is given in Table 6. The value of 4902.87 km^3/sec^2 was obtained in conjunction with the gravitational field solution shown in Table 3. From a combination of Lunar Orbiter solutions, the value 4902.73 km^3/sec^2 has been obtained. This is considered to be the best current estimate, and is in very good agreement with the value of 4902.75 km^3/sec^2 obtained from a combination of analyses from deep space probes.

6. Concluding Remarks

Precise determination of the lunar gravitational potential requires a large number of coefficients in the series representation and combined analyses of tracking data from lunar orbiting spacecraft with a variety of orbital parameters. By means of various approaches in this active research area, considerable progress has been made. Recent results have stimulated discussions and applications to geophyical problems, but the results are not yet definitive and additional research and analyses are required.

References

1. Michael, W. H., Jr., Tolson, R. H.: The Lunar Orbiter Selenodesy Experiment; in: The Use of Artificial Satellites for Geodesy, Vol. II, ed. by G. Veis, National Technical University of Athens, Greece, 1967.

2. Lorell, J.: Lunar Gravity From Tracking Data. In Measure of the Moon, ed. by Z. Kopal and C. Goudas, New York: D. Reidel Co., 1967.
3. Kaula, W. M.: Analysis of Satellite Orbit Perturbations to Determine the Lunar Gravitational Field; in: Measure of the Moon, ed. by Z. Kopal and C. Goudas, New York: D. Reidel Co., 1967.
4. Akim, E. L.: Determination of the Gravitational Field of the Moon by the Motion of AMS Luna 10. Dokl. Akad. Nauk SSSR, Mekh. 170, 1966.
5. Muller, P. M., Sjogren, W. L.: Mascons: Lunar Mass Concentrations. Science 161 (1968) 3842.
6. Muller, P. M., Sjogren, W. L.: Lunar Gravimetrics. Paper presented at 12th Plenary Meeting of COSPAR, Prague, 1969.
7. Blackshear, W. T., Compton, H. R., Schiess, J. R.: Preliminary Results on the Lunar Gravitational Field From Analysis of Long-Period and Secular Effects on Lunar Orbiter I. Presented at NASA Seminar on Guidance Theory and Trajectory Analysis, Electronics Research Center, May 1967.
8. Lorell, J., Sjogren, W. L.: Lunar Gravity: Preliminary Estimates From Lunar Orbiter. Science 159 (1968) 3815.
9. Risdal, R. E.: Development of a Simple Lunar Model for Apollo. Contract Report D2-100819-1, The Boeing Company, Seattle; Washington 1968.
10. Michael, W. H., Jr., Tolson, R. H., Gapcynski, J. P.: Lunar Orbiter: Tracking Data Indicate Properties of Moon's Gravitational Field. Science 153 (1966) 3740.
11. Michael, W. H., Jr., Tolson, R. D., Gapcynski, J. P.: Preliminary Results on the Gravitational Field of the Moon From Analysis of Lunar Orbiter Tracking Data. Paper presented at the American Geophysical Union 48th Annual Meeting, Washington, D.C., April 17–20, 1967.
12. Tolson, R. H., Gapcynski, J. P.: An Analysis of the Lunar Gravitational Field as Obtained From Lunar Orbiter Tracking Data; in: Moon and Planets, Vol. II, ed. by A. Dollfus, Amsterdam: North Holland Publishing Co. 1968.
13. Gapcynski, J. P., Blackshear, W. T., Compton, H. R.: The Lunar Gravitational Field as Determined from the Tracking Data of the Lunar Orbiter Series of Spacecraft. Paper presented at the AIAA/AAS Astrodynamics Specialist Conference, Jackson, Wyoming, September 3–5, 1968.
14. Blackshear, W. T.: Progress on Lunar Gravitational Potential Determination by Analysis of Lunar Orbiter Tracking Data. Paper presented at the American Geophysical Union 50th Annual Meeting, Washington, D.C., April 1969.
15. Koziel, K.: Differences in the Moon's Moments of Inertia. Proc. Roy. Soc. Ser. A, 296 (1967) 1446.
16. Jeffreys, H.: On the Figure of the Moon. M. N. Roy. Astron. Soc. 122 (1961) 5.
17. Michael, W. H. Jr.,: Physical Properties of the Moon as Obtained From Lunar Orbiter Data. Presented at the Fourteenth General Assembly of the International Union of Geodesy and Geophysics, Lucerne, Switzerland, September 1967.
18. Jeffreys, H.: The Earth. Fourth ed., Cambridge University Press 1959.
19. Vegos, C. J., Trask, D. W.: Ranger Combined Analysis, Part II: Determination of the Masses of the Earth and Moon From Radio Tracking Data. NASA Space Programs Summary 37-44, Vol. III, Jet Propulsion Laboratory, Cal. Inst. of Techn., 1967.
20. Anderson, J. D., Hilt, D. E.: Improvement of Astronomical Constants and Ephemerides From Pioneer Radio Tracking Data. AAS Paper No. 68-130, presented at the AAS/AIAA Astrodynamics Specialist Conference, Jackson, Wyoming, September 3–5, 1968.

The Use of Brown's Lunar Theory in Lunar Satellite Perturbations by Sun and Earth

By

A. E. Roy

Department of Astronomy, Glasgow University, Glasgow, UK

Abstract. In order to compute the lunar potential's effect on the orbit of an artificial lunar satellite it is necessary to calculate and subtract perturbations by Earth and Sun to a high degree of accuracy. The present paper describes how the first- and second-order Earth-solar changes in the elements of the selenocentric Keplerian satellite orbit are obtained by making use of Brown's lunar theory. Values are given of the accuracy attained by comparing changes in the orbital elements of a typical lunar satellite (of semi-major axis 1.25 times the Moon's radius, eccentricity 0.17 and orbital inclination 30°) calculated from the first-order analytical theory and from numerical integration of the Gauss-Lagrange planetary equations. It is found for example that during two full orbits, the maximum errors in the semi-major axis and periselenium distance are of order 8 cm at most while angular accuracy is equally satisfactory.

1. Introduction

A number of workers have studied the behaviour of an artificial lunar satellite both analytically and numerically in recent years [1, 2, 3, 4] while the creation of such satellites has enabled an analysis of the lunar gravitational field to be made from lunar orbiter tracking data [5]. In these studies, perturbations of the lunar satellite orbit due to Earth and Sun have to be computed to an accuracy dependent upon the accuracy to which it is desired to determine the lunar gravitational field's harmonics.

The problem is complicated by the fact that a lunar satellite at a height of only 1700 km (a selenocentric distance of order 2 lunar radii) suffers Earth perturbations of the same order of magnitude as those arising from the Moon's oblateness. The Sun itself produces direct perturbations of amplitude some 1/200th that of the Earth. In addition, the Sun has an indirect effect through the changes it produces in the Earth-Moon system and indeed these indirect solar perturbations may be of larger amplitude than the direct ones.

Again, the choice of the reference plane is complicated by the rapid precession of the lunar equator's line of nodes on the plane of the ecliptic (period 18.61 years) and of course the equally rapid precession of the Moon's orbital plane. For a close lunar satellite a suitable reference plane may be that of the lunar equator but for satellites so distant that terrestrial perturbations are of the same order as, or greater than, lunar gravitational disturbances, the ecliptic is the more suitable choice. In the present author's work, the latter plane is chosen. This enables Brown's lunar theory to be used as an explicit description of the Earth-Moon system that includes all significant changes due to the Sun. The Sun's direct effect can be added later as a special and simpler case of the general Earth effect.

2. Construction of the Analytical Theory

The equations of motion of an artificial lunar satellite may be written as [6]

$$\ddot{x} = \frac{\partial U}{\partial x}, \quad \ddot{y} = \frac{\partial U}{\partial y}, \quad \ddot{z} = \frac{\partial U}{\partial z},$$

where the rectangular coordinate system Oxyz is non-rotating but has origin the Moon's centre of mass and

$$U = V + F,$$

where V is the Moon's gravitational potential and F is given by

$$F = -G\left\{m_E\left(\frac{x\,x_E + y\,y_E + z\,z_E}{r_{ME}^3} - \frac{1}{r_E}\right) + m_S\left(\frac{x\,x_S + y\,y_S + z\,z_S}{r_{MS}^3} - \frac{1}{r_S}\right)\right\},$$

the suffixes E and S distinguishing Earth and Sun while r_{ME}, r_{MS}, r_E, r_S are given by

$$r_E^2 = \sum_{xyz} (x_E - x)^2; \qquad r_S^2 = \sum_{xyz} (x_S - x)^2$$

$$r_{ME}^2 = \sum_{xyz} x_E^2; \qquad r_{MS}^2 = \sum_{xyz} x_S^2.$$

For first-order Earth and Sun perturbations, Lagrange planetary equations may then be employed, viz.

$$\dot{a} = A\frac{\partial F}{\partial M}; \quad \dot{e} = E_1\frac{\partial F}{\partial M} + E_2\frac{\partial G}{\partial \omega}; \quad \dot{i} = I_1\frac{\partial F}{\partial \omega} + I_2\frac{\partial F}{\partial \Omega},$$

$$\dot{\Omega} = -I_2\frac{\partial F}{\partial i}; \quad \dot{\omega} = -I_1\frac{\partial F}{\partial i} - E_2\frac{\partial F}{\partial e}; \quad \dot{M} = n - E_1\frac{\partial F}{\partial e} - A\frac{\partial f}{\partial a},$$

where

$$A = 2(a/\mu)^{1/2},$$

$$E_1 = (1-e^2)/e(\mu a)^{1/2}; \qquad E_2 = -(1-e^2)^{1/2}/e(\mu a)^{1/2},$$

$$I_1 = \cos i/[\mu a(1-e^2)]^{1/2}\sin i; \qquad I_2 = -1/[\mu a(1-e^2)]^{1/2}\sin i.$$

The quantities $a, e, i, \Omega, \omega, M$ are the usual elements of the Keplerian undisturbed selenocentric orbit of the satellite.

For first-order perturbations due to the Moon's departure from a point-mass, the same equations are used with U_1 replacing F, where

$$U_1 = V - G m_M / r .$$

To obtain second-order perturbations, cross-products of the first-order effects due to the different causes must be considered.

In this paper, we content ourselves with first-order effects though a second-order theory has been obtained by the present author [7].

3. The First-order Earth's Effect and the Sun's Indirect Effect

Taking the Earth's effect alone, the disturbing function F_E, given by

$$F_E = G m_E \left\{ \frac{1}{r_E} - \frac{x x_E + y y_E + z z_E}{r_{ME}^3} \right\}$$

may be written in the usual way as

$$F_E = \frac{G m_E}{r_{ME}} \left(\frac{r}{r_{ME}}\right)^2 \left\{ \left(\frac{3}{2}\cos^2 S - \frac{1}{2}\right) + \left(\frac{r}{r_{ME}}\right)\left(\frac{5}{2}\cos^3 S - \frac{3}{2}\cos S\right) + \cdots \right\}$$

where

$$\cos S = \frac{x x_E + y y_E + z z_E}{r r_{ME}} .$$

We also have

$$x_E / r_{ME} = \cos\lambda_E \cos\beta_E ,$$

$$y_E / r_{ME} = \sin\lambda_E \cos\beta_E ,$$

$$z_E / r_{ME} = \sin\beta_E ,$$

where r_{ME}, λ_E, β_E, the Earth's selenocentric distance, ecliptic longitude and latitude are related to the Moon's geocentric distance r_{EM}, ecliptic longitude λ_M and latitude β_M by the relations

$$\lambda_E = \lambda_M - 180^\circ, \quad \beta_E = -\beta_M, \quad r_{ME} = r_{EM} .$$

In addition, x/r, y/r, z/r are given by

$$x/r = \cos\Omega \cos u - \sin\Omega \sin u \cos i,$$

$$y/r = \sin\Omega \cos u + \cos\Omega \sin u \cos i,$$

$$z/r = \sin u \sin i,$$

where

$$u = \omega + f, \quad f \text{ being the true anomaly.}$$

Now Brown's lunar theory [8], gives expressions for the Moon's geocentric ecliptic longitude λ_M, latitude β_M, and parallax π_M, where

$$\lambda_M = L_M + \sum_i \sigma_i \sin s_i = L_M + \sigma,$$

$$\beta_M = \sum_i \beta_i \sin v_i,$$

L_M being the Moon's longitude. The σ_i and β_i are slowly varying quantities; the arguments s_i and v_i are linear functions of time with small secular and periodic additions. π_M is of a similar nature to λ_M.

The procedure then is to embody Brown's lunar theory in the expression for F_E, expanding to the desired accuracy. The first-order perturbations for the Earth's effect are then straightforwardly obtained by integrating the Lagrange planetary equations, keeping the elements constant on the right hand sides, but remembering that n, appearing alone in the $\dot{M}$ equation, is of zero order and must therefore have its first-order variation $\Delta_1 n$ inserted before integration from

$$\Delta_1 n = -3n \Delta_1 a/2a,$$

where $\Delta_1 n$ is the first-order variation in a.

It was decided to develop F_E to an accuracy such that observational data from a close lunar orbiter could be used to provide values of the principal moments of inertia of the Moon to 1 part in 20000 with values of the higher harmonics in the Moon's field being capable of being evaluated to a lesser degree of accuracy, depending upon their size and effects, secular or periodic, upon the satellite orbit. To this end, then, F_E was developed to an accuracy of 1 part in 1600, it being found that Legendre polynomials $P_3(\cos S)$ and higher could be neglected. Even if J_4 in the Moon's field, neglecting tesseral and sectorial harmonics, given in the usual notation by

$$U = \frac{G m_M}{r}\left\{1 - \sum_{n=2}^{\infty} J_n \left(\frac{R}{r}\right)^n P_n(\sin\varphi)\right\},$$

is as small as 10^{-6}, its effect will be large compared with that due to any neglected part of F_E for satellites as far out as $r = 1.33 \times$ radius of the Moon (height = 580 km) and of the same order as the neglected part of F_E for $r = 2R$ (height = 1738 km).

The development of F_E is laborious but straightforward. Fortunately, the expressions in Brown's lunar theory may be drastically curtailed without loss of desired accuracy. Functions of the nature of

$$\left(\frac{r}{a}\right)^n \cos m(f+\omega), \qquad \left(\frac{r}{a}\right)^n \sin m(f+\omega)$$

are expressed in the usual way as cosine or sine series, whose arguments are multiples of the mean anomaly and the argument of periselenium. The coefficients, functions of the eccentricity, are the well-known Hansen coefficients.

The final expression for F_E, with numerical values for the arguments arising from Brown's lunar theory, are given in Tables XII, XVI, XVII, XVIII and XIX of [6].

After integration of the Lagrange planetary equations, these tables, plus Table XXV of [7], provide the full first-order theory for the Earth's effect and the Sun's indirect effect through the changes it makes in the Earth-Moon system.

4. The First-order Theory of the Sun's Direct Effect

To the accuracy involved, the Sun may be taken to move in a circular orbit, period one year, about the Moon in the plane of the ecliptic. The expression for the Sun's disturbing function F_S is easily obtained then by putting $\beta_S = \sigma = 0$ in the expression for the Earth's disturbing function.

Table XXI [6] and Table XXVI [7] then give the required first-order theory of the Sun's direct effect.

5. Check on the Accuracy

My colleague, Mr. J. L. M. Ruhlmann, programmed the first-order theory for the Earth's effect in Algol for the University of Glasgow's KDF9 computer. He also programmed the Gauss-Lagrange differential equations of the elements in the same problem. Some idea of the accuracy of the first-order Earth perturbation theory, even without the second order theory, was obtained by considering Earth perturbations of a lunar satellite orbit of elements $a = 1.25 \times$ lunar radius ($a = 2172.5$ km), $e = 0.17$, $i = 30°$, $\Omega = \omega = M = 0$, the Moon being taken to be a point mass.

The Gauss-Lagrange equations were integrated over two orbital periods. Thirty times, well-scattered within this interval of time, were chosen and the analytical theory was used to compute the values of the osculating elements at those times. These were compared with the values given by the numerical integration.

The errors in the semi-major axis and eccentricity (first-order theory minus numerical integration) gave maximum errors in the semi-major axis and periselenium distance of 8 cm and order 10 cm respectively, there being evidence of a long-period second-order term in the eccentricity values. The inclination errors are less than 10^{-5} of a degree in size. The three elements Ω, ω and M had maximum errors of order 4×10^{-4} of a degree, there being evidence again of second order long-period or secular terms not yet included.

It is hoped that future work will enable the principal second-order terms to be taken into account.

References

1. Brumberg, V. A.: General Perturbations of the Elements of Artificial Lunar Satellites. Bull. Inst. Theor. Astron. Leningrad 8 (1962) 705.
2. Kozai, Y.: Motion of a Lunar Orbiter. Publ. Astron. Soc. Japan 15 (1963) 301.
3. Kaula, W. M.: Calculation of Perturbations of Lunar Orbiters. App. III in IMP D and E Feasibility study. Publ. X-672-64-4, NASA Goddard Space Flight Center, 1964.
4. Oesterwinter, C.: Motion of a lunar satellite. Astron. J. 71 (1966) 987.
5. Tolson, R. H., Gapcynski, J. P.: An Analysis of the Lunar Gravitational Field as obtained from Lunar Orbiter Tracking Data. Presented at the IQSY/COSPAR Assemblies, July 1967.
6. Roy, A. E.: The Theory of the Motion of an Artificial Lunar Satellite. I. Development of the Disturbing function. Icarus 9 (1968) 82.
7. Roy, A. E.: The Theory of the Motion of an Artificial Lunar Satellite. II. The First-order and Second-order Theories. Icarus 9 (1968) 133.
8. Brown, E. W.: Tables of the Motion of the Moon. New Haven: Yale Univ. Press 1919.

The Determination of Planetary Masses from Radio Tracking of Space Probes and Planetary Radar

By

W. G. Melbourne

Jet Propulsion Laboratory, Systems Analysis Research Section, Pasadena, CA, USA

Abstract. This paper reviews the progress in recent years in determining the masses of the terrestrial planets through radio tracking information from space probes and from planetary radar measurements. The history over the past decade of radio tracking information from the Mariner, Pioneer and lunar spacecraft, and planetary radar information is reviewed; the progress in these new technologies in terms of the resulting improvements in the accuracy of Doppler and time-delay measurements is emphasized. The use of these new data types to improve the planetary masses, other astronomical constants, and the planetary ephemerides, is described. New values for planetary masses and related constants based on weighted least squares results using information from radio tracking and radar measurements available over the periods 1962–1968 is presented. The internal consistency of these results is reviewed and a discussion of the estimated accuracy of these new determinations is given. The potential accuracy of these results is shown to be about an order of magnitude better than the presently quoted error bounds.

1. Introduction

It has been over seven years since the first U.S. lunar and planetary spacecraft were launched. During this period, 27 spacecrafts through precise Doppler tracking data, (and for recent flights, ranging also) have returned new values of unprecedented accuracy for various astronomical constants. Concurrently, the technology of radio tracking has advanced through better hardware, improved signal processing techniques, and more elaborate computer programs for data processing. The present coherent S-Band Doppler system has a potential measurement accuracy of about 1 part in 10^9 for a spacecraft travelling at planetary velocities. The current ranging system used on Mariners V–VII measures time-delay to 0.1 μsec, which for a spacecraft at 1 A.U. from the earth is equivalent to a range measurement accuracy of 1 part in 10^{10}.

In a related area, the past eight years since the first radar contact with Venus has also been a period of significant improvement in radar technology applied to the study of the surface properties and motions of the planets. It will be possible with the new 450 kW transmitter on the 210 foot antenna at Goldstone, California to measure the time-delay of radar signals bounced from the surface of Venus to an accuracy of 5 μsec over most of the orbit of Venus. These radar observations, particularly those since 1964, when used in conjunction with the meridian circle observations of the planets and the sun, have brought two orders of magnitude improvement in the accuracy of the ephemerides of the terrestrial planets and have allowed a determination of the light time equivalent of the astronomical unit to about 3 parts in 10^8.

This paper reviews the determinations of the mass of Mercury, Venus, Earth, Moon, Mars, and the astronomical unit. Our intent is to briefly describe the radio tracking systems and their inherent accuracy, to discuss the information content in these data types, to discuss the determination of the planetary masses and their theoretical standard deviations, and to assess the true accuracy of these results as limited by unmodelled effects in the tracking data.

2. Radio Tracking Systems

A. Coherent Counted Doppler

The Doppler radio tracking system employed in the JPL Deep Space Network [1] is a coherent loop that includes earth-based transmitting and receiving antennas plus a transponder on the spacecraft that coherently retransmits the received earth-based signal. The two-way coherent counted Doppler system is illustrated schematically in Fig. 1. Here, the *CW* signal f_T referenced to a stable rubidium vapor oscillator is transmitted to the spacecraft (nominally at an *S*-band frequency). The received signal is retransmitted by the transponder on the spacecraft by means of a phase-locked loop at an altered but coherent frequency and is received at the ground receiver as f_R. Thus, f_R has undergone a two-way Doppler shift and a delay of τ seconds relative to the time of transmission of f_T because of the round trip time. The received signal f_R is compared with the current transmitter frequency and the difference is called the Doppler tone f_D. Addition of a 1 MHz bias at this point results in positive numbers for the Doppler shift. The cumulative count of Doppler cycles are stored on paper and magnetic tape for later processing. In practice, the Doppler tone is continuously counted and this continuous count is sampled in a nondestructive manner at fixed time intervals. Typical sample rates are between 1 sec and 10 min. In effect then, the continuously counted Doppler tone (adjusted for the offsets and biases indicated in Fig. 1) is the two-way range difference between the ground-based antenna and the spacecraft that accumulates over a fixed count interval. Dividing by the count interval gives the average range rate (or, approximately twice the radial velocity $\dot{\varrho}$) of the spacecraft relative to the ground antenna.

A simplified first order expression for f_D has been shown in Fig. 1. In reality, the expression used is more complicated involving a Taylor's series expansion in powers of $1/c$ through the third degree and includes relativistic effects, as well. For planetary missions $\dot{\varrho}/c \sim 10^{-4}$, so all $1/c^2$

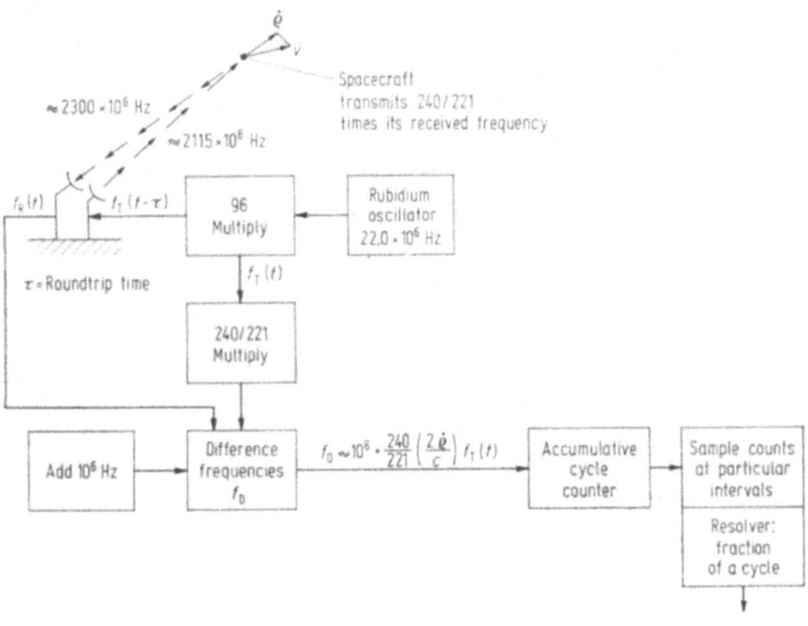

Fig. 1. Simplified coherent Doppler system.

terms are required in view of the accuracy levels presently attainable. With the advent of computers with large core capacity which make double precision computations practical, the computation for the counted Doppler data type can be performed as a range change calculation thereby avoiding the Taylor's series/quadratures approach.

Fundamental to the inherent accuracy of the Doppler system is the stability of the master oscillator. The current rubidium standards have a medium term (i.e. averaged over a count time greater than 100 sec) stability of better than 5 parts in 10^{12} and a short-term (e.g., 1 sec count time) stability of about 2 parts in 10^{11}. With respect to stability there is a simple heuristic analogy between these oscillators and the simple harmonic pendulum. One can think of the pendulum as being affected by two kinds of processes. The first is a kind of Brownian motion phenomenon in which the pendulum bob is bombarded by molecules causing minute but discontinuous changes in the phase of the pendulum's oscillation. Note that this process does not change the true frequency of the pendulum but it does affect the ability of an experimenter to measure the true frequency by averaging the number of cycles over a counting interval. The analog of this is "short-term" stability. The second process is one in which, for example, the length of the pendulum experiences a slow change. Here, the true frequency is affected and the analogy with this is "medium term" stability; it would be detected by an experimenter upon comparing results from longer averaging times where the *RMS* fluctuations due to phase jitter would be small.

It is instructive to derive an expression showing how oscillator instability introduces an error in the measured range change over the count interval. From Fig. 1 it follows that from the Doppler counter one obtains for range change $\Delta\varrho$, the expression

$$\Delta\varrho = \frac{c}{2 f_T} \int_{t_1}^{t_2} \frac{221}{240} [f_D - 10^6]\, dt \tag{1}$$

where t_1 and t_2 mark the beginning and end of the count time, respectively and c is the velocity of light in A.U./A.1 sec. Here, $\Delta\varrho$ is strictly the average of the change in the upleg and downleg distances, but for error analysis purposes, we may ignore this distinction and simply speak of range change. Again from Fig. 1, f_D is

$$f_D = \frac{240}{221} f_T(t) - f_R(t) + 10^6 \tag{2}$$

and

$$f_R(t) = \frac{240}{221} f_T(t-\tau) + \text{Doppler effect} \tag{3}$$

where τ is the round trip light time and $f_T(t-\tau)$ is the frequency transmitted by the ground antenna at a time τ earlier than the time t at which f_R was received. Thus, the error in the range change $\varepsilon_{\Delta\varrho}$ is given by the expression

$$\varepsilon_{\Delta\varrho} = \frac{c}{2 f_T} \int_{t_1}^{t_2} [f_T(t) - f_T(t-\tau)]\, dt\,. \tag{4}$$

It is the medium term oscillator stability with which we are concerned in this paper and $t_2 - t_1$ is typically equivalent to the duration of a pass, namely, tracking from horizon to horizon, or nearly so. In this case, Eq. (4) becomes

$$\varepsilon_{\Delta\varrho} = \frac{c}{2 f_T} \left[\int_{t_2-\tau}^{t_2} f_T(t)\, dt - \int_{t_1-\tau}^{t_1} f_T(t)\, dt \right] \tag{5}$$

which assuming $f_T(t)$ is slowly varying may be written as

$$\varepsilon_{\Delta \varrho} = \frac{c_\tau}{2}\left[\frac{f_T(t_2) - f_T(t_1)}{f_T}\right] = \varrho\left(\frac{\Delta f_T}{f_T}\right). \tag{6}$$

Thus, we see that the error in range change is directly proportional to the product of the range and the oscillator stability. Fig. 2 shows the history of observed oscillator medium-term stability in terms of $\Delta f_T/f_T$

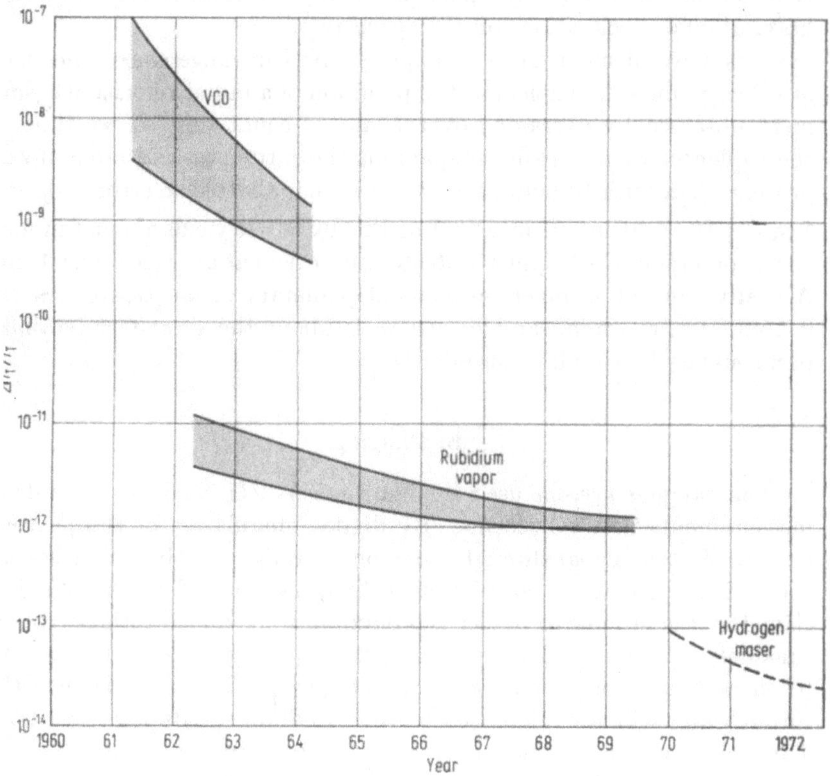

Fig. 2. Observed DSIF Frequency system performance medium-term oscillator stability.

over the past 7 years for the voltage controlled oscillators (VCO) used on earlier flights, the rubidium standards, and a conservative estimate of the performance of the hydrogen maser to be used in the near future. Setting ϱ equal to 1 A.U. in Eq. (6), one sees that for a spacecraft at planetary distances, the present rubidium standards allow a range change measurement precision of 10^{-11} A.U. When this is divided by the distance which a planetary spacecraft typically moves in a pass, i.e., 10^{-2} A.U. per pass, one obtains the measurement precision of 10^{-9} quoted earlier for the Doppler system.

A precision for $\Delta \varrho$ of 10^{-11} is equivalent to about 1 m. If we are to fully exploit this precision, we must take into account all elements affecting the change in the position of the ground antenna and the spacecraft in inertial space over a pass that are of 1 m in magnitude or larger, and all signal propagation effects of this size. Indeed, we shall see that the limitation in our navigational system has not been oscillator instability, but rather, factors such as ephemerides, tracking station locations on the earth, UT.1 and polar motion, propagation effects, and non-gravitational accelerations on the spacecraft.

It is convenient at times to express this 1 m range change precision in velocity units, in particular, the precision of a radial velocity measurement obtained by averaging over 1 min. of count time. If we think of the sequence of 1 min. samples spanning the entire pass as a random walk process, then for a 10-hour pass, each 1 min. is allotted an error budged of 1 m $\div \sqrt{10 \times 60}$ or 0.04 m. Dividing this by 60 sec yields about 1 mm/sec as the precision of a Doppler velocity measurement averaged over 1 min. Actually, current Doppler residuals at planetary distances are closer to 0.5 mm/sec for 1 min. samples, which is about the short-term stability performance for rubidium standards.

B. Ranging

The ranging system used on Mariners V–VII produces time-delay measurements in UTC seconds. By digital modulation of the phase of the transmitted signal through a composite range code involving several pseudonoise sequences, correlation techniques are used on the signal received at the ground to detect the maximum likelihood estimate of the time-delay [2 and 3]. The system upon acquiring the code is then able to operate in a continuous range tracking mode. The delay is measured through an oscillator referenced to the rubidium standard and to UTC. Thus, clock errors in counting round trip time are less than 10 ns for a spacecraft at 1 A.U. The overall performance of the system is in the range of 30–100 ns which for a spacecraft at 1 A.U. distances yields a measurement accuracy of time-delay to about 1 part in 10^{10}.

The radar systems will not be discussed here since considerable literature already exists. (See for example, [4] and [5].)

3. The Status of Error Sources Affecting Tracking Accuracy

Straightforward covariance analyses using the radio tracking data available during the encounter phase show that the gravitational mass value of the principal body may be determined to a precision of the order of 1 part per million (1 ppm) from each flight. To attain planetary mass determinations at an accuracy level of 1 ppm, it is essential that all error sources in the dynamics and in the observations be limited to a level at which their corrupting effect on the determined mass value is no greater than 1 ppm. Here, we briefly review these error sources and assess their effect on the mass determinations.

The various error sources can be grouped into five categories, namely, those error sources that, 1. affect the position of the tracking station, 2. affect the trajectory of the spacecraft, 3. result from propagation effects, 4. result from signal processing and, 5. result from computational deficiencies.

A. Station Positions

Station positions are affected by UT.1, polar motion, precession, nutation, possible crustal motion, tidal deformation, antenna motion, the ephemerides of the earth and the moon, and of course, the assigned values of longitude, and distance from the spin axis. Notice that to the first order, errors in station locations parallel to the spin axis do not affect Doppler tracking data for a distant object. This can readily be verified from an inspection of Fig. 3 where P is a distant spacecraft and 0 is

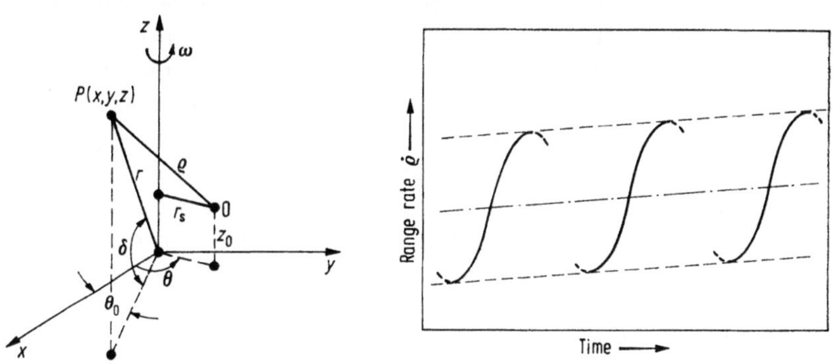

Fig. 3 and 4. Doppler distant-probe.

the topocentric position of the antenna in an equatorial coordinate system with the center of the earth at the origin. A typical pass of two-way Doppler data involves a continuous count of the Doppler frequency from near rise to near set. During this interval the Doppler frequency reflects the motion of the observer about the pole of the earth, and some small projection and cross terms resulting from the observer not being at the center of the earth. The radial velocity $\dot{\varrho}$ of the probe relative to 0 may be expressed as [6]

$$\dot{\varrho} = \dot{r} - \frac{\bar{r}\cdot(\bar{\omega}\times\bar{r}_s)}{r} + 0\left[\frac{r_s}{r},\ \frac{z_0}{r}\right] \tag{7}$$

where ω is the angular frequency of the rotation of the earth and r_s is the perpendicular distance of 0 from the spin axis. For a distant spacecraft, $(r_s/r, z_0/r \ll 1)$, it follows from Fig. 3 and Eq. (7) upon ignoring the small terms that $\dot{\varrho}$ may be expressed as

$$\dot{\varrho} = \dot{r} + \omega r_s \cos\delta \sin(\theta - \theta_0) \tag{8}$$

where δ is the declination of the spacecraft and θ_0 is its right ascension. Expressing θ in terms of universal time t, we have

$$\dot{\varrho} = \dot{r} + \omega r_s \cos\delta \sin\omega(t - t_0)\,. \tag{9}$$

In the actual orbit determination process no such approximations are made; however, the neglected terms here are of the order of 10^{-4}. The observables in Eq. (9) are $\dot{\varrho}$ and the universal time. Note that t_0 corresponds to the time of crossing of the local meridian by the spacecraft, so if t_0 can be determined the right ascension of the spacecraft may be directly inferred from the relation between universal time and sidereal time. Similarly, the amplitude of the diurnal sinusoid in Fig. 4 is directly proportional to $\cos\delta$ so information about $\cos\delta$ is also contained in the diurnal signature (See [6] and [7]). Experience has shown that the Doppler data for a single pass can be represented to 0.5 mm/sec accuracy by as few as 11 coefficients [8] in a series expansion that is basically a continuation of Eq. (7). Eqs. (8) and (9) show that errors in tracking station locations and in the measured value of universal time will affect the Doppler tracking data and cause errors in the values of the right ascension and declination of the spacecraft inferred from the orbid determination process. For example, a 1 meter error in r_s causes an orbit determination error of about 25 csc δ km in the declination direction of the spacecraft at 1 A.U. distance after one pass. The orbit determination

Table 1. *Station Position Error Sources*

Error Source	1-sigma Level in 1969	1-sigma Level Expected in 1972
Tracking Station Locations with Respect to the Crust of the Earth:		
r_s	$1.5\,M$	$0.5\,M$
λ	$3.0\,M$	$1.0\,M$
Pole Motion: Coordinates of Spin Axis with Respect to the Crust of the Earth:		
$\sigma_x = \sigma_y$	$1.5\,M$	$0.7\,M$
Timing (UT.1 – A.1)	5 msec	2.5 msec
Precession, P (arc-sec/century)	0.3″	0.2″

error due to station location or timing errors will be mitigated by combining many days of tracking because of the motion of the spacecraft and the earth about the sun. On the other hand, random accelerations of the spacecraft cause "information loss" in the tracking data and effectively limit the level to which station and timing error effects can be reduced from long arc orbit determination analysis.

Table 1 shows the level of station position errors at this time and their anticipated level in the future. Because of their adverse navigational effects JPL has had a vigorous program [9] of station location improvement resulting in a factor of 10 improvement in the past few years. The polar motion error level is based on IPSM and BIH information; the UT.1 error level is based on information from the USNO. Timing errors within the DSIF are now negligible.

The precession constant affects the position of the tracking station relative to 1950.0 equatorials coordinates. It is well established (for example, cf. [10]) that the adopted precession constant is too small by about 1″/century, which is equivalent to about 0.3 m/year shift in station locations. An absolute shift in station location that accumulates from 1950.0 to some epoch is not serious because it may be absorbed in station location definitions. It is the relative shift over the few years during which there has been accurate tracking that causes difficulty, for example, when one tries to reconcile station locations determined from Mariner II with those from Mariner V.

To summarize, a pass of Doppler data through a continuum of range rate measurements, provides strong determinations of geocentric radial velocity, right ascension and $\cos\delta$ at meridian crossing to a high degree of precision. A 1 m error in station location, pole position, or an equivalent timing error, produces a diurnal sinusoid signature on Doppler data of about 0.1 mm/sec. It can be shown that present station location, timing, and polar motion errors are not significant for planetary mass determination at 1 part in 10^6 accuracies.

B. Spacecraft Trajectory

The trajectory of the spacecraft is influenced by the ephemerides, the masses and gravity harmonics of the gravitational bodies, non-gravitational forces such as maneuvers, radiation pressure and low-thrust forces resulting from unbalanced forces from attitude control jets, and outgassing related to other space science experiments onboard the spacecraft. A related effect is the rotation of the spacecraft resulting in a motion of its antenna about the center of mass.

Table 2 shows the non-gravitational acceleration levels for past planetary spacecraft. There is significant evidence that radiation pressure

Table 2. *Non-gravitational Forces*

Spacecraft	Radiation Pressure Accelerations		Low-thrust Accelerations	Ref.
	Level at 1 A.U.	Percentage Uncertainty	Level	
Mariner II	1×10^{-10} km/sec^2	5%	5×10^{-11} km/sec^2	[10]
Mariner IV	2×10^{-10} km/sec^2	2%	1 to 3×10^{-12} km/sec^2	[11]
Mariner V	2×10^{-10} km/sec^2	1%	$<1 \times 10^{-12}$ km/sec^2	[12]

acceleration (adjusted for the r^{-2} effect) varies during the flight [11–13] by 5–10%, presumably due to a change in the reflectivity of the spacecraft or, occasionally, due to a change in effective area of the spacecraft. The solar flux may also vary but by not more than 1%. Table 2 shows that the low-thrust force level has been reduced considerably since Mariner II. Some of these effects may be partially modelled through telemetered attitude control information, although it is a difficult task [11, 12, 14, 15]. As we shall see, the uncertainty in either of these two effects produces a negligible error in the planetary mass solutions derived from encounter data. Outgassing from other sources aboard the spacecraft has not existed prior to Mariner VI. However, there is an infrared spectrometer aboard Mariners VI and VII which uses gas release to cool the equipment over a period of a few hours near Mars encounter.

Table 3 summarized the accuracy levels of the ephemerides of Venus, Earth, Mars, and the Moon. DE 19 refers to the use of JPL of ephemeris tapes produced from a least-squares fit to source theories by numerical integration [16, 17]. The source theories were Newcomb's theory of the Sun, Mercury, and Venus, and Clemence's second order theory for Mars [18]. The reason for performing least-squares numerical integrations was to obtain smoother velocities than the truncation errors on the theories would allow through numerical differentiation. However, the velocity errors in these early ephemeris tapes were fairly severe; as we shall see an uncompensated radial velocity error of 100 mm/sec of a planet and the related acceleration error during an encounter phase can produce a significant error in the mass determination. The error estimates for these ephemerides are based on corrections that have been determined from planetary radar data [19, 20]. They are also consistent with the estimates given by Kulikov [21].

The ephemeris alluded to as "DE 69" is a currently operational ephemeris at JPL based on gravitationally consistent least squares fit to over 34000 U.S. Naval Observatory meridian circle observations spanning 1912 to 1968, about 800 time-delay planetary radar points from 1964–1968 and 200 Mariner V range points taken during the 10-

Table 3. *Ephemeris Error Bounds*

Planet	DE 19	DE 69 (Currently Used)	
	Heliocentric 1950.0 Equatorial Coordinates		Relatives Coordinate.
Venus			
Position	3×10^{-6} A.U.	1×10^{-6} A.U.	3×10^{-8} A.U.
Velocity	100 mm/sec		0.5 mm/sec
Earth-Moon Barycenter			
Position	3×10^{-6} A.U.	1×10^{-6} A.U.	3×10^{-8} A.U.
Velocity	70 mm/sec		0.3 mm/sec
Mars			
Position	2×10^{-6} A.U.	1×10^{-6} A.U.	2×10^{-7} A.U.
Velocity	50 mm/sec		3 mm/sec
Moon	(ILE)	LE 16 (Integrated Ephemeris)	
Position	1 km	0.1 km	
Velocity	1 mm/sec	0.1 mm/sec	

day encounter interval with Venus,[20]. The radar data have been collected from observations made at the Arecibo Ionospheric Observatory, JPL, and the Lincoln Laboratory of MIT.

It is important to identify two kinds of coordinate systems for these radar ephemerides. Because the radar time-delay data are 1–2 orders of magnitude less accurate than the radio tracking data and because there is rarely more than one normal point of radar time-delay provided per observation day, the right ascension and declination information described earlier is not present in significant strength. Furthermore, the effect of station location errors is greatly attenuated because of the multiple revolutions of the planets during eight years of radar observations. For example, using the actual data set available and shifting λ or r_s of the radar antenna by 25 m, produces only 100 m shift of Venus in the longitudinal direction. In effect then, for error analyses purposes, the radar observations may be considered as geocentric measurements. Since radar measurements are scalars, they do not sense (exept in a second order sense due to the weak coupling with the spin axis of the earth noted above) directions relative to an inertial coordinate system such as the 1950.0 equatorial system. This means that there are at least three angular quantities (which for convenience, we may think of as the mean longitude of the sun, and two quantities defining the orientation of the ecliptic plane) that are only extremely weakly determined by radar. Their solution strength in a combined meridian circle radar data fit depends on the meridian circle observations and, in turn, on the interpretation of the errors in the star catalogs, precession, the definition of the epoch for the longitude of the sun, etc. An uncertainty of 10^{-6} A.U. has been assigned to the planetary coordinates in this system corresponding to about 0.2″ in angular measure, but it is important to remember that this 10^{-6} A.U. error reflects an orientation uncertainty in the 1950.0 equatorial coordinate system, and that any correction of this type applied to one of the planets must be applied to all of them. The "relative" coordinates given in the fourth column of Table 3 are free from this deficiency, and it is the error level in this coordinate system that is important for the accurate determination of planetary masses from encounter data.

The accuracy figures quoted for Venus, Earth and Mars are 1–2 orders of magnitude improved relative to earlier ephemerides. The following simple exercise is informative. Suppose one wishes to calculate the accuracy with which the relative longitude λ_0 between Venus and the earth in the simple circular coplanar equation

$$\tau = \frac{2}{c}\left[1 + a^2 - 2a\cos(n\tau + \lambda_0)\right]^{1/2} \tag{10}$$

may be determined when N radar points are given with observational errors that are uncorrelated, gaussian with mean zero and a standard deviation of σ_τ; the data are equally distributed between greatest elongations. Here, τ is time-delay, a is the semi-major axis, n is the relative mean motion, and $2/c$ has the value of 10^9 μsec. For $N = 100$ and $\sigma_\tau = 10$ μsec the reader may verify that λ_0 may be determined to about 0.0002″ or equivalently, about 1×10^{-9} A.U. along the orbit of Venus. Of course, when one adds the orbital elements of Venus and the Earth, as well as the radius of Venus and the A.U., one would not expect quite as good a result. Nevertheless, the covariance matrix from combined optical and radar fits involving 65 parameters still gives a standard deviation of about 0.001″ for relative longitude and about the same for all other "inplane" elements. The two "out-of-plane" elements describing the orientation of the Venus orbit planet are determined less accurately, but only by about a factor of 2. Of course, formal standard deviations are seldom to be trusted, so a multiplication factor between 5 and 10 has been applied to obtain the accuracy levels quoted in Column 4. Radar ephemerides have been extended for prediction purposes a year beyond the last data point used in the solution. For Venus these radar predictions have consistently been accurate to within 20–30 μsec. Mariner V during the Venus encounter allowed a determination of the geocentric radial distance and velocity of Venus to about 100 m and 0.2 mm/sec, respectively. The corrections to the radar ephemeris in use at that time were $+2.8 \times 10^{-8}$ A.U. and 0.3 mm/sec, [22]. The MIT group [19 and 23] using radar data from MIT and AIO have produced ephemerides with similar reliability in prediction, and which for Venus consistently agree with JPL ephemerides to within the figures given in Column 4. It is important to note that the velocity errors in these new ephemerides have now been reduced to a very low level.

For the moon the abbreviation ILE stands for the Improved Brown Lunar Theory [24] which was evaluated as reported in [17]. The most recent lunar ephemeris at JPL, LE 16, [25] is a numerically integrated least squares fit to a source theory consisting of the Improved Brown Lunar Theory adjusted by the Eckert corrections [26, 27] and by the removal of the aberration term [28]. The constants used in this ephemeris are consistent with the IAU definitions.

It can be shown for the trajectory profiles characteristic of the earth espace, lunar approach, and planetary encounter that uncertainties in the gravity harmonics of these bodies do not hamper mass determinations at 1 ppm accuracy levels.

C. Propagation Effects

The radio signal is affected by charged particles both in the solar system and in the ionosphere of the earth, by the troposphere, by the atmosphere and ionosphere of other planets during an occultation phase, and by gravity effects predicted by general relativity. At S-band frequency, dispersive effects due to electrons in the ionosphere and at times, in the solar system can be significant. Fig. 5 shows the typical ionospheric diurnal effect at S-band in meters for an integration along

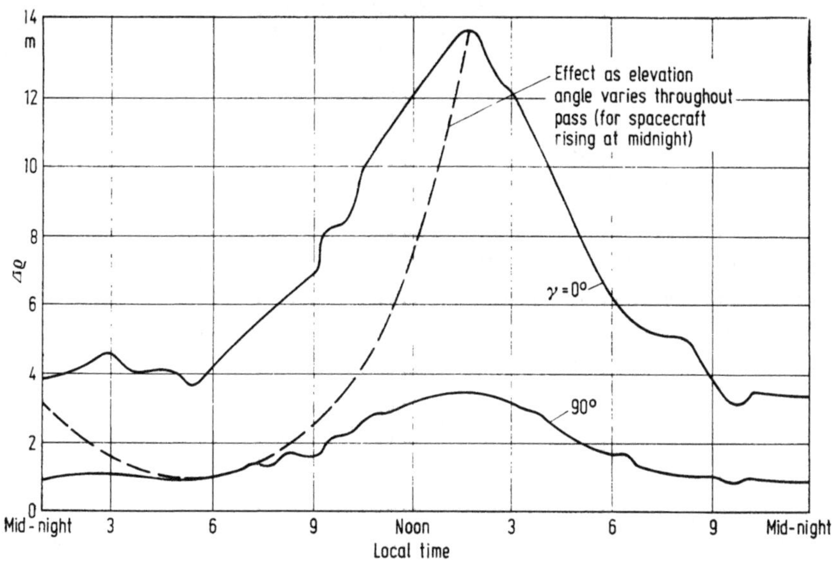

Fig. 5. Ionosphere.

the vertical direction ($\gamma = 90°$) and along the horizontal directions ($\gamma = 0$). Although the ionospheric signature on the tracking data depends on the angle between the sun and the spacecraft, a "worst case" signature is indicated by the dashed curve which corresponds to a spacecraft rising above the horizon at midnight and setting at noon. Note that only the change in integrated electron content along the ray path affects the Doppler data; the absolute content changes very slightly the ground and spacecraft transmission epochs. For the case shown, the one-way Doppler signature reaches a maximum of about 3 mm/sec or a total range change of over 10 m for the pass. Fig. 6 shows a plot of solar

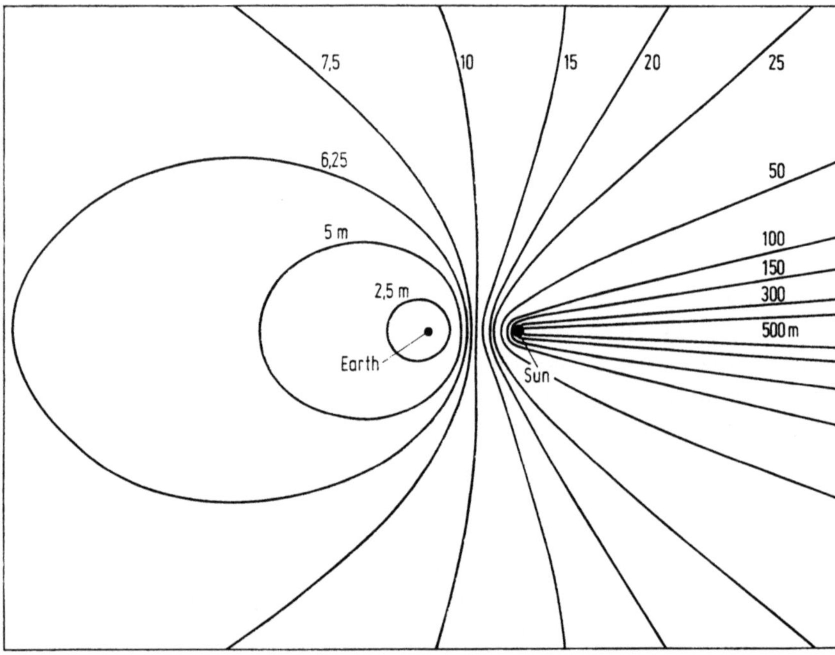

Fig. 6. Contours of geocentric range error at *S*-band.

system electron effects at *S*-Band for an observer on the earth for a steady-state electron distribution. Here, an inverse square solar distance law has been assumed for the number of electrons per c.c., $N_e(r)$, normalized at 6 electrons/c.c. at $r = 1$ A.U. The expression

$$\Delta \varrho_{\text{meters}} = \frac{40.3}{f_T^2} \int_0^{\varrho} N_e(r)\, d\varrho \tag{11}$$

has been used in Fig. 6 to calculate the electron one-way group delay effect. Here, N_e is electrons/M³ and ϱ is in meters. The rate of phase advance on the Doppler is obtained from Eq. (11) by differentiating with respect to time and changing the sign. Note that the Doppler effect due to solar system electrons is small (i.e., $\ll 1$ mm/sec for Venus and Mars encounters) except near the sun. From the Pioneer series, on rate occasions rapid fluctuations in integrated electron content have been observed amounting to as much as 150 m over a period of 4 h, probably due to a wedge effect from a body of electrons associated with a flare or other solar disturbance [28a]. However, the signature of this effect is readily discernible in the residuals being a high frequency phenomenon; it occurs infrequently and consequently, is not considered to be a problen for 1 ppm mass determinations. It is possible to calibrate the ionospheric effects to within about 20% of their total effect through ionosonde and Faraday rotation information. Comparison of the difference between range measurements made at two epochs with the counted Doppler measurement taken over the same time interval enables one to calculate the change in electron content along the ray path. Later flights will probably have dual-frequency (S and X-band) carriers for which electron effects will be eliminated and, in fact, these devices will be uesd to study the distribution of electrons.

D. Signal Processing

Signal processing effects include oscillator stability, electrical path length variations through the equipment at the tracking site and onboard the spacecraft, time synchronization throughout the tracking network and synchronization with UTC, and other miscellaneous hardware errors; all of these are presently negligible.

E. Computational Effects

Computational deficiencies result either from an inability to adequately construct and utilize models describing the set of non-negligible effects taken from those cited above, or from arithmetic or algorithmic shortcomings in the computer programs developed to process the tracking data. The most recent orbit determination program at JPL occupies nearly 500000 locations. Although this program system has been designed to represent Doppler to an accuracy of 0.01 mm/sec and range to 0.1 m, there are many parameters in the models developed for this system whose values have inaccuracies that result in calculated observable errors significantly larger than these precision levels.

Let us summarize the contributions of these diverse error sources on the mass determinations by providing upper bounds for their effects. The mass determinations are derived from the portion of the trajectory lying within the sphere of influence of the principal body. It is informative to consider a very simplified model of the estimation process on ecsape trajectories. Simplified models are not always trustworthy but nevertheless, they sometimes provide insight not readily obtained through computer studies. Consider the 1-dimensional problem with dynamics described by

$$\ddot{x} = -\mu/x^2 \tag{12}$$

and for which the Doppler observable $\dot{\varrho}$ is given by

$$\dot{\varrho} = \dot{x}. \tag{13}$$

The parameters to be estimated are μ, the escape velocity V_∞, and x at a given epoch. The Doppler observations are continuously made over a specified time period; the measurements are uncorrelated and have a precision equivalent to the discrete measurement precision $\sigma_{\dot{\varrho}}$ sampled every 60 sec. The details will be omitted, but it can be shown from the above that the standard deviation for the least squares estimated value of the gravitational mass, μ, is given by

$$\frac{1}{\mu}\sigma_{\hat{\mu}} = \sqrt{2}\,\mu^{-1/2}\,V_\infty^{1/2}\,(60\,\sigma_{\dot{\varrho}}^2)^{1/2}\,F \tag{14}$$

where F is a dimensionless slowly varying function of the observation interval and typically in the range 1–5. Using the actual conditions for the Mariner V earth escape phase, one obtains from Eq. (14) a value for $\sigma_{\hat{\mu}}/\mu$ of about 0.1 ppm for GE, which is somewhat optimistic. However, Eq. (14) does show the behavior of $\sigma_{\hat{\mu}}/\mu$ with μ and V_∞ that is both in accord with one's intuition and in close agreement with results from computer studies using the exact model.

A measure of the sensitivity of $\hat{\mu}$ to unmodelled effects is a more useful by-product of this example and can be obtained from the Gauss-Markov theorem. For example, small acceleration effects may be taken into account by adding the appropriate term to the RHS of Eq. (12). For an unmodelled constant acceleration a, one obtains an expression for the error in the estimate of μ given by

$$\frac{1}{\mu}\delta\hat{\mu} = k V_\infty^{-2}(x_2 - x_1)\,a + \text{small terms}, \quad (V_\infty \geq 1) \tag{15}$$

where k is a dimensionless constant near unity, and $x_2 - x_1$ is the spatial observation interval. Eq. (15), through the $x_2 - x_1$ term, demonstrates the increasing effect of low thrust forces as longer spans of tracking data are included. Related to this is the observation that mass determinations from spacecraft orbiting the earth or the moon are not as reliable as those from escape missions. The quantity V_∞ is in the range 1–5 km/sec for earth escape and target encounter. For lunar missions, the trajectories do not usually have escape conditions in the earth phase. However, the error expressions for this case do not give values appreciably different from those obtained from the escape case with V_∞ equal to 1 km/sec. The quantity $x_2 - x_1$ may be set equal to the radius of the sphere of influence of each body which is about 10^6 km for the earth, Venus, and Mars, and about 10^5 km for the moon. For the uncertainty levels given in Table 2 one sees from Eq. (15) that the non-gravitational accelerations corrupt the mass determinations from escape and encounter data by an amount of the order of 1 ppm or less, a result that has been verified through computer studies using the exact formulation of the dynamics and the observables.

Planetary ephemeris acceleration errors during the encounter phase for Venus and Mars missions can also be estimated with Eq. (15). During the 10-day encounter period of Mariner II with Venus, the error in the geocentric radial velocity of Venus given by DE 19 was observed to change by 100 mm/sec. Using a value for a of 10^{-4} mm/sec² and setting V_∞ equal to 3 km/sec in Eq. (15) gives an error in $\hat{\mu}$ of 10 ppm.

The various error sources mapping directly into a Doppler signature v, can be taken into account by modifying Eq. (13) to read

$$\dot{\varrho} = \dot{x} + v. \tag{16}$$

An expression for the effect of an unmodelled constant velocity is given by

$$\frac{1}{\mu}\,\delta\hat{\mu} = -F_1 V_\infty^{-1} v, \quad V_\infty > 0 \tag{17}$$

where F_1 is a dimensionless function typically near unity. Thus, a 1 mm/sec error source affects the mass estimate by less than 1 ppm, a result that has also been verified by computer studies. A diurnal sinusoidal velocity error of 1 mm/sec in magnitude also has an affect of less than 1 ppm even during the close encounter phase when the gravity gradient is steep.

Uncompensated planetary ephemeris velocity errors of 100 mm/sec, however, can be serious. Eq. (17) suggests that Venus and Mars determinations, where V_∞ is in the range of 3–5 km/sec, can be in error by 20–50 ppm. Computer studies show that this estimate is somewhat pessimistic but that errors of 1–20 ppm can occur depending on the geometry of the hyperbolic trajectory. Thus, early planetary ephemerides through their velocity and acceleration errors have been a significant error source when not properly compensated for in the regression analysis. Position errors in the ephemerides are not important for mass determinations using Doppler data and can be almost totally compensated for by including station location parameters in the regression analysis.

Table 4. *Upper Bound of Effect of Uncompensated Error Sources on Mass Determinations in ppm*

Planet	Error Sources							
	Tracking Station Position¹)	Unmodelled Propagation Effects	Unmodelled Non-gravitational Forces	Gravity Harmonics	Ephemerides		Signal Processing	Computational
					DE 19	DE 69		
Venus	0.1	0.3	2²) to 0.1³)		20	0.5	0.1	0.01
Earth	0.1	0.4⁴)	0.3⁴)	0.1⁵)	—	—	0.1	0.01
Mars	0.1	0.3	0.2		30	1	0.1	0.01
Moon	0.1	0.4	0.3⁴)	1⁶)	1	0.5	0.1	0.01

¹) Present Uncertainties. – ²) Based on Mariner II. – ³) Based on Mariner V. – ⁴) Based on lunar spacecraft. – ⁵) Based on uncertainty in J_2 given in [10]. – ⁶) Based on uncertainty in the triaxal model for the moon given in [28b].

Table 4 gives upper bound error values for mass determinations in ppm due to the effect of these error sources when not taken into account by including in the regression analysis. We conclude that mass determinations from radio tracking data during encounter at an accuracy level of 1 ppm are not limited by known error sources at their present level of uncertainty.

4. The Planetary Masses from Radio and Radar Tracking

In this section we review the progress in determining the masses of the inner planets, the moon, and the astronomical unit through radio and radar experiments. A survey of the best available determinations by all methods of the masses of the principal planets up to about 1963 has been given by Clemence [10, 29]. In addition, a weighted average of these determinations was given for each planet. Since then significant progress has been made on the mass of the earth-moon $(E + M)/S$ from Eros observations, through the work of Lieske [30], and through a revision and updating of Rabe's earlier determination [31, 32, 33].

The strongest determinations of $S/(E + M)$ through radio tracking information are derived from individual determinations of GE, GM, E/M, and with GS through the defining relation

$$GS = k^2 (c\,\tau_A)^3 \tag{18}$$

where k is the Gaussian constant and τ_A is the light-time in A.1 sec for 1 A.U.

Table 5 gives the determination of GE and GM in units of km³/sec², E/M, standard deviations, and related information. The GM and GE

Table 5. *The Mass of the Earth and the Moon*

Launch Date	Source	GE ²) km³/sec² 398000+	Std. Dev. km³/sec²	GM ²) km³/sec² 4902+	Std. Dev. km³/sec²	E/M ¹) 81.0+	Std. Dev.	Ref.
Jan. '62	Ranger III	601.63	2.5	—	—	—		[34]
April '62	Ranger IV	601.87	13.3	—	—	—		]34]
Oct. '62	Ranger V	599.20	13.2	—	—	—		[34]
Jan. '64	Ranger VI	600.69	1.1	0.66	0.19	0.3030		[34]
July '64	Ranger VII	601.34	1.5	0.54	0.17	0.3050		[34]
Feb. '65	Ranger VIII	601.14	0.7	0.63	0.12	0.3035		[34]
Mar. '65	Ranger IX	601.42	0.6	0.71	0.30	0.3022		[34]
	Combined Ranger VI–IX	601.22	0.4	0.63	0.07	0.3035		[35]
May '66	Surveyor I	601.27	0.8	0.65	0.24	0.3032		[36]
Apr. '67	Surveyor III	601.11	0.8	0.64	0.25	0.3034		[36]
July '67	Surveyor IV	601.19	1.0	0.63	0.25	0.3035		[36]
Sept. '67	Surveyor V	601.10	0.6	0.63	0.24	0.3035		[36]
Nov. '67	Surveyor VI	601.11	0.5	0.64	0.24	0.3034		[36]
Jan. '68	Surveyor VII	601.11	0.8	0.64	0.24	0.3034		[36]
Nov. '66	Lunar Orbiter II	600.88	2.1	0.66	0.19	0.3030		[37]
Aug. '62	Mariner II	—	—	—	—	0.3001	0.0013	[10]
Nov. '64	Mariner IV	601.83	1.4	—	—	0.3015	0.0017	[11]
June '67	Mariner V	601.49	0.4	—	—	0.3013	0.0017	[38, 39]
Dec. '65	Pioneer VI	—	—	—	—	0.3005	0.0007	[39]
Aug. '66	Pioneer VII	601.86	1.8	—	—	0.3021	0.0004	[39]
Feb. '69	Mariner VI			—	—	—	—	
Mar. '69	Mariner VII			—	—	—	—	
'64–'68	Radar					0.302	0.002	[23]

¹) For the Ranger, Surveyor, and Lunar Orbiter flights, a value of 398601.2 has been used for GE.

²) Although the basic lengths unit in radio tracking and radar is the light-sec, these values are expressed in metric units for convenience. The IAU value of 299792.5 km/sec for c has been used to convert from light-seconds to kilometers.

information from the escape and approach phases of the remaining Lunar Orbiter missions has not yet been reduced. The reader is reminded that the expression of GM and GE in metric units is for convenience only. The basic length unit in radio and radar experiments is the light-second and at no place is c in m/sec required. The equations for the observables and the dynamics, and all the constants form a complete and consistent system when length, time, and mass are expressed in units of light-seconds, A.1 sec, and solar masses, or alternately, in A.U.'s, A.1 sec, and solar masses with τ_A defining the scale between the A.U. and the light second. The metric value of c used for conversion is the IAU value of 299792.5 km/sec.

The ratio E/M is derived for the lunar flights using the individual determinations of GM and a fixed value of GE of 398601.2 km³/sec². For the planetary mission the ratio E/M is obtained directly through a measurement of the lunar inequality derived from the 12 m/sec sinusoidal motion of the center of mass of the earth about the earth-moon barycenter [11, 40]. To a good approximation, this signature on the Doppler is given by

$$\dot{\varrho} = n_{☾} a_{☾} \left(\frac{M}{E + M}\right) \cos\beta \sin(\lambda - \lambda_{☾}) \tag{19}$$

where $n_{☾}$ and $a_{☾}$ are the mean motion and perturbed mean distance of the moon, and λ and β are the longitude and latitude of the spacecraft relative to the lunar orbit plane. β varies over the flight and is small, so the correlation between the estimates of E/M and β is small.

The consistency of the GE and GM determinations from the lunar flights is impressive, more so than the assigned statistics would suggest. However, the formal statistics listed are not to be taken too literally, since the various investigators, in attempts to account for unmodelled effects, generally introduced effective data noise figures that were between 3 and 10 times the observed Doppler noise. The Ranger IV and V flights were of short duration, hence their large standard deviations for GE. Because of high escape speeds and constraints due to the flight operations sequence on planetary spacecraft, planetary missions do not provide tracking as close to the earth as did the lunar missions. It can be shown for the planetary missions that during the first four hours after injection time, the partial derivative $\partial\dot{\varrho}/\partial\mu$, decreases by a factor of 4.

At the time of the Ranger mass determinations, there was some concern that the totally unmodelled ionosphere signature on the L-Band Doppler tracking from the Rangers might corrupt the mass determinations. However, the subsequent lunar flights at S-Band where the ionospheric effect is diminished by about a factor of 6 do not show a significant difference. For the lunar mass determinations a correction for the geocentric lunar radial distance at lunar encounter was included in the solution set. These corrections have been shown to be similar to the correction between the ILE and LE 16. It has been shown that uncompensated lunar ephemeris velocity errors of 1 mm/sec affect the GM values by less than 1 ppm. The same upper bound holds true for the velocity effects resulting from 10 m level station location and time effects, which was about the error level for these sources when these mass determinations were finished.

There is a curious and as yet, unexplained systematic difference amounting to about 0.002 (20 ppm) between the E/M ratios determined from lunar inequality measurements with the Mariner and Pioneer series and the direct GE/GM determinations from the lunar flights. The Doppler signature of the lunar inequality coefficient shown in Eq. (19) is nearly sinusoidal and leads to a good separation from the other parameters in the least squares solution. The precision with which the amplitude of a sinusoid may be determined from N equally distributed samples of measurement precision $\sigma_{\dot{\varrho}}$ is $\sqrt{2/N}\,\sigma_{\dot{\varrho}}$. For $\sigma_{\dot{\varrho}} = 1$ mm/sec, and a data point obtained every 10 min for 200 days, one may easily show that the 12 m/sec amplitude in Eq. (19) leads to a measurement precision of E/M of about 0.0003 which is about half that obtained from the actual covariance matrix for $\sigma_{\dot{\varrho}} = 1$ mm/sec. It can be shown that the defects in the ILE affect the Mariner and Pioneer E/M determinations by no more than 0.0005. Non-gravitational forces on planetary spacecraft are a possible source. There is some slight evidence from temperature measurement variations on Mariner V of a 28-day cycle, presumably associated with local solar activity returning after a solar rotation period. The reader may easily verify that a 28-day sinusoidal solar flux variation of 0.1% and of appropriate phase can produce an error in E/M determined from the lunar inequality measurement of about 0.001. On the other hand, it is unlikely that all determinations from planetary flights would be biased in the same direction due to this cause. Most recent error analyses indicate that the Mariner and Pioneer determinations are the most reliable and that a standard deviation of 0.001 is realistic.

Returning to the mass determinations, the light-time equivalent of the A.U., τ_A, is one of the most precisely determined constants in astronomy. It is currently considered to be known to at least 0.03 ppm [19, 20], although the formal standard deviation from the combined radar and optical observations is about 0.002 ppm. Experience has shown it to be a very stable parameter in a least squares sense with respect to reasonable changes in the data set, parameters solved-for, solution rank used, etc. The current value is

$$\tau_A = 499.004788 \pm 0.000015 \text{ A.1 sec.} \tag{20}$$

Through Eq. (18) and using for GE and GM the values of 398601.2 km^3/sec^2 and 4902.64 km^3/sec^2, the values for $S/(E + M)$ so indicated in Table 6 is obtained. The last entry in Table 6 is obtained from the theoretical relationship between the A.U. and $S/(E + M)$ given in [41] using a perturbed mean lunar distance value of 384400.2 km deduced from lunar radar bounce measurements.

We now turn to the other planetary masses. Table 7 summarizes recent determinations of the mass of Mercury. None of these recent determinations is significantly different from Newcomb's value of 6000000. The radar solutions are based on the perturbations by Mercury on Venus and to a lesser extent by the other planets. They are obtained while fixing the mass values of the other inner planets at the spacecraft determined values. Here, the radar does not have the solution power exhibited earlier, although, with time the additional observations of increased precision should improve this determination.

Table 6. *Recent Mass Values for the Earth-Moon*

Source	$S/(E + M)$
IAU (1964)	328912
$GS/(GE + GM)$ from Table 5	328900.2 ± 0.4
Lieske [30]	328915 ± 4
Rabe and Francis [33]	328939 ± 40
Astronomical Unit	328898 ± 3*

* This standard deviation is based on the uncertainty in the perturbed mean distance of the moon inferred from lunar radar bounce measurements.

Table 7. *Recent Determinations of the Mass of Mercury*

Source	S/M
IAU	6000000
Duncombe [43]	5970000 ± 455000
Planetary Radar [42]	5983000 ± 10000*
Planetary Radar [23]	6039000 ± 15000*

* Formal Standard Deviations. Based on the stability of the S/M value in the radar solutions, a more representative value for the probable error is 25000.

For Venus, only the Mariners II and V determinations are shown in Table 8. The results for Venus (and Mars) are obtained from the encounter phase of the Mariner trajectories having a duration of about 10 days. The mass is directly measured in $(\text{light-sec})^3/\text{sec}^2$, and the expressions in Eq. (18) and (20) have been used to obtain the mass ratios. The history of the Mariner II determinations as our technology advanced is interesting and the results are included in Table 8. The assigned probable error on the early Mariner II mass determinations was set an order of magnitude higher than the formal statistics to allow for several deficiencies known to exist at that time. Again, it has been shown earlier that the non-gravitational acceleration uncertainties had an effect of the Venus mass of at almost a few ppm. On the other hand, ephemeris errors have had a significant effect. The result reported in [44] used this

Table 8. *Mass of Venus from Mariners II and V*

Source	GS/GM
Mariner II	
Ref. [44] (1963)	408526 ± 30
Ref. [10] (1966)	408508 ± 6
Ref. [45] (1968)	408522 ± 5
Mariner V	
Doppler: Refs. [38, 45] (1968)	408521.8 ± 1
Range: (1968)	408522.7 ± 2

Table 9. *Mass of Mars from Mariner IV*

Source	GS/GM
Mariner IV	
Ref. [46] (1965)	3098600 ± 600
Ref. [47] (1968)	3098710 ± 50

Newcomb ephemeris. The result reported in [10] also used this ephemeris but an attempt was made to correct the ephemeris error in the regression analysis. However, a constraint was applied in order to conserve the mean motion of Venus. Unfortunately, this constraint was introduced in a form that did not allow an adequate adjustment in the velocity of Venus. Finally, the result given in [45] is 10 ppm displaced from the original determination and used a current radar-based ephemeris. Note that the most recent value from Mariner II is consistent with the Mariner V results which rely on the same ephemeris.

The most recent result from Mariner V Doppler given in [45] also included a determination of J_2. The result given for J_2 is $(-5.6 \pm 8) \times 10^{-6}$ which is not significantly different from zero. A mass determination from ranging measurements is also given in Table 8. It was obtained from combining the Mariner V ranging data during encounter with all available planetary radar and solar and planetary meridian circle observation in a 65 parameter least squares fit. It should still be considered provisional; accordingly, a standard deviation of 2 has been assigned based on the stability of this parameter, even though the formal error is about 0.3. The consistency of this determination with the other results should be noted.

The Mariner IV results for Mars are shown in Table 9. Null [46] gave considerable attention to the corrupting effects of unmodelled parameters through extensive computer studies and finally assigned an upper error bound on the mass result of 600 reciprocal solar masses. However, Null also presented in [46] a figure showing the quadratic variation of the weighted sum of the squares of the Doppler residuals after convergence versus various fixed values for the mass of Mars. From curvature of this quadratic at its minimum one obtains a value of 43 for the formal standard deviation using a Doppler accuracy level of 30 mm/sec. The actual Doppler accuracy at the time was nearer to 1 mm/sec, so under the assumption of unmodelled parameters of negligible effect, one obtains a formal standard deviation of about 1.5 (0.5 ppm) which can be considered to be the potential accuracy of the Mariner IV determination of the mass of Mars. NULL has recently [47] reprocessed the Mariner IV data with updated parameters and using a current radar-based ephemeris. The result is now 3098710 with a still somewhat cautious assigned standard deviation of 50.

In summary, the mass determinations for the inner bodies from radio and radar information show a high degree of internal consistency. The present accuracy of these determinations is impressive; the potential accuracy of these results is about an order of magnitude better than the presently assigned error bounds. To perform mass determinations at the 1 ppm level requires an unprecedented degree of knowledge and modelling of those elements affecting the dynamics and the observations. The recent improvements in the planetary ephemerides has been a significant factor in strengthening our confidence in these results. Work is in progress toward reprocessing most of the data from which the present results were obtained using the most recent models and computer programs. The completed analyses should yield results, not significantly different from the present values, but with accuracies approaching the inherent precision of the data.

References

1. DSIF S-Band Tracking and Communication System, Vol. II. EPD-256, Jet Propulsion Laboratory 1965.
2. TAUSWORTHE, R. C.: Digital Communication and Tracking: Ranging Measurement, SPS 37-42, Vol. II, p. 52. Jet Propulsion Laboratory 1968.
3. TAUSWORTHE, R. C.: Ranging the 1967 Mariner to Venus. Paper No. 36.4, 1967, IEEE Intern. Conf. Record, 20–23 March, New York.
4. GOLDSTEIN, R. M.: Radar Exploration of Venus. TR 32-396, Jet Propulsion Laboratory 1965.
5. EVANS, J. V., HAGFORS, T., ed.: Radar Astronomy. New York: McGraw-Hill 1968.
6. HAMILTON, T. W., MELBOURNE, W. G.: Information Content of a Single Pass of Doppler Data, SPS 37-39, Vol. III, p. 18. Jet Propulsion Laboratory 1966.
7. CURKENDALL, D. W.: Six-Parameter Model Characterizing Doppler Information Received While Tracking a Distant Spacecraft, SPS 37-47, Vol. II, p. 15. Jet Propulsion Laboratory 1967.
8. DYER, P., MCREYNOLDS, S. R.: Data Compression and its Application to Orbit Determination, SPS 37-53, Vol. II, p. 39. Jet Propulsion Laboratory 1968.
9. MOTTINGER, N. A.: Breaking The 10 Meter Level in Obtaining Consistent Station Location Solutions from the Reduction of Deep Space Probe Data. Jet Propulsion Laboratory, April 23, 1969.
10. CLEMENCE, G. M.: Annual Review of Astronomy and Astrophysics (ed. by L. Goldberg). Vol. III, p. 93, Palo Alto, California 1965.
11. ANDERSON, J. D.: Doctoral Thesis, UCLA, 1967. Also: Determination of the Masses of the Moon and Venus and the Astronomical Unit from Radio Tracking Data of the Mariner II Spacecraft. TR 32-816. Jet Propulsion Laboratory 1967.

12. Null, G. W., Gordon, G. J., Tito, D. A.: The Mariner IV Flight Path and Its Determination from Tracking Data. TR 32-1108, Jet Propulsion Laboratory 1967.
13. Pease, G. E., et al.: Mariner V Trajectory and Its Determination from Tracking Data. TR 32-1363, Jet Propulsion Laboratory.
14. Bourke, R. D., McReynolds, S. R., Thuleen, K. L.: Translation Forces on the Mariner V Spacecraft Stemming from the Attitude Control System. J. Spacecraft and Rockets, in press.
15. Bourke, R. D., Curkendall, D. W., McReynolds, S. R.: A Technique for the Reduction of Random Attitude-Control-System-Generated Non-Gravitational Forces. SPS 37-48, Vol. II, p. 27, Jet Propulsion Laboratory 1967.
16. Peabody, P. R., Scott, J. F., Orozco, E. G.: Users Description of JPL Ephemeris Tapes. TR 32-580, Jet Propulsion Laboratory 1964.
17. Peabody, P. R., Scott, J. F., Oroczo, E. G.: JPL Ephemeris Tape E9510, E9511, and E9512. TM 33-167, Jet Propulsion Laboratory 1964.
18. Clemence, G. M.: Theory of Mars. Completion. Astron. Papers 16 (1961) Part 2.
19. Ash, M. E., Shapiro, I. I., Smith, W. B.: Astron. J. 72 (1967) 338.
20. Melbourne, W. G., O'Handley, D. A.: A Consistent Ephemeris of the Major Planets and the Solar System. SPS 37-51, Vol. III, p. 4, Jet Propulsion Laboratory 1968.
21. Kulikov, D. K.: Accuracy of the Inner Planet Coordinates, in: IAU Symposium No. 21, Le Systeme De Constantes Astronomique, ed. by J. Kovalevsky, Paris: Gauthier-Villars and Cie. 1965.
22. Anderson, J. D., Cain, D. L., Efron, L., Goldstein, R. M., Melbourne, W. G., O'Handley, D. A., Pease, G. E., Tausworthe, R. C.: J. of Atmospheric Sciences 5 (1968) 1171.
23. Ash, M. E., Shapiro, I. I., Smith, W. B.: Private Correspondence, 1969.
24. Eckert, W. J., Jones, R., Clark, H. K.: Improved Lunar Ephemeris, 1952 to 1959, Washington, D.C.: U.S. Government Printing Office 1954.
25. Mulholland, J. D.: Application of a Transformation of Equinox in the Lunar Theory: Lunar Ephemeris Number 16. SPS 37-57, Vol. II, Jet Propulsion Laboratory 1969.
26. Eckert, W. D., Smith, H. F., Jr.: Astron. Papers of the American Ephemeris, Vol. 19, Part II. U.S. Nautical Almanac Office, GPO, Washington, D.C.
27. Clemence, G. M., Porter, J. G., Sadler, D. H.: Aberration in the Lunar Ephemeris. Astron. J. 57 (1952).
28. Clemence, G. M.: Remarks on Current Lunar Theory, in: Proceedings of the JPL Seminar on Uncertainties in the Lunar Ephemeris, ed. by J. D. Mulholland. Technical Report 32-1247, Jet Propulsion Laboratory 1968.
28a. Koehler, R. L.: J. of Geophysical Research, Space Physics 73 (1968) 4883.
28b. Melbourne, W. G., Mulholland, J. D., Sjogren, W. L., Sturms, F. M., Jr.: Constants and Related Information for Astrodynamic Calculations, 1968. Technical Report 32-1306, Jet Propulsion Laboratory 1968.
29. Clemence, G. M.: Masses of the Principal Planet, Trans. IAU 12B (1968) 609.
30. Lieske, J. H.: Astron. J. 73 (1968) 628.
31. Rabe, E.: Astron. J. 55 (1950) 112.
32. Schubart, J., Zech, G.: Nature 214 (1967) 900.
33. Rabe, E., Francis, M.: Astron. J. 72 (1967) 856.
34. Sjogren, W. L., Trask, D. W., Vegos, C. J., Wollenhaupt, W. R.: Physical Constants as Determined from Radio Tracking of the Ranger Lunar Probes. TR 32-1057, Jet Propulsion Laboratory 1966.
35. Vegos, C. J., Trask, D. W.: Ranger Combined Analysis, Part II: Determi-

nation of the Masses of the Earth and Moon from Radio Tracking Data. SPS 37-44, Vol. III, p. 11, Jet Propulsion Laboratory 1967.
36. Wong, S. K.: Deep Space Station Locations and Physical Constants Solutions of Surveyor Missions. SPS 37-52, Vol. II, p. 12, Jet Propulsion Laboratory 1968.
37. Mottinger, N. A., Sjogren, W. L.: Consistency of Lunar Orbiter II Ranging and Doppler Data. SPS 37-46, Vol. III, p. 19, Jet Propulsion Laboratory 1967.
38. Anderson, J. D., Pease, G. E., Efron, L., Tausworthe, R. C.: Astron. J., Vol. 73, No. 2, Part II (1968) 52.
39. Anderson, J. D., Hilt, D. E.: Improvement of Astronomical Constants and Ephemerides from Pioneer Radio Tracking Data. Astronaut. Soc. Paper No. 68-130, Jackson, Wyoming, 1698.
40. Hamilton, T. W.: Proceedings of the IAU Symposium No. 21: Le Système De Constants Astronomique, ed. by J. Kovalevsky, Paris: Gauthier-Villars and Cie. 1965.
41. Brouwer, D.: Relations Among Some Important Astronomical Constants, IAU Symposium No. 21, ibid.
42. Melbourne, W. G., O'Handley, D. A.: Recent Development Ephemerides and the Mass of Mercury. SPS 37-53, Vol. III, p. 1, Jet Propulsion Laboratory 1968.
43. Duncombe, R. L.: Motion of Venus, 1750–1949, Astronomical Papers, Vol. XVI, Part. I (1958) 39.
44. Anderson, J. D., Null, G. W., Thornton, C. T.: The Evaluation of Certain Astronomical Constants from the Radio Tracking of Mariner II in: Progress in Astronautics and Aeronautics, Vol. IV, p. 311, New York: Academic Press 1964.
45. Anderson, J. D., Efron, L.: The Mass and Dynamical Oblateness of Venus. Presented at the 129th Meeting of the American Astronomical Society, University of Hawaii, Honolulu, Hawaii, March 30 – April 2, 1969.
46. Null, G. W.: Astron. J. 72 (1967) 1292.
47. Null, G. W.: Private Communications 1968.

Satellite Trajectory Determination and their Expected Errors OGO-IV, GEOS-I*

By

F. O. Vonbun

National Aeronautics and Space Administration – Goddard Space Flight Center, Greenbelt, MD/USA

Abstract. Precision trajectory determination, as needed for some of NASA's spacecraft, requires a close analysis of the errors of the state vector. Particularly important is the determination and separation of the errors caused by the tracking systems from those caused by the trajectory determination process. Only a very good understanding of this effect will lead to the required decrease of trajectory errors.

In this paper, orbits from GEOS-A and OGO spacecraft are analyzed and their trajectory errors estimated using our precision Definitive Orbit Determination System. Then, a tracking systems error study is performed to determine those errors of the state vector which are due only to the tracking systems. From these it is concluded that the major contribution to the total errors in the state is due to the uncertainties in the trajectory analysis (geopotential, integration procedures, etc.) rather than from the errors of our present worldwide distributed tracking stations. Therefore, future work to improve our orbit determination capability will be directed to improve our software capability.

Orbital Uncertainty Estimates

General

One important quantity to know in addition to the six dimensional "orbit vector" ($[x_1\ x_2\ x_3,\ \dot{x}_1\ \dot{x}_2\ \dot{x}_3]^T$ or $[a, e, i, \Omega, \omega, M]^T$)** which in the essence of orbit determination, is the knowledge of its accuracy. If an orbit is determined, the question immediately arises: How good is the orbit or how accurate or what are the errors associated with that calculated orbit? This is a rather difficult question to answer and is subject to a lot of controversy. An orbit, being a six dimensional quantity, cannot be "measured" in the physical sense. Only parameters such as range, range rate, and angles can be determined via actual physical measurements. The orbit itself has to be calculated and also the "Orbital Uncertainty Estimates" (OUE) which may be a better description than orbital errors.

---

* Abbreviated version of the paper submittered to COSPAR.

** T – means transposed.

Orbital Uncertainty Estimate Using Real Data

The Orbital Uncertainty Estimate as originally introduced by J. SIRY and D. STEWART [1] are based upon practical orbit determination and include all errors known and unknown which do occur when an orbit is calculated. The principle is rather simple and is depicted in Fig. 1, in a

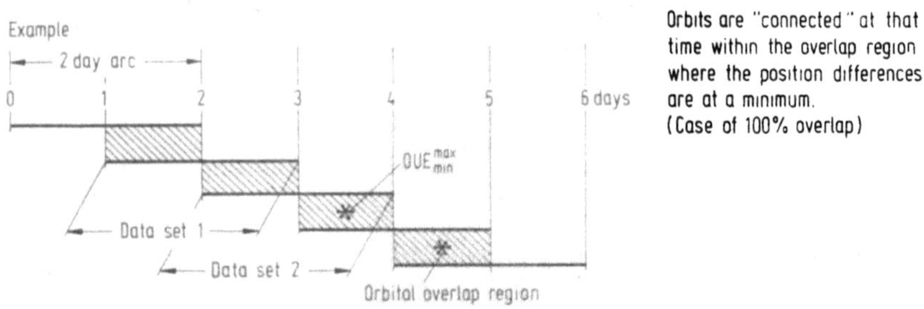

Fig. 1. Orbital uncertainty estimate (OUE) (schematic).

schematic way. Orbits are calculated, say for a two-day period where one day is always "overlapped". The maximum and minimum position differences in the overlap region is recorded as a measure of the uncertainty associated with the orbit. These arcs have different data sets, different tracking stations and thus do, as mentioned, describe reality. The orbit errors not only depend on the errors in the orbit computation itself, the tracking systems errors, but also to a large extent on the tracking geometry and the distribution and availability of tracking data "around" the world and thus the orbit. Gaps in information due to the lack of coverage for a particular orbit or temporary instrument or communications failures, which do occur occasionally, all influence the final accuracy with which an orbit can be determined. These factors are, of course, all considered when these orbital uncertainty estimates using the overlap method are determined [1].

Orbital Errors Using Error Theory

In many cases, particularly during project planning phases, "Error Analyses" have to be performed on a theoretical basis only since the tracking system to be used is not even built and/or the spacecraft not designed. These techniques have been used for the basic design of the Goddard Range and Range Rate System and the Apollo Tracking Network as well as for plans of a new "Orbiting Tracking System" which is under investigation at present, ([2] through [8]).

Figs. 2, 3 and 4 show, in graph form, the Orbital Uncertainty Estimates which have been obtained for the OGO-IV and GEOS-I. The

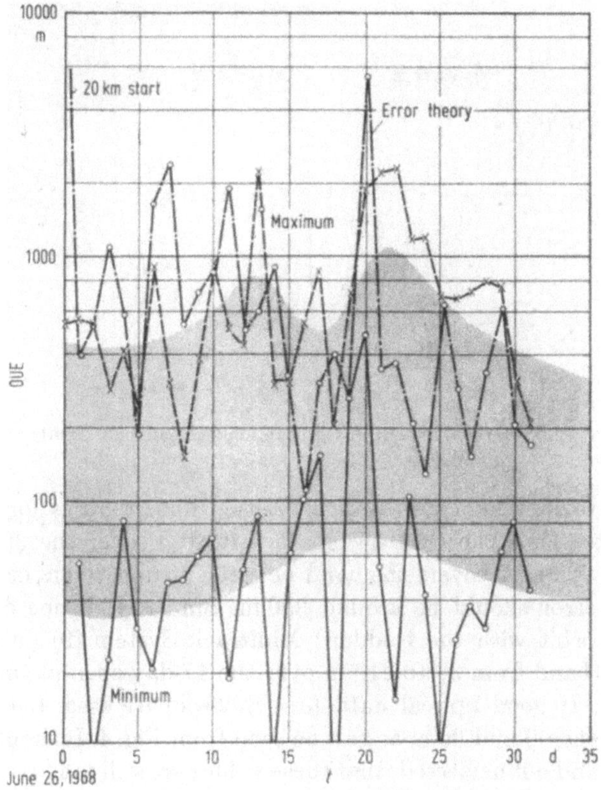

Fig. 2. OGO-IV Orbital uncertainty estimates (daily variations only).

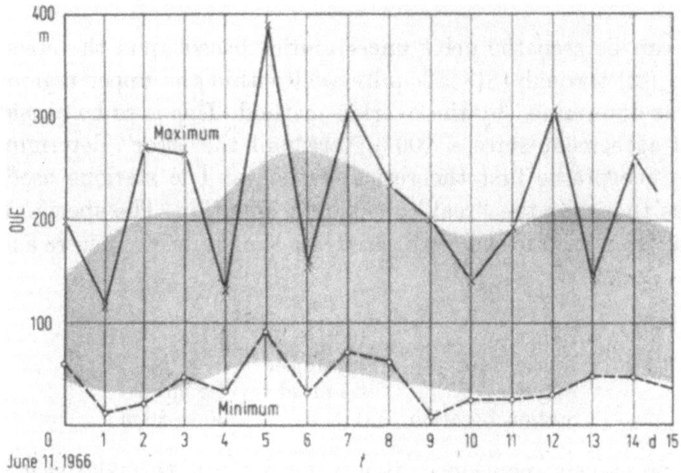

Fig. 3. GEOS-I orbital uncertainty estimates minitrack data (daily variations only).

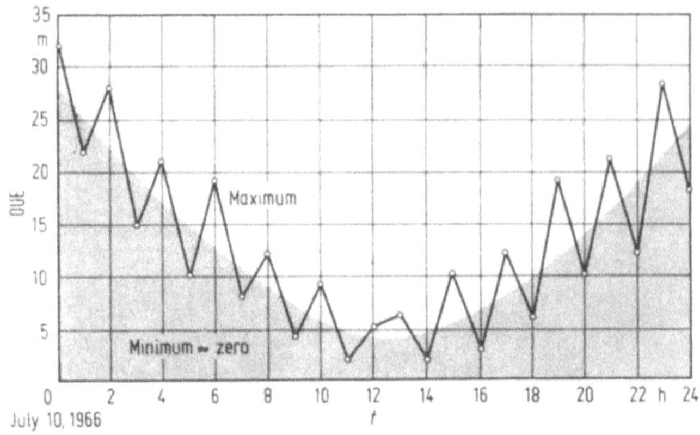

Fig. 4. GEOS-I orbital uncertainty estimates optical data only (hourly variations).

hatched regions indicate average position uncertainties for these satellite orbits. Thus the OGO-IV spacecraft OUE over the 30-day period depicted in Fig. 2 covers the band between 50 and 700 m or the average position error should be around 350 m. Similarly, if one determines a GEOS-I orbit with the Goddard Minitrack System [9] only, the OUE covers a band from 30 to 210 m over the 17-day span of time as shown in Fig. 3. If good optical data ($\sigma \measuredangle \leq 2\,\widehat{\sec}$) are used, the OUE varies only between 0 and 30 m as can be seen from Fig. 4. It should, however, be noted and not neglected, that these values are still to a "certain extent" relative values rather than absolute ones.

An error analysis ([2] through [8]) for the 30-day OGO-IV orbit has been performed and the result is superimposed on the OUE shown in Fig. 2.

As can be seen the orbit uncertainties based upon the error theory applied ([2] through [8]) fall quite well within the upper regions of the errors as determined by the overlap method. This is to be considered as a good agreement since a 100% "error of the error" determination is still an acceptable first theoretical estimate. The stations used are the same as those for the "real" orbit determination. The specific tracking system errors used for the error analyses shown in Fig. 2 were as follows:

	Noise	Bias
Range	10 m	10 m
Range Rate	1.2 cm/sec	0
Angles	1.5 mrad	1.5 mrad
Station Location	0	30 to 80 m

As previously mentioned, it is very difficult to make accurate and pure theoretical predictions of expected orbit errors. Tracking systems error parameters [7, 9 through 13] have to be determined and chosen from previous orbit analyses. As can be seen, these "theoretical" errors are on the high side as shown in Fig. 1. The reason is that rather large system "bias" error values were used in the error theory. It is those biases which contribute most to the orbit errors. Noise contributions reduce statistically inversely proportional to the square root of the number of measurements used where bias contributions stay, by their nature, constant [2, 3, 4, 5], no matter how many measurements were taken and actually used.

From a tracking systems design point of view, this is important to take into consideration since it prevents an underestimation of the problem. In brief, if a mission can be flown with pressimistic assumed tracking system accuracies, its final success is then assured in a more adequate fashion from a ground tracking and navigation point of view (4, 6, 7, 13, through 15).

Orbit Projection

Meaning and Purpose

Orbit projection is the mathematical extension of an orbit or six dimensional state vector ($X_{6\times 1}$) into the future. An orbit is determined over a certain time interval, say two days as shown in Fig. 5 as an example, and then projected.

The purpose of such projections is to make station predictions and thus station scheduling for the network possible. The necessary needed prediction accuracy depends on the acquisition capability of the tracking and communications systems used in the network. For our Minitrack System, this accuracy is rather low since the antenna beam is a steady fan beam, (spacecraft passes through). The time of passage varies then only for a near-earth spacecraft by one second for an along track error of 7000 m to 8000 m. For systems using narrow beams such as radars, range and range rate or laser systems, the accuracy must be such that the spacecraft can be predicted within the antenna beamwidth.

In one of our recent experiments, orbit predictions for the GEOS-I spacecraft were made within $\pm$ 200 to 400 m over one week in advance. Thus we were able to perform, for the first time, daytime tracking with our Goddard laser which was sent to Carnarvon (Australia) for a tracking system intercomparison experiment. Results of this specific experiment will be published in the future.

Fig. 5 depicts, in some detail, the projection of orbits and gives the meaning of the average projection differences $\overline{\Delta P_{ik}}$ in position and $\overline{\Delta V_{ik}}$ in velocity. These values are obtained by differencing the orbit projection and the definitive orbit obtained later over the same time interval. These numbers thus present an estimate of the accuracy with which orbits can be projected. Examples are given for the GEOS-I and OGO-IV orbits.

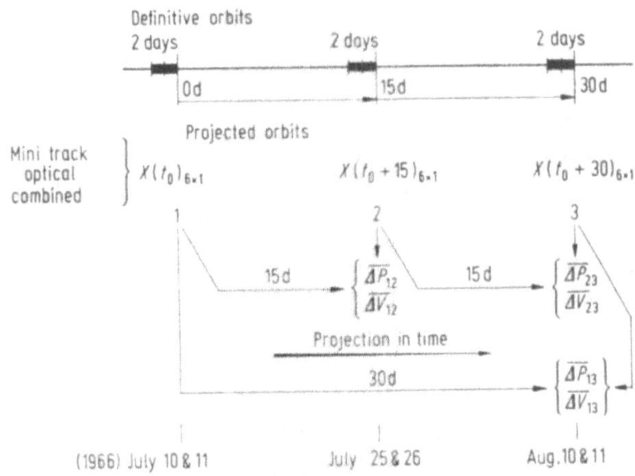

Fig. 5. Projected minus definitive orbit differences (schematic).

Projection of the Orbit of GEOS-I

Table 1 shows the projection results obtained for the GEOS-I spacecraft when different kinds of tracking data were used for the two-day

Table 1. *GEOS-I Projected Minus Definitive Orbit Difference Table* (m, m/sec)

Average Position and Velocity Diff.	Days	A		B		C	
		Minitrack Data Only	$\overline{\Delta P}$/Tot. Dist.	Optical Data Only	$\overline{\Delta P}$/Tot. Dist.	Minitrack Optical $(R+\dot{R})$ Data	$\overline{\Delta P}$/Tot. Dist.
$\overline{\Delta P}_{12}$	15	408	2×10^{-8}	110	6×10^{-9}	286	$1.5\times10^{8-}$
$\overline{\Delta V}_{12}$	15	0.30	—	0.10	—	0.24	—
$\overline{\Delta P}_{23}$	15	1880	9×10^{-8}	746	4×10^{-8}	425	2×10^{-8}
$\overline{\Delta V}_{23}$	15	1.63	—	0.65	—	0.40	—
$\overline{\Delta P}_{13}$	30	1070	6×10^{-8}	766	4×10^{-8}	1457	8×10^{-8}
$\overline{\Delta V}_{13}$	30	0.94	—	0.70	—	1.30	—

$$\frac{\Delta\,\text{Position}}{\text{Total distance traveled by Satellite}} \doteq \begin{array}{ll} A: & (2\ \text{to}\ 9)\times10^{-8} \\ B: & (0.6\ \text{to}\ 2)\times10^{-8} \\ C: & (1\ \text{to}\ 8)\times10^{-8} \end{array}$$

starting orbit (see Fig. 5). As can be seen, the differences are rather independent of the original tracking data used. One would, in general, assume that the value of $\overline{\Delta P_{13}} \doteq 1070$ m for a Minitrack starting orbit should be larger than $\overline{\Delta P_{13}} \doteq 1480$ m where all tracking data combined were used to calculate the original starting orbit. This last orbit should be a better one and thus the projection difference should be smaller. The reason that this is not so in this particular case seems to be that the projection sensitivity overrides the "better" orbit. In brief, the initial conditions, or the last set of osculating elements which are projected determine the real differences.

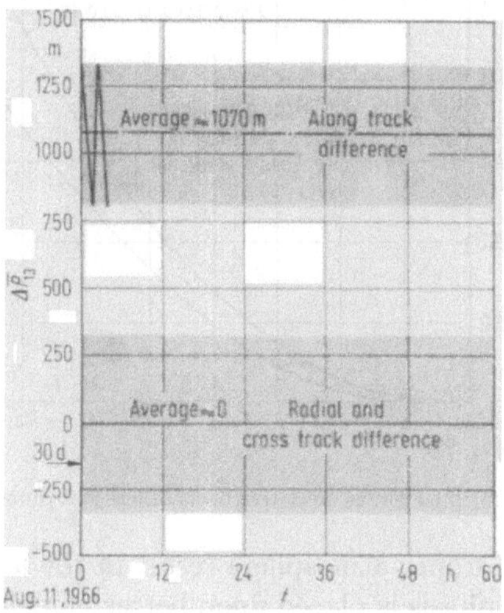

Fig. 6. GEOS-I projected minus definitive orbit difference (total position).

Fig. 6 depicts the projected orbital differences for the GEOS-I orbit over 30 days and then continues these over the next 48 h. The along track error varies approximately sinusoidally over the range of ± 260 m as indicated by the hatched area with an average error of 1070 m.

The radial and cross track components vary over an approximate range of ± 330 m and average out to zero over a time larger than the orbit period which has been theoretically proven by Bonavito [5].

Projection of the Orbit of OGO-IV

The OGO-IV orbit as shown in Table 1 was also projected 15 days in advance starting June 26, 1968. The position difference of the definitive and the projected orbit is shown in Fig. 7 as a function of time. These position differences increase in a rather non-linear and almost quadratic fashion. After 15 days they are between 400 km and 470 km depending on the atmospheric model used in the orbit program. This is an extremely large difference as compared to the GEOS-I predictions discussed earlier. The main reason is that the GEOS orbit is out of the drag region (height $\geq$ 1000 km) where the OGO-IV orbit is within the drag region ($\leq$ 500 km). In brief, the large and rather crude unpredictable variations of

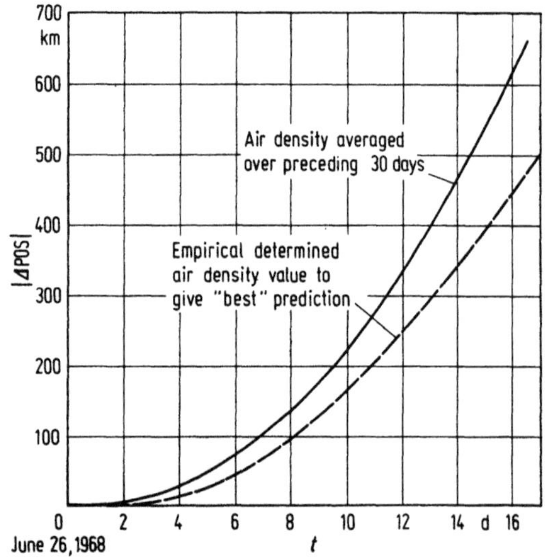

Fig. 7. OGO-IV projected minus definitive orbit differences [minitrack, $(R+\dot{R})$ data].

the density in the upper atmosphere result in the large prediction differences shown. Emphasis is placed upon better modeling in order to bring these values down. Good prediction over large periods of time make network operations more effective because of a better tracking station scheduling. In addition, prediction can be and is used in orbital error studies in general. Using for instance a laser system during daytime, as mentioned before, does need good and accurate orbit predictions. This is one way "to measure to a certain extent" a six dimensional state vector. This can and is therefore being used in our orbital accuracy studies.

Acknowledgement

The author would like to take this opportunity to thank the following members of the Mission and Trajectory Analysis Division. Mr. P. Schmid for the data analyses; Messrs. G. Marechek and E. Doll for their valuable help in performing the presented orbital computations and Mr. T. Cooley and Mrs. A. Marlow for their help in the orbit error analyses performed.

References

1. Siry, J. W., Stewart, D. J.: Goddard Orbit Information, GSFC X-550-69-136, February 1969.
2. Vonbun, F. O., Kahn, W. D.: Tracking Systems, Their Mathematical Models and Their Errors, Part I – Theory. NASA TN D-1471, October 1962.
3. Kahn, W. D., Vonbun, F. O.: Tracking Systems, Their Mathematical Models and Their Errors, Part II – Least Squares Treatment. NASA TN D-3776, December 1966.
4. Schmid, S. F.: Users Manual, Mark II Error Propagation Program. NASA Contract No. NAS5-9700, Philco. Palo Alto, California, February 15, 1966.
5. Bonavito, N. L.: An Analysis of Satellite Position Uncertainty by Statistical Mechanics, GSFC X-552-68-465, December 1968.
6. Vonbun, F. O., Mengel, J. T.: Tracking and Communications for Planetary Manned Mission, J. Spacecraft and Rockets, Vol. 5, No. 5, pp. 863–865, July 1968.
7. MSC/GSFC: Apollo Mission and Navigation Systems Characteristics, NASA ANGW Report, AN-1.3, December 15, 1967.
8. Cooley, J. L., Marlow, A.: Orbital Error Studies – Tracking from a Synchronous Spacecraft, GSFC X-551-67-7, January 1969.
9. Schroeder, C. A.: Electronic Interferometer Tracking, 5th National Symposium on Space Electronics and Telemetry, September 19–21, 1966, Washington, D.C.
10. Kruger, B.: The Range Rate Error Due to the Averaging Techniques of Doppler Measurements. GSFC X-513-65-100, March 5, 1965.
11. Manned Spacecraft Center/Goddard Space Flight Center, Apollo Navigation Ground and Onboard Capabilities, ANWG Technical Report AN-2.1, September 1, 1966.
12. Kruger, B.: Effects of Correlated Noise with Applications to Apollo Tracking Problems, NASA TN D-4121, February 1968.
13. Marsh, J. G., et al.: Intercomparison of the Minitrack and Optical Tracking Networks Using GEOS-I Long Arc Orbital Solutions, Part I, GSFC X-552-68-105, December 1967. – Schmid, P. E.: Atmospheric Tracking Errors at S- and C-Band Frequencies, GSFX X-507-65-398, October 5, 1968.
14. Wolf Research: Coordinate and Time Reference Systems Used in Astrodynamics, NASA Contract No. NAS5-9756-124, September 1968.
15. Schmid, S. F.: Application of State-Space Methods to Navigation Problems, in: Advances in Control Systems, Vol. 3. New York, London: Academic Press 1966.

Appendix A. Tracking Stations used for Orbit Determination OGO-IV

	Station Code	Teletype Code	Station Coordinates: Geodetic Latitude [°]	[′]	[″]	East Longitude [°]	[′]	[″]	Spheroid Height [m]
A. Minitrack Station									
1. Fort Myers (Florida, USA)	FTMYRS	GYRS	26	32	53.080	278	8	3.806	−32.482
2. Quito (Ecuador)	QUITOE	GQUI	0	−37	−22.639	281	25	15.238	3564.619
3. Lima (Peru)	LIMAPU	GAPU	−11	−46	−37.569	282	50	58.864	44.049
4. Santiago (Chile)	SNTAGO	GAGO	−33	−8	−58.769	289	19	52.591	715.300
5. Newfoundland (Canada)	NEWFLD	GFLD	47	44	28.738	307	16	46.676	68.465
6. Winkfield (England)	WNKFLD	LWNK	51	26	40.670	359	18	8.355	86.000
7. Johannesburg (South Africa)	JOBURG	GBUR	−25	−53	−2.706	27	42	25.416	1556.250
8. Tananarive (Madagascar)	MADGAR	LTAN	−19	0	−25.21	47	18	0.46	1361.4465.3
9. Orroral (Australia)	ORORAL	AORR	−35	37	37.51	148	57	10.71	947.3107.0
10. Fairbanks (Alaska)	ALASKA	GULA	64	58	37.516	212	28	33.882	1024.8
B. Range and Range Rate									
1. Rosman (North Carolina, USA)	ROSRAN	GROS	−35	11	41.192	277	7	26.020	833.530
2. Tananarive (Madagascar)	MADGAR	LTAN	−19	−1	−23.352	47	18	7.964	1392.560
3. Carnarvon (Australia)	CARVON	ACRO	−24	−54	−16.254	113	42	58.584	20.545

Appendix B. Tracking Stations used for Orbit Determination GEOS-I

	Station Code	Teletype Code	Station Coordinates						
			Geodetic Latitude			East Longitude			Spheroid Height
			[°]	[′]	[″]	[°]	[′]	[″]	[m]
A. Minitrack Station									
1. Blossom Point (Maryland, USA) (Deactivated 9/30/66)			38	25	49.448	282	54	48.659	−40.366
2. Fort Myers (Florida, USA)	FTMYRS	GYRS	26	32	53.080	278	8	3.806	−32.482
3. Woomera (Australien)	OOMERA	AWOM	−31	−23	−26.965	136	52	14.254	158.784
4. Quito (Ecuador)	QUITOE	GQUI	0	−37	−22.639	281	25	15.238	3564.619
5. Lima (Peru)	LIMPAU	GAPU	−11	−46	−37.569	282	50	58.864	44.049
6. Santiago (Chile)	SNTAGO	GAGO	−33	−8	−58.769	289	19	52.591	715.300
7. Mojave (California, USA)	MOJAVE	GAVE	35	19	47.579	243	5	59.186	884.934
8. Johannesburg (South Africa)	JOBURG	GBUR	−25	−53	−2.706	27	42	25.416	1556.250
9. St. Johns (Newfoundland)	NEWFLD	GFLD	47	44	28.738	307	16	46.676	68.465
10. College (Alaska)	ALASKA	GULA	64	52	17.783	212	9	37.297	149.113
11. Winkfield (England)	WNKFLD	LWNK	51	26	40.670	359	18	8.355	86.000
B. Range and Range Rate									
1. Madagascar	MADGAR	LTAN	−19	−1	−23.352	47	18	7.964	1392.560
2. Rosman (North Carolina, USA)	ROSRAN	GROS	35	11	41.192	277	7	26.020	833.530
3. Carnarvon (Australia)	CARVON	ACRO	−24	−54	−16.254	113	42	58.548	20.545
C. Stadan – Optical									
1. Blossom Point (Maryland, USA)	1BPOIN		38	25	49.536	282	45	48.720	−38.695
2. Fort Myers (Florida, USA)	1FTMYR	GYRS	26	32	53.048	278	8	3.709	−31.844
3. Woomera (Australia)	1OOMER	AWOM							
4. Mojave (California, USA)	1MOJAV	GAVE	35	19	47.583	243	5	58.897	887.388
5. Johannesburg (South Africa)	1JOBUR	GBUR	−25	−53	2.656	27	42	25.372	1557.154

Appendix B – (Continued)

	Station Code	Teletype Code	Station Coordinates: Geodetic Latitude [°]	[′]	[″]	East Longitude [°]	[′]	[″]	Spheroid Height [m]
6. East Grand Forks (Minnesota, USA)	1GFORK		48	1	20.930	262	59	19.400	212.916
7. Rosman (North Carolina, USA)	1ROSMA	GROS	35	12	7.030	277	7	40.810	867.000
8. Tananarive (Madagascar)	1TANAN	LTAN	−19	0	−33.261	47	17	58.893	1355.870
D. Special Optical (SPEOPT)									
1. Edinburg (Texas, UAS)	1EDINB		26	22	46.350	261	40	7.350	25.000
2. Columbia (Missouri, USA)	1COLBA		38	53	35.894	267	47	40.789	299.981
3. Bermuda Island	1BERMD		32	21	48.995	295	20	33.733	−16.757
4. San Juan (Puerto Rico)	1PURIO		18	15	28.284	294	0	23.580	16.682
5. Sudbury (Ontario)	1SUDBR		46	27	20.520	279	3	10.350	234.000
6. Jamaica (British West Indies)	1JAMAC		18	4	34.200	283	11	27.030	433.000
7. Denver (Colorado, USA)			39	38	47.590	255	23	38.494	1763.565
E. Smithsonian Astrophysical Laboratory (SAO)									
1. Organ Pass (New Mexico, USA)	1ORGAN		32	25	24.710	253	26	48.300	1621.000
2. Olifantsfontein (South Africa)	1OLFAN		−25	−57	−37.690	28	14	51.470	1571.000
3. San Fernando (Spain)	1SPAIN		36	27	46.660	353	47	36.550	66.000
4. Arequipa (Peru)	1QUIPA		−16	−27	−58.04	288	30	24.02	2489.0
5. Curaçao (Lesser Antilles)	1CURAC		12	05	24.92	291	09	43.96	−24.0
6. Jupiter (Florida, USA)	1JURTR		27	01	14.21	279	53	12.95	−26.0
7. Villa Dolores (Argentina)	1VILDO		−31	−56	−36.39	294	53	36.10	647.0
8. Maui (Hawaii)	1MAUIO		20	42	25.64	203	44	33.22	3038.0
9. Woomera (Australia)	AUSBAK		31	23	27.74	136	52	42.29	148.0
10. Edwards Air Force Base (California, USA)	EDWAFB		34	57	50.14	242	05	07.71	767.0

Appendix C.

Earth Gravity Models used for Orbit Determination

Parameters for the Earth's Ellipsoid

Gravitational Constant (GM) [km³/sec²]	3.986032×10^5
Semi-Major Axis (a) [km]	6378.165
Flattening f	1/298.25

Normalized Spherical Harmonic Coefficients

		SAO M1		NWL 5E-6	
n	m	$\bar{C} \times 10^6$	$\bar{S} \times 10^6$	$\bar{C} \times 10^6$	$\bar{S} \times 10^6$
2	0	−484.1735		−484.194	
2	1			0.016	0.062
2	2	2.379	−1.351	2.446	−1.519
3	0	0.9623		0.984	
3	1	1.936	0.266	2.148	0.275
3	2	0.734	−0.538	0.978	−0.906
3	3	0.561	1.620	0.585	1.625
4	0	0.5497		0.507	
4	1	−0.572	−0.469	−0.495	−0.575
4	2	0.330	0.661	0.274	0.671
4	3	0.851	−0.190	1.030	−0.247
4	4	−0.053	0.230	−0.413	0.336
5	0	0.0633		0.045	
5	1	−0.079	−0.103	0.032	−0.119
5	2	0.631	−0.232	0.637	−0.328
5	3	−0.520	0.007	−0.389	−0.124
5	4	−0.265	0.064	−0.549	0.148
5	5	0.156	−0.592	0.215	−0.594
6	0	−0.1792		−0.219	
6	1	−0.047	−0.027	−0.085	0.192
6	2	0.069	−0.366	0.129	−0.457
6	3	−0.054	0.031	−0.020	−0.134
6	4	−0.044	−0.518	−0.193	−0.316
6	5	−0.313	−0.458	−0.093	−0.786
6	6	−0.040	−0.155	−0.324	−0.360
7	0	−0.0860		0.105	0.083
7	1	0.197	0.156	0.331	−0.195
7	2	0.364	0.163	0.350	0.045

Normalized Spherical Harmonic Coefficients (Continued)

		SAO M1		NWL 5E-6	
n	m	$\bar{C} \times 10^6$	$\bar{S} \times 10^6$	$\bar{C} \times 10^6$	$\bar{S} \times 10^6$
7	3	0.250	0.018	0.323	−0.244
7	4	−0.152	−0.102	−0.467	0.021
7	5	0.076	0.054	0.055	−0.244
7	6	−0.328	0.063	−0.477	
7	7	0.055	0.096		
8	0	0.0655			
8	1	−0.075	0.065		
8	2	0.026	0.039		
8	3	−0.037	0.004		
8	4	−0.212	−0.012		
8	5	−0.053	0.118		
8	6	−0.017	0.318		
8	7	−0.0087	0.031		
8	8	−0.248	0.102		
9	0	0.0122			
9	1	0.117	0.012		
10	00	0.0118			
10	01	0.105	−0.126		
10	02	−0.105	−0.042		
10	03	−0.065	0.030		
10	04	−0.074	−0.111		
11	00	−0.0630			
11	01	−0.053	0.015		
12	00	0.0714			
12	01	−0.163	−0.071		
12	02	−0.103	−0.0051		
12	12	−0.031	0.0008		
13	00	0.0219			
13	12	−0.06769	0.06245		
13	13	−0.059	0.077	−0.03	0.11
14	00	−0.0332			
14	01	−0.015	0.0053		
14	11	0.0002	−0.0001		
14	12	0.00261	−0.02457		
14	14	−0.014	−0.003		
15	09	−0.0009	−0.0018		
15	12	−0.07473	−0.01026		
15	13	−0.058	−0.046	−0.06	−0.06
15	14	0.0043	−0.0211	0.01	−0.03

Appendix D. Gravitational Field Description

$$U = \frac{GM}{r}\left\{1 + \sum_{n=2}^{k}\sum_{m=0}^{n}\left(\frac{a}{r}\right)^n p_n^m(\sin\Phi)\,[C_{nm}\cos m\lambda + S_{nm}\sin m\lambda]\right\},$$

where

G is the universal gravitational constant,
M is the mass of the earth,
r is the geocentric satellite distance,
a is the earth's mean equatorial radius,
φ is the sub-satellite geocentric latitude,
λ is the sub-satellite east longitude,
$p_n^m(\sin\Phi)$ are the associated Legendre polynomials of degree n and order m, and
C_{nm}, S_{nm} are the gravitational coefficients.

The gravitational coefficients are related to the normalized gravitational coefficients ($\overline{C}_{nm}$, $\overline{S}_{nm}$) as indicated below:

$$C(n,m) = [(n-m)!\,(2n+1)\,K/(n+m)!]^{1/2}\,\overline{C}(n,m),$$

$$S(n,m) = [(n-m)!\,(2n+1)\,K/(n+m)!]^{1/2}\,\overline{S}(n,m),$$

where,

$K = 1$ when $m = 0$
$K = 2$ when $m \neq 0$.

Revised Values for Coefficients of Zonal Spherical Harmonics in the Geopotential

By

Y. Kozai

Tokyo Astronomical Observatory, Mitaka, Tokyo, Japan and
Smithsonian Astrophysical Observatory, Cambridge, MA, USA

Abstract. From precisely reduced Baker-Nunn observations for 12 artificial satellites with inclinations between 28° and 96°, coefficients of zonal spherical harmonics up to the 21st order in the expression of the gravitational potential of the earth are derived.

Since Kozai's (1968) determination of the coefficients of zonal spherical harmonics for the geopotential, E. M. GAPOSCHKIN and his colleagues at the Smithsonian Astrophysical Observatory have obtained, from reduced Baker-Nunn observations, orbital elements of very high accuracy for several satellites. Consequently, complete analyses have been made for eight satellites in order to determine $(O–C)$ for secular motions and amplitudes of long-periodic terms, where the computed values are based on Kozai's coefficients determined in 1964. These values (KOZAI, 1964) are the following:

$$\left.\begin{aligned} J_2 &= 1082.639, & J_3 &= -2.546, & J_4 &= -1.649, & J_5 &= -0.210,\\ J_6 &= 0.646, & J_7 &= -0.333, & J_8 &= -0.270, & J_9 &= -0.053,\\ J_{10} &= -0.054, & J_{11} &= 0.302, & J_{12} &= -0.357, & J_{13} &= -0.114,\\ J_{14} &= 0.179, \end{aligned}\right\} \quad (1)$$

where the unit is 10^{-6} and the following values are used for the geocentric gravitation constant GM and the equatorial radius of the earth a_e:

$$GM = 3.98601 \times 10^{20}\ \text{cm}^3\,\text{sec}^{-2}, \quad a_e = 6.37816 \times 10^8\ \text{cm}. \quad (2)$$

In the present determination, 12 satellites are chosen. Their names as well as their anomalistic mean motions in revolutions per day, inclinations, eccentricities, and periods of observations used, are given in Table 1, which also presents approximate values of the secular motions

Table 1. *Satellites Chosen with Their Mean Orbital Elements, Secular Motions, and Amplitudes of* $\left.\begin{matrix}\sin\\ \cos\end{matrix}\right\}\omega$ *Terms*

Satellite	n [rev/day]	i	e	Periods [MJD]	$\dot{\omega}$ [deg/day]	$\dot{\Omega}$ [deg/day]	A_ω	$A_\Omega \times 10^2$	$A_i \times 10^3$	$A_e \times 10^4$
60 ν1	13.454	28.33°	0.0166	38924–39062	8.265°	−5.065°	1.656°	0.10°	−0.85°	4.80
59 α1	11.442	32.88	0.1660	36620–38530	5.258	−3.498	0.156	0.75	−6.98	4.62
	11.480	32.88	0.1642		5.293	−3.521	0.158	0.73	−6.92	4.62
62 $\alpha\varepsilon$1	9.126	44.80	0.2428	37870–38606	1.987	−1.860	0.112	1.64	−7.79	5.23
60 ι2	12.197	47.23	0.0114	37192–38576	2.978	−3.101	3.330	1.17	−0.40	6.63
63 26A	14.099	49.74	0.0614	38222–38988	3.505	−4.168	0.735	1.18	−2.38	7.94
	14.116	49.74	0.0607		3.514	−4.179	0.744	1.17	−2.35	7.95
62 $\beta\mu$1	13.345	50.14	0.0070	37974–38574	2.963	−3.609	6.363	1.15	−0.26	7.84
65 89A	11.968	59.38	0.0717	39074–39574	0.653	−2.247	1.008	3.76	−3.05	12.50
61 15A	13.870	66.82	0.0080	37548–38390	−0.695	−2.425	1.886	0.78	−0.05	2.64
64 01A	13.920	69.91	0.0015	38550–38866	−1.276	−2.133	29.000	0.04	−0.02	7.31
64 64A	13.746	79.70	0.0129	38698–39132	−2.535	−1.078	4.850	0.07	−0.15	10.99
65 81A	13.797	87.37	0.0747	39090–39472	−3.042	−0.282	0.910	0.08	−0.23	11.75
	13.814	87.37	0.0739		−3.050	−0.283	0.930	0.08	−0.23	11.76
61 $\alpha\delta$1	8.677	95.85	0.0121	38428–38972	−0.978	−0.210	3.760	−0.02	0.06	7.91

Table 2. *(O–C) and Residuals for Secular Motions*

	$\dot{\omega} \times 10^6$		$\dot{\Omega} \times 10^6$	
	(O–C)	v	(O–C)	v
60 ν1	170° ± 100°	40° ± 100°	−125° ± 5°	−1° ± 5°
59 α1	32 ± 1	1 ± 10	−9 ± 1	0 ± 5
62 αε1	40 ± 6	2 ± 6	7 ± 3	2 ± 3
60 ι2	220 ± 50	47 ± 50	−11 ± 1	4 ± 10
63 26A	920 ± 10	−52 ± 80	1 ± 1	19 ± 40
62 βμ1	600 ± 60	60 ± 100	−42 ± 1	8 ± 15
65 89A	−110 ± 20	−26 ± 20	−70 ± 10	−7 ± 10
61 15A	−300 ± 80	65 ± 80	22 ± 1	−3 ± 10
64 01A	600 ± 800	620 ± 800	56 ± 8	9 ± 8
64 64A	−400 ± 100	−110 ± 100	90 ± 10	15 ± 10
65 81A	620 ± 30	−8 ± 80	50 ± 1	−27 ± 18
61 αδ1	−35 ± 50	−47 ± 50	−2.9 ± 0.5	0.6 ± 0.5

and amplitudes of long-periodic terms due to spherical harmonics of odd orders.

Table 2 gives the values of (O–C) based on my previous coefficients (1) and the standard deviations mentioned there are from the analyses of observations. In the computation of the weights of the equations of condition, however, the standard deviations derived from the observations are not used, since the equations of condition include coefficients only up to 20th order. Neglect of higher order terms causes some errors in computed values. Therefore, the standard deviations assigned are increased for some data and are given as those of the residuals v. For determination of even-order harmonics amplitudes of long-periodic terms with argument 2ω are also used, although they are not given in Table 2.

Normal equations to determine corrections of coefficients of even-order harmonics are constructed by 52 equations of condition, and are solved at first with seven unknowns, for which Σv^2 is 216. Then the equations are solved with 8, 9, and 10 unknowns; Σv^2 for these cases are, respectively, 216, 33, and 33. The residuals based on the last solution for each equation is given in Table 2. The residuals for some data, especially for the secular motion of the node for 1965 81A, are considerably larger than their standard deviations. This suggests that it may be necessary to take into account more terms to compute their secular motions accurately.

In order to determine corrections to coefficients of odd-order harmonics, 46 equations of condition for the amplitudes of the long-periodic terms with argument ω are used. The standard deviations assigned to (O–C) are derived from observations, and those to the residuals, if they are given, are computed by taking into account effects of higher order

Table 3. *(O–C) and Residuals for Amplitudes of Long-Periodic Terms*

	$A_e \times 10^7$		$A_i \times 10^5$		$A_\omega \times 10^4$		$A_\Omega \times 10^4$	
	(O–C)	v	(O–C)	v	(O–C)	v	(O–C)	v
60 ν1	16 ± 10	6 ± 10	0° ± 30°	0°	40° ± 10°	−1° ± 10°	0° ± 3°	0°
59 α1	−31 ± 5	−1 ± 5	1 ± 5	−4	−17 ± 3	0 ± 6	−2 ± 3	2
62 αε1	−15 ± 8	2 ± 8	−2 ± 3	−4	−1 ± 2	−1 ± 2	2 ± 3	3
60 ι2	−26 ± 6	3 ± 12	−6 ± 6	6	−240 ± 30	+100 ± 100	0 ± 1	0
63 26A	−120 ± 10	20 ± 30	14 ± 15	10	−170 ± 20	−10 ± 40	−2 ± 4	1
62 βμ1	−80 ± 8	−10 ± 20	0 ± 10	0	−590 ± 40	0 ± 150	−2 ± 2	−2
65 89A	−40 ± 10	−20 ± 20	−8 ± 8	−7	30 ± 15	0 ± 20	10 ± 2	2
61 15A	−110 ± 10	40 ± 50	0 ± 5	0	−1900 ± 200	−800 ± 800	−3 ± 4	0
64 01A	−580 ± 10	−90 ± 90			−20000 ± 1000	100 ± 1000		
64 64A	−340 ± 50	−20 ± 50	0 ± 8	0	−1100 ± 200	300 ± 200	6 ± 3	1
65 81A	600 ± 30	−20 ± 30	−10 ± 10	−10	600 ± 20	30 ± 50	20 ± 2	2
61 αδ1	30 ± 15	0 ± 15	−6 ± 7	−6	−300 ± 500	−400 ± 500	−2 ± 2	2

harmonics. The equations are solved with 7, 8, 9, and 10 unknowns, and the values of Σv^2 are, respectively, 223, 38, 38, and 22.

The solutions with 10 unknowns for these two cases are,

$$\left.\begin{array}{llll}
dJ_2 = -0.011, & dJ_4 = 0.056, & dJ_6 = -0.144, \\
\pm 2 & \pm 7 & \pm 14 \\
dJ_8 = 0.152, & dJ_{10} = -0.300, & dJ_{12} = 0.315, \\
\pm 20 & \pm 25 & \pm 27 \\
dJ_{14} = -0.252, & dJ_{16} = 0.187, & \\
\pm 28 & \pm 26 & \\
dJ_{18} = -0.231, & dJ_{20} = -0.005, & \\
\pm 22 & \pm 22 & \\
dJ_3 = 0.008, & dJ_5 = -0.020, & dJ_7 = -0.028, \\
\pm 4 & \pm 7 & \pm 15 \\
dJ_9 = -0.047, & dJ_{11} = -0.100, & dJ_{13} = -0.009, \\
\pm 23 & \pm 35 & \pm 49 \\
dJ_{15} = -0,174, & dJ_{17} = 0.085, & \\
\pm 61 & \pm 65 & \\
dJ_{19} = -0.216, & dJ_{21} = 0.145, & \\
\pm 52 & \pm 29 &
\end{array}\right\} \quad (3)$$

and the coefficients thus determined are

$$\left.\begin{array}{llll}
J_2 = 1082.628, & J_3 = -2.538, & J_4 = -1.593, & J_5 = -0.230, \\
J_6 = 0.502, & J_7 = -0.361, & J_8 = -0.118, & J_9 = -0.100, \\
J_{10} = -0.354, & J_{11} = 0.202, & J_{12} = -0.042, & J_{13} = -0.123, \\
J_{14} = -0.073, & J_{15} = -0.174, & J_{16} = 0.187, & J_{17} = 0.085, \\
J_{18} = -0.231, & J_{19} = -0.216, & J_{20} = -0.005, & J_{21} = 0.145,
\end{array}\right\} (4)$$

where the unit is in 10^{-6}.

To reduce standard deviations for the solutions, more data and the inclusion of much higher order harmonics are necessary. There are gaps in the inclination around 40° and below 25°. Particularly, satellites with inclinations less than 25° are needed to reduce correlation among coefficients in the equations of condition.

The full text of this paper will appear in the Special Report No. 295 of the Smithsonian Astrophysical Observatory.

References

Kozai, Y. 1964. New Determination of Zonal Spherical Harmonics Coefficients of the Earth's Gravitational Potential. Smithsonian Astrophys. Obs. Spec. Rep. No. 165.

Kozai, Y. 1968. Improved Values for Coefficients of Zonal Spherical Harmonics in the Geopotential. In: Geodetic Satellite Results during 1967, ed. by C. A. Lundquist. Smithsonian Astrophys. Obs. Spec. Rep. No. 264, pp. 43–56.

Improved Values for the Tesseral Harmonics of the Geopotential and Station Coordinates

By

E. M. Gaposchkin

Smithsonian Institution, Astrophysical Observatory, Cambridge, MA, USA

Abstract. Baker-Nunn data on 24 close earth satellites, laser data from the five satellites with retroreflectors, and range and range-rate data from the GRARR system are combined in a global solution for gravity-field parameters and station coordinates. The data base represents a significant improvement on previous work, with increased accuracy and better distribution. The gravity field is given complete to 14th degree and order, and more than 29 station nets are determined.

1. Introduction

During 1966, the Smithsonian Astrophysical Observatory (SAO) published numerical parameters for the gravity field and station coordinates of the earth [1, 2], and the fundamental results were presented at the COSPAR meeting of that year.

A major revision of the process and programs has since been undertaken, and the results given here are the first from this effort. The basic method for determining the geodetic parameters is unchanged. However, the process can now deal with range and velocity measurements combined with optical data. The improved accuracy of these results is in part a consequence of the use of accurate range data from laser tracking equipment.

In addition to a significant change in the computer programs, many new data are available from observing programs organized specifically for this analysis. Also, there are some additional participating observatories, and several of the Baker-Nunn cameras comprising the basic network have been moved (9028, 9029, 9031). There are now 29 stations or station nets to be determined in the fundamental system.

2. Data Types Available

2.1. *Baker-Nunn and other optical data.* Baker-Nunn data were used exclusively in 1966. The accuracy was quoted as 4 arcsec. Subsequent analysis indicates that the accuracy of recently aquired data approaches 2 arcsec. The principal source of error now seems to be atmospheric turbulence.

2.2. *Select files*. A new mode in data taking and reduction results in observations of significantly higher accuracy. Using long-range forecasts, we choose those intervals, 30 days or longer, during which most of the stations can observe the satellite. More predictions are sent (i.e., 3 per pass), and the satellite is placed on a high-priority status. For select files, five frames are measured. The multiple-frame reductions are used to obtain synthetic observations, and many files are made up entirely of these data. The accuracy of synthetic observations is between 1 and 3 arcsec.

2.3. *Other optical data*. Optical data from the other Baker-Nunn cameras and from a group of European cameras (Haute Provence, Zimmerwald, Malvern), and some K-50 camera data taken by SAO are used.

2.4. *Laser data*. The laser systems at Organ Pass, New Mexico (7901), those operated by the Centre National d'Etudes Spatiales (CNES), Paris, France, at Haute Provence, France (7815), Colomb-Bechar, Algeria (7818), and Stephanion, Greece (7816), and that operated by the Goddard Space Flight Center (GSFC) at Greenbelt, Maryland (7050), have contributed data. The current laser systems require cornercube reflectors mounted on the satellite; five satellites are so equipped: 1964 64A, 1965 32A, 1965 89A, 1967 11A, and 1967 14A. Laser data have a nominal accuracy of 1.5 m [3].

2.5. *Goddard Range and Range Rate (GRARR) data*. GSFC operates four range and range-rate systems at Rosman, North Carolina (4713), Tananarive, Madagascar (4714), Carnarvon, Australia (4715), and Fairbanks, Alaska (4716). Data on three satellites, 1965 81A, 1965 89A, and 1968 2A, are available from this network.

3. Data Available

Table 1, summarizing the data chosen for the analysis, gives the satellite designation, other identification, and nominal orbital elements. There are a total of 244 arcs, although only one solution used all of them. Of the 24 satellites, 12 are objects not used in the 1966 solution. There is also new material on several of the original 12, the most important being the select files. The number of days per arc is given. In 1966, the arc lengths were typically 14 days. Now satellites with lower perigee heights (i.e., < 600 km) are used in 2-week arcs, but all other files can be given in 1-month arcs.

Table 1. *Summary of Dynamical Data*

Satellite	Name or Other Designation	Inclination [deg.]	Eccentricity	Semi-major Axis [km]	Perigee Height [km]	Number of Arcs	Days/arc	New Satellite	Select files	Laser Data	GRARR Data
1960 13A	Courier 1B (1960 $\nu 1$)	28	0.016	7465	965	6	30	×	×		
1959 1A	Vanguard 2 (1959 $\alpha 1$)	33	0.165	8300	557	33	30		×		
1959 1B	(1959 $\alpha 2$)	33	0.183	8502	567	8	14				
1959 7A	(1959 $\eta 1$)	33	0.188	8493	515	19	14				
1967 14A	D1 D	39	0.053	7337	569	4	17	×	×	×	
1967 11A	D1 C	40	0.052	7336	579	4	17	×	×	×	
1965 32A	Explorer 27 (BE-C)	41	0.026	7311	941	10	30	×	×	×	
1962 29A	Telstar 1 (1962) $\alpha\varepsilon 1$)	44	0.241	9672	962	13	30				
1960 9B	(1960 $\iota 2$)	47	0.011	7971	1512	46	30				
1962 68A	Relay 1 (1962 $\beta\gamma 1$)	48	0.284	10757	1316	6	30				
1962 60A	Anna 1B (1962 $\beta\mu 1$)	50	0.007	7508	1077	25	30		×		
1963 26A	Geophysical Research	50	0.062	7237	424	15	14				
1965 89A	Explorer 29 (GEOS 1)	59	0.073	8074	1121	10	30	×	×	×	×
1961 15A	Transit 4A (1961 01)	67	0.008	7318	885	8	30				
1961 15B	Injun 1 (1961 02)	67	0.008	7316	896	4	30				
1965 63A	Secor 5	69	0.079	8159	1137	2	30	×	×		
1964 1A		70	0.002	7301	921	4	30	×			
1964 64A	Explorer 22 (BE-B)	80	0.012	7362	912	2	30	×		×	
1965 81A	OGO 2	87	0.075	7344	420	9	14	×			×
1966 5A	Oscar 07	89	0.023	7417	868	1	30	×	×		
1963 49B	5 BN-2	90	0.005	7473	1070	5	30	×			
1961 28A	Midas 4 (1961 $\alpha\beta 1$)	96	0.013	10005	3503	3	50				
1968 2A	Explorer 36 (GEOS 2)	106	0.031	7709	1101	6	11	×			×
1965 78A	OV1 2	144	0.182	8306	416	2	30	×			
						244					

4. Theoretical Considerations

4.1. *Time system.* SAO has its own master clock and through VLF maintains in principle its own coordinated time system called A.S. It is within a few microseconds of A1. The principal reduction in time was to convert from UTC to A.S. The French laser data were given with reference to A3, and a simple table was used to convert this to A.S.

The UT1 data used in 1966 were the final values of the United States Naval Observatory (USNO). This year, values of the Bureau International de l'Heure (BIH) were adopted. The polar motion data were taken from the International Polar Motion Service (IPMS). The IPMS data are referred to the mean pole of 1900–1905.

4.2. *Resonance.* In the dynamical theory for the tesseral harmonics, divisors of the form $(l - 2p)\dot{\omega} + (l - 2p + q)n + m(\dot{\Omega} - \dot{\theta})$ appear in the formulas. With typical values of the orbital parameters, certain combinations of l, m, p, q give rise to small numerical values of these divisors, leading to large perturbations. The occurrence of resonant terms is a general property of a linearized solution. There is an extensive literature on the subject, but as yet no completely satisfactory treatment exists.

The occurrence of resonances between the gravity field of the earth and a satellite has been viewed as an opportunity to determine particular harmonics with extreme accuracy. Resonant harmonics were determined concurrently with the full gravity field, as was the case in [1].

4.3. *Initial values.* As in 1966, the determination of the zonal harmonics was a precursor to this analysis. Kozai [4] presents the results of the zonal-harmonic determination to $J(21)$, used here as starting values. The coefficients are listed for reference:

Zonal harmonics to J(21)

$J(2)$ =	1.08262800E-03	$J(3)$ =	−2.5380E-06
$J(4)$ =	−1.5930E-06	$J(5)$ =	−2.3000E-07
$J(6)$ =	5.0200E-07	$J(7)$ =	−3.6200E-07
$J(8)$ =	−1.1800E-07	$J(9)$ =	−1.0000E-07
$J(10)$ =	−3.5400E-07	$J(11)$ =	2.0200B-07
$J(12)$ =	−4.2000E-08	$J(13)$ =	−1.2300E-07
$J(14)$ =	−7.3000E-08	$J(15)$ =	−1.7400E-07
$J(16)$ =	1.8700E-07	$J(17)$ =	8.5000E-08
$J(18)$ =	−2.3100E-07	$J(19)$ =	−2.1600E-07
$J(20)$ =	−5.0000E-09	$J(21)$ =	1.4400E-07

The tesseral harmonics and station coordinates used as initial values are those given in 1966 and updated by GAPOSCHKIN and VEIS [5] and by GAPOSCHKIN [6].

The fundamental values for GM, a_e, and c adopted are unchanged:

$$GM = 3.986013 \times 10^{20}\ \mathrm{cm^3\,sec^{-2}},$$
$$a_e = 6.378155 \times 10^{6}\ \mathrm{m},$$
$$c = 2.997925 \times 10^{10}\ \mathrm{cm\,sec^{-1}}.$$

5. Analysis

The principal objective of the first solution in the present analysis was the determination of station coordinates. A network of 29 stations or station nets was chosen, along with 202 harmonic coefficients. These coefficients represent the geopotential $l = 12$, $m = 12$ with higher terms. The complete data set of 244 arcs was used, resulting in 94377 observation equations. A combination with simultaneous-observation data was also performed. The results of the combination solution were adopted for the initial coordinates for subsequent analysis.

The resonant harmonics were not very well determined, and a series of small iterations was performed to improve them before embarking on another large iteration. Forty-three files were used to improve 56 gravity-field parameters. Since for resonance there is a very important interaction between orbital elements and gravity-field parameters, the normal equations were reduced; i.e., the effects of the orbital elements were rigorously removed. Reducing the normal equation requires a great deal of computer time, and this step was not taken in the other analysis.

A second large iteration solving for the same combination of station coordinates and tesseral harmonics was performed. Somewhat fewer arcs were employed, and the final solution used 74045 observed quantities. The combination solution was again made, and this solution adopted.

Table 2 lists the corrections to the SAO station coordinates, the dynamical solution, and the adopted combination solution for the two iterations. LAMBECK [7] will discuss combination solutions and the intercomparison of the results.

A third and a fourth iteration were performed, the station coordinates being held fixed and gravity-field parameters being determined complete through $l = 14$, $m = 14$ with many higher terms, a total of 260 parameters. For these solutions the distribution of satellites is significantly different. The laser satellites and the select files formed the base, and other files were added to improve the distribution and to ensure that all resonant harmonics were sampled. The numerical results of the final iteration are given in Table 3.

Table 2. *Corrections to SAO Station Coordinates for Two Iterations, in m (Dynamical and Combinations Solution)*

Station Number	Location		Iteration 1 Dyn.	Iteration 1 Comb.	Iteration 2 Dyn.	Iteration 2 Comb.
9001	Organ Pass,	ΔX	8	6	3	1
7901	New Mexico	ΔY	−2	−2	4	−1
		ΔZ	6	5	5	0
9002	Pretoria,	ΔX	5	6	−3	−2
	South Africa	ΔY	13	13	8	6
		ΔZ	9	8	8	8
9003	Woomera,	ΔX	−5	−5	−8	−13
	Australia	ΔY	−5	−5	7	6
9023	Islans Lagoon, Australia	ΔZ	4	4	−7	−7
9004	San Fernando,	ΔX	0	2	−1	−4
	Spain	ΔY	4	6	9	4
		ΔZ	1	1	9	7
9005	Tokyo, Japan	ΔX	0	−2	4	−3
		ΔY	−5	2	4	9
		ΔZ	0	−4	−4	−9
9006	Naini Tal, India	ΔX	−6	−5	−6	−7
		ΔY	4	5	4	6
		ΔZ	3	5	2	3
9007	Arequipa, Peru	ΔX	−4	4	5	11
		ΔY	3	−1	2	−3
		ΔZ	3	8	14	20
9008	Shiraz, Iran	ΔX	−3	0	8	−1
		ΔY	8	3	5	−6
		ΔZ	2	−5	7	−12
9009	Curaçao,	ΔX	1	3	6	8
	Antilles	ΔY	3	−1	8	1
		ΔZ	−2	2	−15	−4
9010	Jupiter, Florida	ΔX	4	3	13	6
		ΔY	−10	−10	−1	−4
		ΔZ	2	2	5	1
9011	Villa Dolores,	ΔX	−4	2	0	13
	Argentina	ΔY	−4	−1	−1	5
		ΔZ	6	7	13	18
9012	Maui, Hawaii	ΔX	2	2	16	13
		ΔY	−2	−2	−8	−8
		ΔZ	−2	−1	−8	−3
9028	Addis Ababa,	ΔX	4	−2	−8	−29
	Ethiopia	ΔY	33	29	−8	−3
		ΔZ	−9	−8	−29	−12
9029	Natal, Brazil	ΔX	32	48	−3	42
		ΔY	9	20	−18	11
		ΔZ	17	19	3	28
9031	Comodore	ΔX	−49	−62	11	5
	Rivadavia,	ΔY	−53	−89	−3	−22
	Argentina	ΔZ	102	107	31	26
9091	Athens, Greece	ΔX	39	18	−21	−23
7816	Stephanion,	ΔY	−53	−52	3	3
	Greece	ΔZ	−36	−12	3	15

Table 3. *Final Iteration of Gravity-field Coefficients*

C 0202	2.4254 E-06	S 0202	−1.3886 E-06	C 0301	1.8890 E-06	S 0301	2.4283 E-07
C 0302	7.7442 E-07	S 0302	−6.9232 E-07	C 0303	6.8595 E-07	S 0303	1.4175 E-06
C 0401	−6.2146 E-07	S 0401	−4.5668 E-07	C 0402	3.2637 E-07	S 0402	6.6172 E-07
C 0403	8.9192 E-07	S 0403	−1.5576 E-07	C 0404	−1.3290 E-07	S 0404	3.6881 E-07
C 0501	−7.3825 E-08	S 0501	−6.0636 E-08	C 0502	5.7520 E-07	S 0502	−3.3186 E-07
C 0503	−4.0553 E-07	S 0503	−7.4684 E-08	C 0504	−2.8969 E-07	S 0504	2.1412 E-07
C 0505	1.0303 E-07	S 0505	−6.1585 E-07	C 0601	−5.8792 E-08	S 0601	1.6939 E-08
C 0602	1.4110 E-08	S 0602	−4.0484 E-07	C 0603	−4.6458 E-08	S 0603	1.0964 E-08
C 0604	1.2897 E-08	S 0604	−4.5315 E-07	C 0605	−3.0454 E-07	S 0605	−4.5616 E-07
C 0606	−1.1110 E-07	S 0606	−2.1435 E-07	C 0701	1.4744 E-07	S 0701	8.2090 E-08
C 0702	2.6101 E-07	S 0702	1.0599 E-07	C 0703	2.8865 E-07	S 0703	−1.7021 E-07
C 0704	−2.4043 E-07	S 0704	−5.8888 E-08	C 0705	−7.1596 E-08	S 0705	2.0261 E-08
C 0706	−9.6911 E-08	S 0706	−5.3172 E-08	C 0707	1.0161 E-08	S 0707	2.2874 E-08
C 0801	−9.0227 E-08	S 0801	9.9734 E-08	C 0802	1.6532 E-08	S 0802	1.3586 E-09
C 0803	−4.1668 E-08	S 0803	−4.4190 E-10	C 0804	−1.7103 E-07	S 0804	−2.7316 E-08
C 0805	−2.0515 E-07	S 0805	2.0780 E-09	C 0806	−2.9284 E-07	S 0806	1.9113 E-07
C 0807	5.4118 E-08	S 0807	−4.5432 E-08	C 0808	−1.8889 E-07	S 0808	2.6442 E-07
C 0901	1.5856 E-07	S 0901	6.6933 E-08	C 0902	2.6853 E-08	S 0902	−4.3179 E-08
C 0903	6.4882 E-08	S 0903	−1.9638 E-07	C 0904	−5.8112 E-08	S 0904	1.6871 E-07
C 0905	−2.7483 E-07	S 0905	−2.3172 E-07	C 0906	1.8756 E-07	S 0906	−7.0800 E-08
C 0907	−3.7641 E-07	S 0907	−1.1281 E-07	C 0908	2.1854 E-07	S 0908	1.2990 E-07
C 0909	−1.8079 E-08	S 0909	−4.3952 E-08	C 1001	1.7978 E-07	S 1001	−1.2145 E-07
C 1002	−1.8764 E-07	S 1002	−3.3856 E-08	C 1003	3.8713 E-08	S 1003	−3.1222 E-08
C 1004	−8.4700 E-08	S 1004	−3.4137 E-07	C 1005	−1.6657 E-07	S 1005	−8.0049 E-08
C 1006	−2.1078 E-07	S 1006	−1.8677 E-07	C 1007	1.2873 E-07	S 1007	−9.5396 E-09
C 1008	1.7516 E-07	S 1008	5.3480 E-09	C 1009	−2.5212 E-07	S 1009	−2.2043 E-07
C 1010	−4.5272 E-08	S 1010	−1.5199 E-07	C 1101	−5.0593 E-08	S 1101	−1.0891 E-07
C 1102	9.4266 E-08	S 1102	−1.8045 E-07	C 1103	−9.6861 E-08	S 1103	−1.7612 E-07
C 1104	3.5759 E-08	S 1104	−2.5963 E-07	C 1105	−1.2095 E-07	S 1105	1.7795 E-07
C 1106	−1.8964 E-07	S 1106	−2.2929 E-07	C 1107	−1.9079 E-07	S 1107	−1.2939 E-07
C 1108	2.7889 E-07	S 1108	−7.8948 E-09	C 1109	−2.2805 E-07	S 1109	−1.1865 E-07
C 1110	6.0993 E-08	S 1110	−2.9601 E-07	C 1111	−3.5274 E-07	S 1111	5.5789 E-08
C 1201	−1.4617 E-07	S 1201	−1.4706 E-08	C 1202	1.0119 E-07	S 1202	2.7808 E-08
C 1203	−5.7710 E-09	S 1203	−4.8572 E-08	C 1204	−1.2421 E-07	S 1204	−3.6757 E-07
C 1205	−1.0407 E-07	S 1205	2.5547 E-07	C 1206	−2.0828 E-07	S 1206	1.3037 E-08
C 1207	2.0469 E-07	S 1207	−9.2668 E-08	C 1208	1.3588 E-07	S 1208	−5.5234 E-08
C 1209	−3.7225 E-07	S 1209	−5.5694 E-08	C 1210	−3.5682 E-08	S 1210	3.5200 E-08
C 1211	−2.2488 E-07	S 1211	−5.4819 E-07	C 1212	−3.9027 E-08	S 1212	−1.2305 E-08
C 1301	3.9726 E-08	S 1301	5.0526 E-08	C 1302	7.9672 E-08	S 1302	−1.3046 E-07
C 1303	−1.3005 E-07	S 1303	−3.4983 E-7	C 1304	5.2045 E-08	S 1304	2.6359 E-07
C 1305	1.2045 E-07	S 1305	−3.0647 E-07	C 1306	5.7250 E-09	S 1306	2.1019 E-07
C 1307	−7.3660 E-08	S 1307	2.1052 E-07	C 1308	2.7212 E-07	S 1308	−1.9716 E-07
C 1309	−1.6228 E-07	S 1309	−2.4882 E-08	C 1310	−1.3343 E-07	S 1310	−1.9381 E-07
C 1313	−6.2448 E-08	S 1313	6.1823 E-08	C 1312	−6.0412 E-08	S 1312	1.0918 E-07
C 1402	−3.0512 E-08	S 1402	4.8635 E-08	C 1401	3.7690 E-08	S 1401	4.5836 E-08
C 1404	3.4391 E-08	S 1404	−1.3212 E-07	C 1403	−4.7246 E-08	S 1403	−6.1314 E-08
C 1406	4.6893 E-08	S 1406	2.8975 E-07	C 1405	5.9280 E-08	S 1405	−1.3595 E-08
C 1408	3.1004 E-07	S 1408	−4.2585 E-07	C 1407	3.9419 E-07	S 1407	4.4420 E-09
C 1410	−1.7151 E-07	S 1410	−7.9050 E-08	C 1409	−2.0911 E-07	S 1409	−5.0210 E-08
C 1412	6.8150 E-11	S 1412	−3.5491 E-08	C 1413	5.5284 E-08	S 1413	7.1516 E-08
C 1414	−4.7041 E-08	S 1414	−2.9202 E-08	C 1501	5.8095 E-08	S 1501	−7.7068 E-08
C 1502	−2.5850 E-08	S 1502	−2.4968 E-07	C 1509	−1.7344 E-07	S 1509	9.4020 E-08
C 1513	−3.7670 E-08	S 1513	−1.3087 E-08	C 1512	−6.6781 E-08	S 1512	3.6896 E-08
C 1515	−2.6090 E-09	S 1515	−4.0450 E-09	C 1514	1.0871 E-08	S 1514	−2.1258 E-08
C 1602	2.4628 E-07	S 1602	−2.6962 E-07	C 1601	9.2640 E-09	S 1601	3.7310 E-08
C 1612	1.3688 E-08	S 1612	−2.3639 E-08	C 1613	5.1342 E-08	S 1613	6.2217 E-08
C 1614	2.4130 E-09	S 1614	−1.8274 E-08	C 1712	8.5690 E-09	S 1712	2.3175 E-08
C 1713	1.4390 E-08	S 1713	−5.5430 E-09	C 1714	−4.8829 E-08	S 1714	2.4795 E-08
C 1813	8.5167 E-08	S 1813	5.3735 E-08	C 1814	−2.2983 E-08	S 1814	−3.6881 E-08
C 1912	−9.9970 E-09	S 1912	−4.8020 E-09	C 1913	1.2940 E-08	S 1913	−5.3811 E-08
C 1914	−3.2309 E-08	S 1914	−2.2662E-08	C 2013	1.1264 E-07	S 2013	9.4412 E-08
C 2014	−3.5536 E-08	S 2014	−3.1711 E-09	C 2113	−3.0953 E-08	S 2113	−1.4629 E-08
C 2114	1.3960 E-07	S 2114	1.6852 E-08	C 2214	−3.5670 E-09	S 2214	2.0455 E-08

Fig. 1. Level curves of geoid height at 10-m intervals, including terms through C/S 14,14. f = 1/298.257.

The numerical results have been superseded by further analysis. The final results are to be found in the Journal of Geophysical Research.

6. Discussion

As a first solution with revised programs, this analysis has improved the detailed knowledge of the gravity field to $l = 14$, $m = 14$ and higher order terms to $l = 22$, $m = 14$. The accuracy of the station coordinates for the 12 stations of the basic Baker-Nunn network seems to be 10 m; the other sites, with fewer data, have a larger uncertainty. The successful combining of optical, range, and velocity measurements demonstrates that each type of observation is useful for satellite geodesy.

It still remains to determine a geopotential that can give a 10-m satellite position, especially for a satellite with low perigee height, where errors in the gravity-field model are magnified. For satellite 1961 28A ($q = 3500$ km), it is possible to determine orbits to the accuracy of the data (2 to 3 arcsec). For a geodetic satellite such as 1965 89A ($q = 1100$ km), residuals for laser observations approach 10 m.

Future analysis will improve both the station coordinates and gravity-field model and the determination of *GM*. There is nothing to prevent the ultimate accuracy of site locations from reaching the limit imposed by UT1 measurements (5 m). The gravity field can be improved by continued analysis of existing data, but more important improvements will come with greater amounts of more accurate data, e.g., laser data, or with better data distribution, such as could be available from Doppler tracking on a polar satellite and from a low-inclination geodetic satellite. With more laser data, the scale constant (*GM*) can be improved. This improvement could be attempted now, but station positions should first be known to 5 m in order to ensure an accuracy in *GM* comparable to that determined by the Jet Propulsion Laboratory (JPL) using deep-space probes.

References

1. Gaposchkin, E. M. (p. 685), Köhnlein, W. J. (p. 994), Veis, G. (pp. 776, 778), Whipple, F. L. (p. 675): Space Research VII, Amsterdam: North Holland 1966.
2. Lundquist, C. A., Veis, G. (eds.): Geodetic Parameters for a 1966 Smithsonian Institution Standard Earth. Smithsonian Astrophys. Obs. Spec. Rep. No. 200 (1966).
3. Lehr, C. G.: Geodetic and Geophysical Applications of Laser Satellite Ranging. Presented at the IEEE Symposium on Geoscience Electronics, April 1969.
4. Kozai, Y.: Revised Values for Coefficients of Zonal Spherical Harmonics in the Geopotential. Smithsonian Astrophys. Obs. Spec. Rep. No. 295 (1969).

5. Gaposchkin, E. M., Veis, G.: Space Research VIII, p. 42, Amsterdam: North Holland 1967.
6. Gaposchkin, E. M.: Dynamical Determination of Station Location using Geos 1 Data. Smithsonian Astrophys. Obs. Spec. Rep. No. 264 (1967) 25.
7. Lambeck, K.: Comparisons and Combinations of Geodetic Parameters estimated from Dynamic and Geometric Satellite Solutions and from Mariner Flights. Presented at the 12th Meeting of COSPAR, Prague, May 1969.

On the Determination of Parameters of the Earth's Figure from Satellite Orbit Dynamics

By

M. Burša

Research Institute of Geodesy, Prague, ČSSR

Abstract. From observations of perturbations of Earth satellite orbits, various sets of harmonic coefficients have been derived in the development of the gravitational potential of the Earth. If the geocentric constant is given, it is possible, using these quantities, to determine the shape of the equipotential surfaces of the external gravitational potential or of the gravity potential. To determine the dimensions of the surfaces, however, further data would be necessary: the value of the potential on the surface investigated, or the radius-vector of the surface at least in one special case. The paper will be limited to the basic shape characteristics – to the polar and equatorial flattening and the orientation of the equatorial ellipse.

Different ways of determining the "best fitting" parameters are discussed; e.g., under minimum condition of the integral of the squares of Gaussian anomalous curvatures over the whole reference surface, using large sets of harmonics available.

Numerical values obtained: Potential on geoidal surface $W_0 = 62637.23 \times 10^3 \, m^2 s^{-2}$, scale factor for lengths $R_0 = 6363675$ m, $a = 6378138$ (mean), $\alpha = 1:298.286$ (mean polar), $\alpha_1 = 1:89000$ (mean equatorial).

Published in: Studia Geophysica et Geodaetica 13 (1969) 337.

Accuracy of Satellite Orbits Obtainable by Synchronous Satellite Tracking

By

C. F. Martin

Wolf Research and Development Corporation, Applied Sciences Division, Space and Planetary Sciences Department, Riverdale, MD, USA

Abstract. A comprehensive error analysis is made of the accuracy of satellite orbits (both low altitude and highly eccentric) which can be obtained from range and range rate measurements by synchronous satellites. Included in the investigation are the effects of measurements noise and biases, earth based station location errors, and geopotential coefficient errors. Results are presented for each source in the form of satellite epoch element errors and position errors propagated along the trajectory for a nominal three day tracking period. All correlations are fully accounted for between the effects of the various error sources on epoch elements and on trajectory. All analyses are performed for tracking by both a single synchronous satellite and by three synchronous satellites spaced 120° apart.

I. Introduction

The possibility of obtaining satellite orbits by tracking by synchronous, or near synchronous, satellites is intriguing for a number of reasons. Among these are the capability of near continuous tracking and the fact that the tracking satellite moves, in general, completely around the tracked satellite orbital plane each day. The synchronous satellite thus has some of the advantages of a series of ground based stations distributed around the equator. Against these advantages must be balanced the large distance to the synchronous satellite and the difficulty of accurately positioning the synchronous satellite. It is the subject of this paper to balance these factors against each other and estimate the orbital accuracy which can be obtained by synchronous satellite tracking of both a low and a high eccentricity satellite.

The basis upon which the satellite orbital accuracy estimates are based are briefly outlined in Sec. II. Sec. III discusses the accuracy with which the position of the synchronous satellites can be obtained. Sec. IV then presents the main results of this paper, namely the orbital accuracy estimates for tracking by one and by three synchronous satellites spaced

120° around the equator. Finally, Sec. V presents conclusions which may be drawn from this study as to the general capabilities of satellite to satellite tracking, the major sources of error, and projections as to its future capabilities.

II. Accuracy Estimation Technique

A theoretical orbital accuracy estimate can be viewed as nothing more than the transformation or propagation of the variance-covariance matrix of the errors in the parameters upon which the orbit depends. The problem lies in obtaining this variance-covariance matrix. The most common procedure for obtaining definitive orbits is to iterate a Bayesian least squares solution for increments to the set of orbital parameters defining the orbit. The measurement model for such a solution may be expressed as

$$m = m_0 + A\,\delta a + K\,\delta\gamma + \varepsilon \tag{1}$$

where

m is the actual measurement set,
m_0 is the measurement set as computed from the previous iteration's solution,
$A = \partial m/\partial a$ is the matrix of partial derivatives of the measurements with respect to the orbital element parameters,
δa is the adjustment of the orbital elements to be solved for,
γ is a set of corrections which should be applied to parameters affecting the actual or calculated measurements. Some error, $\delta\gamma$, is assumed to exist in these corrections, but no adjustments are to be made,
$K = \partial m/\partial\gamma$ is the matrix of partial derivatives of the measurements with respect to the unadjusted γ parameters,
ε is the vector of measurement noise.

Ignoring any errors which might exist in the γ parameters, the standard weighted least squares solution of (1) for adjustments to the orbital elements is

$$\delta\hat{a} = (A^T W A)^{-1} A^T W (m - m_0), \tag{2}$$

where W is the measurement weight matrix. In orbit determination, the matrix is generally assumed to be based upon an estimate of the measurement noise. Specifically, assuming that W is the inverse of the variance of ε,

$$W^{-1} = E(\varepsilon\,\varepsilon^T). \tag{3}$$

The variance of the estimate provided by (2) can now be calculated from the definition

$$\operatorname{Var} \delta \hat{a} = E[(\delta \hat{a} - \delta a)(\delta \hat{a} - \delta a)^T]. \tag{4}$$

From Eqs. (1) and (2),

$$\begin{aligned} \delta \hat{a} - \delta a &= (A^T W A)^{-1} A^T W (A\, \delta a + K\, \delta \gamma + \varepsilon) - \delta a \\ &= (A^T W A)^{-1} A^T W (K\, \delta \gamma + \varepsilon). \end{aligned} \tag{5}$$

Substituting into (4), and assuming that the errors in the γ parameters are uncorrelated with the measurement noise,

$$E(\gamma\, \varepsilon^T) = 0, \tag{6}$$

it follows that

$$\operatorname{Var}(\delta \hat{a}) = (A^T W A)^{-1} + [(A^T W A)^{-1} A^T W K] \operatorname{Var} \gamma\, [(A^T W A)^{-1} A^T W K]^T. \tag{7}$$

The total variance is thus essentially decomposed into a noise contribution and an unadjusted parameter contribution. In most practical situations, the noise contribution is negligible compared to the unadjusted parameter effects.

The orbital accuracies quoted in this paper are based upon a computer implementation of Eq. (7) and its propagation along the satellite orbit. The computations are made in such a manner that the contributions of the effect of each γ parameter error is made separately so that the most deleterious error sources are immediately recognizable.

III. Accuracy of Synchronous Satellite Orbit

To obtain an estimate of the accuracy with which a synchronous satellite could be positioned, the orbit was assumed to be obtained by continuous tracking for 2 weeks (one measurement/30 min) by 3 radars making range and range rate measurements. The stations used, along with the measurement and station position errors assumed, are shown in Table 1. The satellite itself was located over the Atlantic, so a reason-

Table 1. *Stations used to Estimate Synchronous Satellite Position Accuracy*

Station Location	Position Error			Range Noise	Range Rate Noise	Range Bias
	σ_{Lat} [m]	σ_{Lon} [m]	σ_{Height} [m]	[m]	[cm/sec]	[m]
Winkfield, England	15	0	15	1	.2	5
Pretoria, South Africa	30	30	15	1	.2	5
Merritt Island, Fla., USA	15	15	15	1	.2	5

ably good tracking geometry was realized. In addition to the station position and measurement noise errors, geopotential coefficient errors were also propagated. The model for geopotential coefficient errors is described in the next section.

The dominant source of error in positioning the synchronous satellite was found to be the set of station position errors. By comparison, measurement noise and geopotential coefficient errors have negligible effects. The 5 m range biases have effects comparable to the station position errors. The minimum total position accuracy, including all the error sources, is shown in Table 2. This minimum occurs near the middle of

Table 2. *Estimated Synchronous Satellite Position Accuracy*

σ_{Latitude}	5 m
$\sigma_{\text{Longitude}}$	125 m
σ_{Height}	6 m

the tracking period. By far the largest component of error is in the along track direction. For low altitude satellites this is, fortunately, a direction nearly perpendicular to all range measurements and would thus be expected to make a minimal contribution to satellite orbit determination error.

IV. Satellite Determined Satellite Orbits

To arrive at an assessment of the capabilities for determining satellite orbits by synchronous satellite tracking, two classes of tracked satellites were investigated. Keplerian orbital elements used as examples of these two classes are shown in Table 3. Existing satellites with these elements are GEOS-II for the low eccentricity satellite and IMP-IV for the high eccentricity satellite.

Table 3. *Keplerian Elements for Satellites Tracked by Synchronous Satellites*

	GEOS-II	IMP-IV
Semi-major axis (meters)	7716993	111932770
Eccentricity	.031	.907
Inclination (degrees)	105.8	67.2
Right Ascension of Ascending Node (degrees)	230.3	165.2
Argument of Perigee (degrees)	314.6	188.4

The synchronous satellites were simulated by fixed tracking stations making range and range-rate measurements from synchronous satellite altitude. Position errors for these stations were taken to be those appropriate for a synchronous satellite as given in Table 2. Measurements were assumed to have the accuracies shown in Table 4 and to exist every ten minutes for GEOS-II and every 15 min for IMP-IV. The duration of track was taken to be 3 days (about 36 satellite revolutions) for GEOS-II and 4 days (one satellite revolution) for IMP-IV.

Table 4. *Measurement Accuracy for Satellite to Satellite Tracking*

Measurement	Noise	Bias
Range	5 m	10 m
Range-Rate	2 cm/sec	None

The accuracy with which a satellite orbit can be generated, starting from a given position and velocity at some epoch, is limited primarily (for satellites not significantly affected by atmospheric drag) by uncertainties in the earth's geopotential coefficients. In general, published sets of geopotential coefficients do not have an accompanying variance-covariance matrix to express the uncertainty in the set. Even if they did, they would probably be of little use for the very reason that the first term in Eq. (7) is not a valid measure of the true uncertainty in estimated satellite orbital elements. That is, unadjusted parameter errors would predominate over effects of measurement noise.

Sets of geopotential coefficients developed by different individuals or organizations would be expected to have somewhat different unmodeled errors in their computation. Consequently, gravity models developed independently using data of comparable accuracy should be able to produce orbits of comparable accuracy. Further, differences between orbits which they produce should be a valid measure of the accuracy of the orbits. The SAO M1 gravity model [1] is presently considered to be one of the most accurate widely available geopotential models. The APL 3.5 gravity model [2] is also widely available and is essentialy complete through eighth degree and order tesserals and twelfth degree zonals. The procedure adopted for obtaining errors in low degree and order coefficients was then to difference the SAO and APL tesserals through (8,8) and zonals through J_{12}. These differences are shown in Table 5. Since the SAO model is considered to be more accurate than the APL model [3], only one fourth of these differences were then propagated as geopotential errors.

It should be emphasized that the complete set of coefficients shown in Table 5 constitutes a single error model term. The net effect that is propagated is one fourth of the differences that would result between performing the data reduction using the SAO M1 gravity model and using the APL 3.5 gravity model.

Table 5. *SAO-APL Gravity Model Differences. One-fourth of these Differences is the Assumed Error in Low Degree and Order Coefficients*

Index		Value		Index		Value	
N	*M*	*C*	*S*	*N*	*M*	*C*	*S*
2	0	$-.547 \times 10^{-7}$		3	0	$.13 \times 10^{-6}$	
4	0	$-.249 \times 10^{-6}$		5	0	$-.182 \times 10^{-6}$	
6	0	$.276 \times 10^{-6}$		7	0	$-.26 \times 10^{-6}$	
8	0	$-.34 \times 10^{-6}$		9	0	$-.23 \times 10^{-6}$	
10	0	$-.54 \times 10^{-7}$		11	0	$.802 \times 10^{-6}$	
12	0	$-.667 \times 10^{-6}$		2	2	$.93 \times 10^{-9}$	$.987 \times 10^{-7}$
3	1	$-.104 \times 10^{-6}$	$-.548 \times 10^{-7}$	3	2	$.165 \times 10^{-6}$	$-.48 \times 10^{-7}$
3	3	$.140 \times 10^{-7}$	$-.894 \times 10^{-7}$	4	1	$.946 \times 10^{-8}$	$.273 \times 10^{-7}$
4	2	$.210 \times 10^{-7}$	$-.488 \times 10^{-7}$	4	3	$-.318 \times 10^{-9}$	$.118 \times 10^{-7}$
4	4	$-.333 \times 10^{-8}$	$-.840 \times 10^{-9}$	5	1	$.185 \times 10^{-6}$	$-.547 \times 10^{-7}$
5	2	$-.586 \times 10^{-7}$	$-.172 \times 10^{-7}$	5	3	$.202 \times 10^{-7}$	$.319 \times 10^{-8}$
5	4	$-.174 \times 10^{-8}$	$-.252 \times 10^{-8}$	5	5	$-.467 \times 10^{-9}$	$-.186 \times 10^{-9}$
6	1	$.368 \times 10^{-7}$	$.100 \times 10^{-6}$	6	2	$-.286 \times 10^{-7}$	$.261 \times 10^{-7}$
6	3	$.121 \times 10^{-7}$	$.417 \times 10^{-9}$	6	4	$-.995 \times 10^{-9}$	$.343 \times 10^{-10}$
6	5	$.108 \times 10^{-9}$	$-.409 \times 10^{-10}$	6	6	$.127 \times 10^{-10}$	$-.178 \times 10^{-10}$
7	1	$-.517 \times 10^{-7}$	$-.455 \times 10^{-7}$	7	2	$.938 \times 10^{-8}$	$-.102 \times 10^{-7}$
7	3	$.203 \times 10^{-8}$	$-.316 \times 10^{-8}$	7	4	$.325 \times 10^{-10}$	$.218 \times 10^{-9}$
7	5	$-.470 \times 10^{-10}$	$-.853 \times 10^{-10}$	7	6	$-.171 \times 10^{-10}$	$.482 \times 10^{-10}$
7	7	$.670 \times 10^{-12}$	$-.448 \times 10^{-11}$	8	1	$-.498 \times 10^{-7}$	$.783 \times 10^{-7}$
8	2	$.564 \times 10^{-8}$	$-.629 \times 10^{-8}$	8	3	$-.182 \times 10^{-9}$	$.215 \times 10^{-8}$
8	4	$.187 \times 10^{-9}$	$.649 \times 10^{-10}$	8	5	$.241 \times 10^{-10}$	$-.218 \times 10^{-10}$
8	6	$-.173 \times 10^{-12}$	$.971 \times 10^{-11}$	8	7	$.910 \times 10^{-12}$	$-.515 \times 10^{-12}$
8	8	$.130 \times 10^{-12}$	$-.987 \times 10^{-14}$				

In addition to errors in low degree and order geopotential coefficients, errors in resonant coefficients were also propagated for the GEOS-II satellite. Since the resonant coefficients for GEOS-II all have similar effects (i.e., approximately the same period), errors were assumed for only a single pair of coefficients. The 13th degree and order pair of coefficients have the largest resonance effect, so an error of 5% of their nominal values was chosen as a reasonable upper bound for the present or near future uncertainty in resonant coefficients.

The effects of the various error sources on the GEOS-II orbital parameters are shown in Table 6 for tracking by one and by three synchronous satellites. (For tracking by 3 satellites, station errors are similar for all stations, so only a typical station's errors are listed.) The error source having the largest effects on the orbital elements for both cases is the set of low degree and order geopotential coefficients. Further, the effects are not appreciably different between the one and three satellite tracking situations.

Table 6. *Effects of Various Error Sources on GEOS-II Keplerian Epoch Elements. Effects on Angles are in Arc Seconds*

Number of Tracking Satellites	Orbital Parameter	Meas Noise	Range Bias	Stat. Long.	Stat. Lat.	Stat. Ht	Low Order Gravity	Resonance Coefficients	
								C	S
One	a	.007655 m	−.003 m	.000	.000	.002 m	−.585 m	.093 m	−.106 m
	e	$.7 \times 10^{-7}$	$-.14 \times 10^{-6}$	.0	$-.7 \times 10^{-7}$	$-.8 \times 10^{-7}$	$.5 \times 10^{-6}$	$.2 \times 10^{-7}$	$.4 \times 10^{-7}$
	i	.014532	−.082	.591	−.000	.048	.109	−.015	.013
	Ω	.016191	−.009	.001	−.001	−.005	−.381	.003	−.007
	ω	.297104	−.488	−.001	−.224	−.287	8.118	−.002	.159
	M	.301654	−.458	−.001	.214	.270	−8.523	.212	−.284
Three	a	.00417 m	.002 m	.001 m	.001 m	.001 m	−.586 m	.086 m	−.103 m
	e	$.4 \times 10^{-7}$	$-.5 \times 10^{-7}$	$-.25 \times 10^{-7}$	$-.25 \times 10^{-7}$	$-.3 \times 10^{-7}$	$.59 \times 10^{-6}$	$.5 \times 10^{-8}$	$.2 \times 10^{-7}$
	i	.008348	.025	−.000	−.000	.015	.143	−.014	.014
	Ω	.00935	−.002	.198	−.000	−.001	−.360	−.003	−.005
	ω	.171409	−.159	−.008	−.076	−.094	7.858	−.001	.803
	M	.172409	.154	.011	.074	.091	−8.218	.188	−.198

The propagation of the total error set into the first 9 h of the three day GEOS-II orbit is shown in Fig. 1. The strong variations in the accuracy shown are due to the particular choice of the geopotential coefficient error model and are not to be construed as true accuracy variations. The main conclusion which we can draw from Fig. 1 is that the overall orbital accuracy is about 30 m. In addition, it is approximately

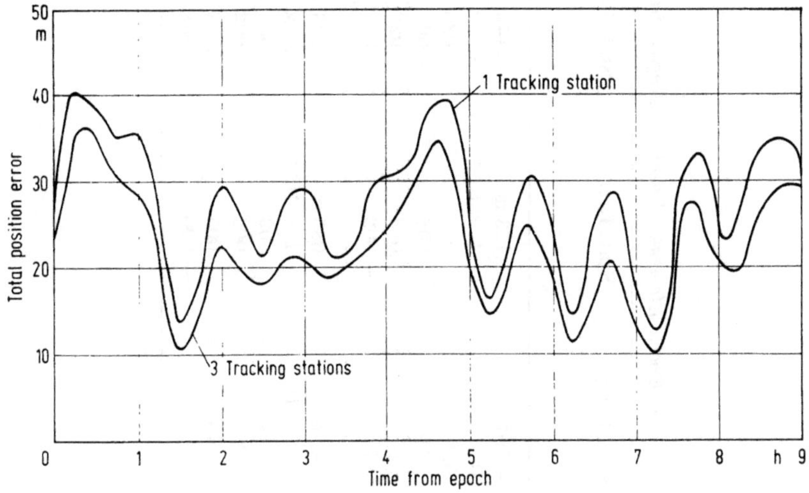

Fig. 1. Orbital accuracy for GEOS-II.

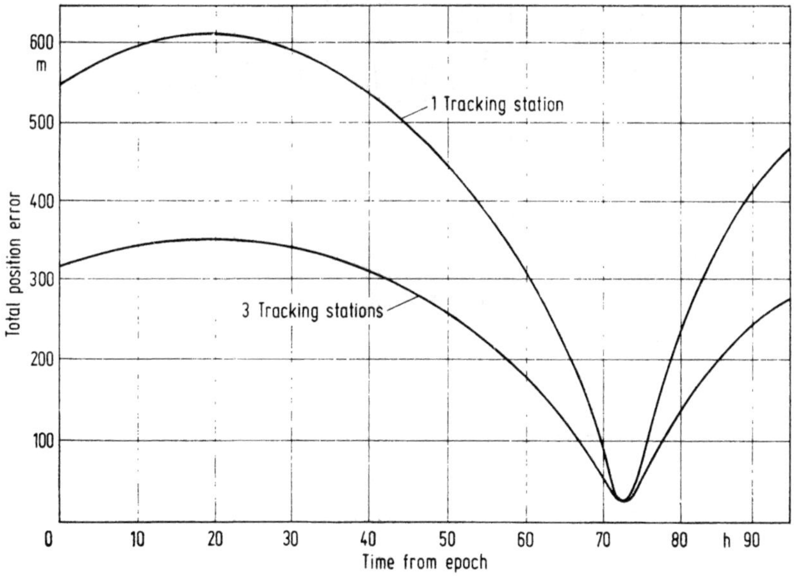

Fig. 2. Orbital accuracy for IMP-IV.

Table 7. *Effects of Various Error Sources on IMP-IV Keplerian Epoch Elements. Effects on Angles are in Arc Seconds*

Satellite	Orbital Parameter	Meas Noise	Range Bias	Stat. Long.	Stat. Lat.	Stat. Ht.	Low Order Gravity
	a	.243 m	−.1.391 m	−.001 m	.173 m	.142 m	.057 m
	e	$.3 \times 10^{-8}$	$-.63 \times 10^{-7}$	0.0	$.4 \times 10^{-8}$	$-.8 \times 10^{-8}$	$.2 \times 10^{-8}$
One	i	.006	.054	0.0	0	.033	.017
Tracker	Ω	.003	.026	.59	−.005	−.003	.004
	ω	.006	−.064	0	.018	.022	.005
	M	.003	−.017	0	−.002	0	−.001
	a	.136 m	−.138 m	.996 m	.085 m	.123 m	−.267 m
	e	$.2 \times 10^{-8}$	$-.3 \times 10^{-7}$	$.8 \times 10^{-8}$	$.1 \times 10^{-8}$	$-.6 \times 10^{-8}$	$.4 \times 10^{-8}$
Three	i	.00342	−.010	−.014	.002	−.001	.011
Trackers	Ω	.001 607	.002	.203	−.002	−.003	.000
	ω	.007 976	−.014	−.014	.005	.008	.017
	M	.001 667	−.002	.020	−.000	.001	−.004

constant throughout the three day tracking period and is only marginally improved by the use of more than one tracking satellite.

For the high eccentricity satellite, the errors in the orbital elements are shown in Table 7. The dominant source of error now is not geopotential errors, but station position error. For this case, the use of more than one tracker does substantially improve the overall orbit accuracy, which is shown in Fig. 2. The somewhat peculiar shape of the accuracy curves for the IMP satellite is due to geometry. The sharp minimum occurs at the time of the satellite's closest approach to the earth. It may be noted also from Table 7 that the major orbit error is one of orientation, in particular, the right ascension of the ascending node.

V. Conclusions

Based on the results presented in the preceding sections, we can conclude that:

1. Particularly for low altitude satellites, a single synchronous satellite is capable of producing satellite orbits almost as good as several such trackers.

2. An accuracy on the order of 30 m can be achieved for low altitude satellites.

3. An accuracy of about 500 m can be achieved for a satellite which has a semi-major axis of 10^8 m or more.

4. The accuracy of satellite orbits determined by synchronous satellites can be improved in the same way as can orbits determined by earth based trackers, viz., more accurate geopotential coefficients and more accurate station positions.

5. Tracking by synchronous satellites offers a promising means of maintaining definitive orbits for both low and high eccentricity satellites.

References

1. Lundquist, C. A., Veis, G.: Geodetic Parameters for a 1966 Smithsonian Institution Standard Earth. Smithsonian Astrophys. Obs. Spec. Rep. No. 200 (1966).
2. Guier, W. H., Newton, R. R.: The Earth's Gravitational Field as Deduced from the Doppler Tracking of Five Satellites. J. Geophys. Res. 70 (18), September 1965.
3. Marsh, J. G.: Orbital Analysis and Prediction for the GEOS Satellites Using Different Sets of Geopotential Coefficients. Trans. Am. Geophys. Union 50 (4), April 1969.

Resonant Satellite Geodesy by High Speed Analysis of Mean Kepler Elements

By

C. A. Wagner

National Aeronautics and Space Administration – Goddard Space Flight Center, Greenbelt, MD, USA

and

B. C. Douglas

Wolf Research and Development Corporation, Riverdale, MD, USA

Abstract. A general resonant orbit and gravity constant determining program has been developed accepting short arc mean Kepler elements as data. The evolution of these elements is calculated by numerical integration of their long period and secular variations.

With only slowly-changing mean element coordinates being integrated, a step size of the order of an orbit revolution or more is achieved. Satellite ephemerides over 5000 revolutions are calculated in about a minute on an IBM/360 computer. Partial derivatives of the evolved mean elements with respect to initial values and gravity constants are readily evaluated from numerically generated variant trajectories.

To test the method, simulated and actual orbit data from long resonant twelve-hour trajectories have been processed for gravity information by a least-squares technique. The gravity recovery is in excellent agreement with the model for the simulated data, and with previous analytic results from the actual data on Cosmos 41.

Extensive data from many resonant 12 and 24 h satellites is currently being processed by the mean element program. Preliminary determinations from this data are given. The final result should be definitive information on more than a quarter of the longitude harmonics through 8,8.

Introduction

Resonant satellite geodesy has undergone considerable elaboration since 1961 when A. H. Cook of England wrote (to our knowledge) the first theoretical treatise on the subject [1]. At present we think the theoretical aspects are understood for most orbit classes, although a general analytic theory of satellite orbital resonance is still an unsolved problem of celestial mechanics. However, we have developed a general semi-analytic theory that should be adequate for the purposes of resonant satellite geodesy. This theory and its application to the solution for the geopotential is the main subject of our paper.

Since 1965, there has been a startling increase of the data base for resonant satellite geodesy. What used to be regarded as a rather esoteric theoretical subject with only a few isolated examples has multiplied to such an extent that we should now speak of the nonresonant orbit as the isolated exception rather than the rule. In April 1968, 600 objects were being tracked in the United States. Ninety seven of these objects had primary resonant beat periods greater than 10 days (see Fig. 1). The primary resonant orbits are those with nearly repeating ground tracks in one day. Eighteen of these had beat periods greater than one-hundred days.

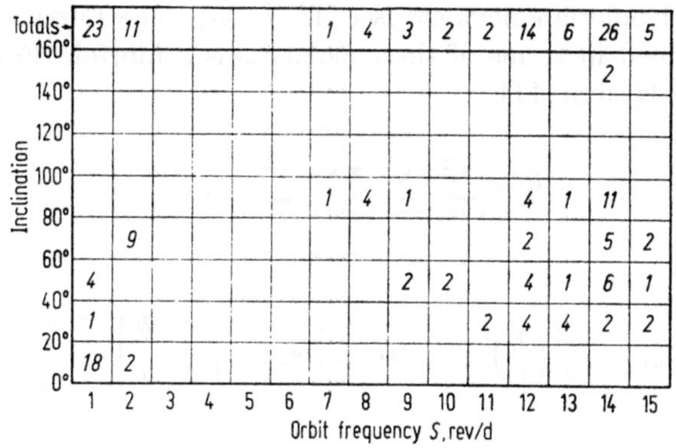

Fig. 1. Characteristics of 97 existing resonant orbits with beat periods greater than 10 days.

Most remarkably, six objects were found with nearly commensurate orbits in two days [2]. These also had greater than 100 day beat periods. When analyzed, they should provide the first really good information from satellites on many constants of odd order between $m = 17$ and 29.

Many orbits in shallow resonance (i.e., with less than about 10 day beat period) have already been used by the Smithsonian Astrophysical Observatory (SAO) and others for geodetic solutions [3, 4, 5, 6]. In addition to these, many deeply resonant orbits [7] should provide definitive solutions for a large number of longitude harmonics by analyses of mean elements similar to those of Kozai [8] and King-Hele's group [9, 10] on the zonals.

Referring to Fig. 1 there appears to be a sufficient number of existing orbits resonant with terms of order 12 to 15 with a good distribution of inclinations and eccentricities to allow the definitive determination of many of these constants. For 8th to 12th order terms, there are fewer resonant orbits with a more restricted distribution of elements, but possibly still enough to give definitive determinations of a few constants of these orders. At the low frequency end of the spectrum the situation is very favorable for definitive geodetic solutions because of the large number of deliberately commensurate communications satellites.

Integration of Resonant Orbits

Previous analytic solutions for the evolution of resonant orbits have been restricted in one or more ways which preclude universal application. This general problem includes not only all relevant resonance effects from rapidly circulating to librating orbits, but also should not be restricted by orbit inclination or eccentricity. We achieve a uniform solution by numerically integrating the coupled Lagrange planetary equations for the Keplerian elements (see [11, p. 23]). The Earth's disturbing potential defined in terms of these elements is a fourfold Fourier series due to W. M. Kaula [11].

$$R = \sum_{l=2}^{\infty} \sum_{m=0}^{l} \sum_{p=0}^{l} \sum_{q=-\infty}^{\infty} V_{lmpq},$$

where:

$$V_{lmpq} = \frac{\mu}{a}\left(\frac{a_e}{a}\right)^l F_{lmp}(i)\, G_{lpq}(e)\, J_{lm} \begin{bmatrix} \cos\Psi \\ \sin\Psi \end{bmatrix}_{l-m \text{ odd}}^{l-m \text{ even}},$$

and:

$$\Psi = [(l-2p)\,\omega + (l-2p+q)\,M + m\,(\Omega - \theta_g - \lambda_{lm})]\,.$$

The C_{lm}, S_{lm} and J_{lm}, λ_{lm} sets of gravity coefficients are related by:

$$C_{lm} = J_{lm} \cos m\,\lambda_{lm}$$

$$S_{lm} = J_{lm} \sin m\,\lambda_{lm}\,.$$

In this representation a, e, i, M, Ω, and ω are the conventional Kepler elements, semi-major axis, eccentricity, inclination, mean anomaly, right ascension of the ascending node and argument of perigee. The constants μ, a_e and θ_g are the Earth's Gaussian gravity constant, mean equatorial radius, and Greenwich hour angle.

This integration technique would offer no advantages over the straight numerical integration of the equations of motion in cartesian coordinates except that the Fourier analysis of the disturbing potential enables us to ignore the great majority of terms which have only short period effects. Only the resonant tesseral and secular and long period zonal (oblateness) terms are calculated for the geopotential effects. These terms are those for which:

$$l - 2p + q = 0, \quad m = 0 \text{ (zonals)}$$

and

$$l = 2p + q = m/s, \quad m \neq 0 \text{ (longitude terms)},$$

where s is the resonant orbit's commensurability ratio expressed as its nearest rational mean motion in revolutions per day.

It can be shown that integrating the long period and secular variations of the elements is equivalent to a 1st order integration of canonical variables without short period terms. These variables are exactly those which would be obtained by a Von Zeipel transformation (following Brouwer [12]) or through the method of Brown and Shook [13].

The long period sun and moon effects of W. M. Kaula [14] are also included in the integration. Drag is incorporated by inclusion of the long period variations of the elements given by formulae similar to those of King-Hele [15].

Integration of the slowly evolving mean element coordinates permits time steps of about one day, in contrast to about 1/100 revolution when short period effects are included.

Differential Correction Program

A differential correction program using independently determined mean Kepler elements as the data type has been built around the numerical integrator described in the previous section. The great efficiency which can be achieved by integrating mean elements permits us to solve numerically for the partials of the Kepler element updates with respect to initial mean elements and gravity constants in reasonable computing time from variant trajectory differencing. Using these partials and the mean element residuals from a reference trajectory, we form and solve the normal equations for a least squares estimation (of the corrections to the initial elements and gravity constants) in the usual way [11]. The differential correction process is iterated until the sum of the squared residuals reaches a minimum.

Gravity Recovery with Cosmos 41 Data

Fig. 2 shows the variation in a set of North American Air Defense Command (NORAD) mean semi-major axes for Cosmos 41 (1964-49D) in 1965. Actually this data represents a conversion from the NORAD mean-mean motion to a semi-major axis ($\bar{a}$) averaged with respect to mean anomaly. This gives a semi-major axis compatible with the integration described previously. The conversion is:

$$\bar{a} = \left[\frac{\mu^{1/2}}{\bar{\bar{n}}} + \frac{3}{2} J_2 \left(\frac{a_e}{a}\right)^2 a^{3/2} \left(1 - \frac{3}{2}\sin^2 i\right)(1 - e^2)^{-3/2}\right]^{2/3},$$

where $J_2 \doteq 1083 \times 10^{-6}$, and $\bar{\bar{n}}$ is the mean-mean motion. This conversion was necessary because the NORAD semi-major axes were defined from $\bar{\bar{n}}$ according to KOZAI (1959) [16] and do not represent mean elements averaged over mean anomaly.

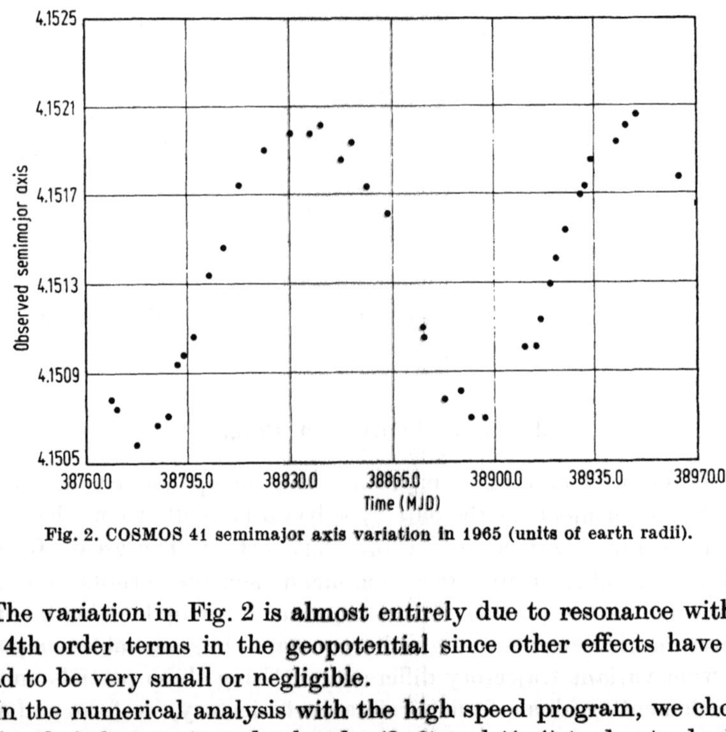

Fig. 2. COSMOS 41 semimajor axis variation in 1965 (units of earth radii).

The variation in Fig. 2 is almost entirely due to resonance with 2nd and 4th order terms in the geopotential since other effects have been found to be very small or negligible.

In the numerical analysis with the high speed program, we chose to fix the (2, 2) harmonic and solve for (3, 2) and (4, 4) to dry to duplicate the analytic solution over the same data obtained by WAGNER [17]. Starting with (3, 2) and (4, 4) values quite far from the final ones, the differential correction program using this $\bar{a}$ data converged to an improved initial semi-major axis, (3, 2) and (4, 4) constants in less than a minute of running time on an International Business Machines (IBM) 360/91 electronic computer. The trajectories were computed in this run at about 5000 revolutions per minute. The final residuals after convergence are shown in Fig. 3 and appear to be quite random. The scatter is typical of that seen for NORAD elements. The root mean square residual is slightly lower than in the analytic solution.

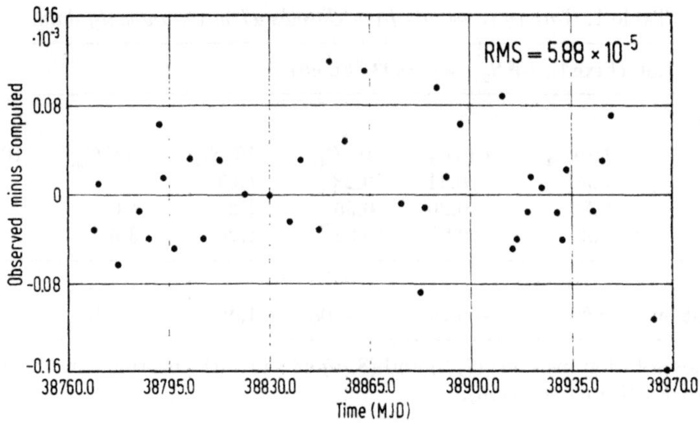

Fig. 3. COSMOS 41 semimajor axis residuals after improved gravity solution (units of earth radii).

Fig. 4 shows a comparison of the corrected gravity constants with the previous analytic results, and is also encouraging. However, the residuals in mean anomaly from the corrected solution show small perio-

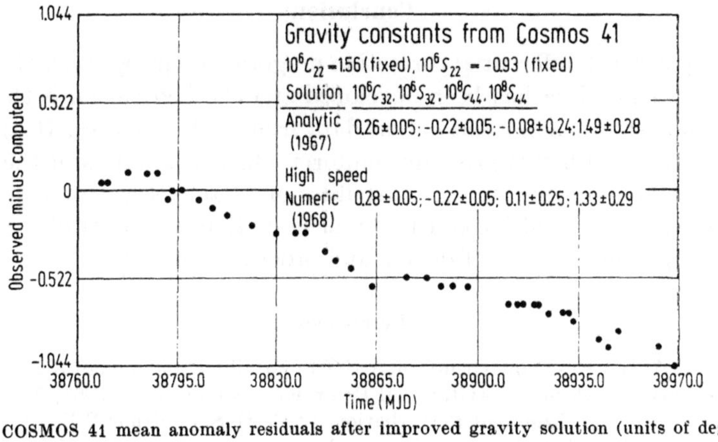

Fig. 4. COSMOS 41 mean anomaly residuals after improved gravity solution (units of degrees).

dicities consistent with 6th order resonant effects superimposed on a secular trend which is most probably due to a slight incompatibility remaining in the semi-major axis definition.

Table 1 shows the results of differential corrections of this orbit using mean anomaly data as the observation type. The solution for (3, 2) and (4, 4) is compatible with the previous results on the semi-major axis according to the random variances allowed by the data for the constants. The solutions for (6, 6) are intriguing when compared to the SAO-66 M1 [3] nonresonant solution. But they [as well as (3, 2) and (4, 4)] are not

Table 1. *Gravity constants from Cosmos 41 mean anomaly data*

$10^6 C_{22} = 1.56$ (Fixed), $10^6 S_{22} = -0.93$ (Fixed)

Solution σ(°)	$10^6 C_{32}$	$10^6 S_{32}$	$10^8 C_{44}$	$10^8 S_{44}$	$10^{11} C_{66}$	$10^{11} S_{66}$
0.038	0.32	−0.21	0.28	1.27		
0.036	0.33	−0.20	0.35	1.26	4.0	−1.2
0.036*	0.33	−0.20	0.35	1.26	3.6	−2.8
SAO 1966, M1 Solution	0.26	−0.22	−0.08	1.49	−0.9	−3.6

* This solution includes (7, 6) and (8, 6) effects with constants from SAO 1966 Standard Earth (M1 solution).

to be considered definitive because, as the last corrected solution indicates, there is still a considerable influence on this data from other resonant harmonics. Definitive determinations will require the use of a number of resonant orbits.

Conclusion

A high speed differential correction program accepting mean elements as observation data has been developed and checked out on both simulated (not shown here) and actual data from 12 h satellites. It appears possible that with this program, uniform, efficient analysis of the more than a score of deeply resonant satellites of 12 and 24 h period will ultimately produce definitive constants for more than a quarter of the longitude geopotential terms of degree and order less than 8.

References

1. Cook, A. H.: Resonant Orbits of Artificial Satellites and Longitude Terms in the Earth's External Gravitational Potential. Geophys. J. 4 (1961) 53–72.
2. Wagner, C. A.: Resonant Perturbations of Earth Satellites in 2 Day Commensurable Orbits. NASA-Goddard Space Flight Center Document X-643-68-373, Greenbelt, Md. 1968.
3. Gaposchkin, E. M.: Tesseral Harmonic Coefficients and Station Coordinates from the Dynamic Method. In: Geodetic Parameters for a 1966 Smithsonian Institution Standard Earth, Vol. 2; Smithsonian Astrophysical Observatory, Cambridge, Mass. 1966.
4. Anderle, R. J.: Observations of Resonance Effects on Satellite Orbits Arising from the Thirteenth and Fourteenth Order Tesseral Gravitational Coefficients, J. Geophys. Res. 70/10 (1965) 2453.
5. Yionoulis, S. M.: A Study of the Resonance Effects Due to the Earth's Potential Function, J. Geophys. Res. 70 (1965) 5991–5996.
6. Kaula, W. M.: Analysis of Geodetic Satellite Tracking to Determine Tesseral Harmonics of the Earth's Gravitational Field. Publication No. 656, Institute

of Geophysics and Planetary Physics, University of California, Los Angeles, Calif. 1968.

7. Wagner, C. A., Douglas, B. C.: Perturbations of Existing Resonant Satellites. Planetary Space Sc. 17 (1969) 1507–1517.
8. Kozai, Y.: The Zonal Harmonic Coefficients. In: Geodetic Parameters for a 1966 Smithsonian Institution Standard Earth, Vol. 2, Smithsonian Astrophysical Observatory, Cambridge, Mass. 1966.
9. King-Hele, D. G., Cook, G. E., Scott, D. W.: Even Zonal Harmonics in the Earth's Gravitational Potential – A Comparison of Recent Determinations. Planetary Space Sc. 14 (1966) 42–52.
10. King-Hele, D. G., Cook, G. E., Scott, D. W.: Odd Zonal Harmonics in the Geopotential, Determined from 14 Well Distributed Satellite Orbits. Planetary Space Sc. 15 (April 1967) 741.
11. Kaula, W. M.: Theory of Satellite Geodesy, Waltham, Mass.: Blaisdell Press 1966.
12. Brouwer, D.: Solution of the Problem of Artificial Satellite Theory without Drag. Astron. J. 64 (1959) 378.
13. Brown, E. W., Shook, C. A.: Planetary Theory, New York City: Dover Press 1964.
14. Kaula, W. M.: Development of the Lunar and Solar Disturbing Functions for a Close Earth Satellite. Astron. J. 67 (1962) 300.
15. King-Hele, D. G.: Theory of Satellite Orbits in an Atmosphere, London: Butterworth's Press 1964.
16. Kozai, Y.: The Motion of a Close Earth Satellite. Astron. J. 64 (1959) 367.
17. Wagner, C. A.: Determination of Low Order Resonant Gravity Harmonics from the Drift of two Russian 12-Hour Satellites. J. Geophys. Res. 73, No. 14, July 1968.

Résultats du Dépouillement des Mesures Laser en Géodésie Semi-Dynamique

par

B. Lago

Centre National D'Etudes Spatiales, Brétigny-sur-Orge, France

Abstract. The topic of this paper is to give a survey of the methods used in Space Geodesy (non-geometrical geodesy) with 3 stations located around the Mediterranean Sea when accurate tracking data such as Laser or Doppler measurements are available.

By using the best value of the perturbations, we can determine the dimensions of the triangle, its location with regards to the center of masses and its orientation.

Here enclosed are the first results obtained by elementary processes of Laser measurements on D1-D. These measurements were made on the 8, 9 and 10 of May 1967.

1. Définition

Si on dispose d'un réseau de stations de poursuite, on peut suivre, pendant des durées allant de plusieurs jours à plusieurs mois, un même satellite et restituer la trajectoire réellement suivie. Les déformations que présente cette trajectoire par rapport à une orbite Képlérienne sont révélatrices des forces qui s'exercent sur le satellite, en plus de l'attraction newtonienne centrale, en particulier des forces d'origine gravitationnelle terrestre.

Si le réseau des stations est bien distribué à la surface de la Terre, notamment en longitude, la connaissance des déformations de l'orbite sera suffisante pour remonter des effets aux causes, des déformations à l'évaluation de certains des termes qui les produisent. Lorsqu'un traitement de données de poursuite remettra en cause, au moins partiellement, les caractéristiques du potential terrestre, il sera dit de «géodésie dynamique».

Mais, pour des raisons de nombre trop faible de stations, d'insuffisance de diversité en longitude (c'est le cas de la campagne DIADEME où les stations étaient confinées en bordure de la Mer Méditerranée), cet objectif peut être estimé trop ambitieux. Un objectif plus modeste permet toutefois de s'intéresser aux équations dynamiques sans remettre en cause le modèle de potentiel de gravitation terrestre, de restituer au mieux la trajectoire suivie et de préciser simultanément la position des stations par rapport à cette trajectoire. Un tel objectif, intermédiaire entre géodésie géométrique et géodésie dynamique, sera désigné sous le nom de géodésie semi-dynamique.

Pour poursuivre un tel objectif, il sera nécessaire de disposer d'un modèle suffisamment correct du potentiel de gravitation terrestre, et, plus généralement, d'un modèle des différentes forces qui agissent sur le satellite.

2. Traitement Elémentaire

Il est intéressant de grouper les mesures faites en un ensemble sur lequel portera le traitement élémentaire. La loi de groupement est importante à définir. Dans le cas de la campagne DIADEME, les stations, toutes situées dans la même région de la Terre à une latitude Nord voisine de l'inclinaison des satellites concernés (DIADEME I et II), voyaient chaque jour in groupe de trois à quatre orbites successives, puis, à nouveau après une éclipse de $18H$ environ, un nouveau groupe.

La loi de groupement peut se ramener au nombre de groupes successif de trois à quatre orbites pris en compte en même temps; ce peut être 1, 2, 3, 4 etc. correspondant à des mesures portant sur des durées totales de 6, 30, 54, 78 h. La loi de groupement qui s'est révélée la meilleure dans cet objectif de géodésie semi-dynamique est celle qui correspond à deux groupe portant sur un temps total de 30 h.

Chaque traitement élémentaire consiste en un ajustement des valeurs théoretiques des mesures et valeurs mesurées grâce à un critère moindre carré.

Dans les paramètres intérieurs (on désignera ainsi les paramètres qui n'ont de sens que relativement au traitement élémentaire par opposition aux paramètres extérieurs qui ont un sens indépendant du cas actuel) vont être d'abord six conditions initiales à savoir les coordonnées point et vitesse à un instant origine ou encore les paramètres Képlériens osculateurs à ce même instant origine ou tout système équivalent.

Comme paramètres extérieurs, nous considérerons les coordonnées des stations dans un trièdre géocentrique classique d'origine le centre de masse de la Terre. Dans le cas de trois stations, il y aura donc 9 tels paramètres. Toutefois, ces 9 paramètres ne peuvent pas être laissés absolument libres car il paraît hors de question de revenir sur la définition de leur longitude moyenne. Il y a en fait impossibilité à déterminer simultanément l'ascension droite du nœud à l'instant origine et la longitude moyenne des stations et, tout naturellement, c'est cette dernière que l'on renonce à modifier; ce qui revient à imposer la condition:

$$\sum_{i=1}^{i} (X_i \, d\, Y_i - Y_i \, d\, X_i) = 0 \; (C)$$

(X_i, Y_i, Z_i étant les coordonnées de la ième station dans le trièdre géocentrique précédemment défini).

Pour chaque jeu de valeurs des paramètres intérieurs ou extérieurs, il faut être capable de calculer la position instantanée du satellite avec, pour conséquence, le calcul des valeurs théoriques des mesures suivi de l'évaluation des dérivées partielles de ces quantités par rapport aux différents paramètres.

Cette facilité (calcul de la position instantanée du satellite) peut être recherchée dans une méthode de Mécanique Céleste mais aussi dans une intégration numérique par la méthode de Cowell classiquement très bien adaptée à un tel cas. Ceci est réalisé grâce à un programme arithmétique sur IBM 360/65 réalisé par J. P. Chassaing et nommé GIN. Malgré l'inconvénient* d'un coût plus élevé en temps ordinateur, cette méthode a l'avantage de la simplicité. Les hypothèse faites se traduisent dans les expressions admises pour les seconds membres des équations différentielles, c'est-à-dire les composantes cartésiennes de l'accélération. Chaque effet (effet de gravitation de la Terre, de la Lune et du Soleil, effet de frottement atmosphérique, effet de la pression de radiation) donnera sa contribution à ces composantes grâde à un modèle soigneusement choisi qu'il n'est pas du propos de cet exposé de présenter davantage. En ce qui concerne l'effet de la gravitation de la Terre, on a choisi essentiellement l'ensemble de coefficients publiés par le S.A.O. sous le nom de *Standard Earth*.

Le formalisme numérique permet donc à chaque application du critère des moindres carrés linéaires ou à chaque itération du cas non linéaire, d'expliciter les matrices désignées par X (matrice des différentielles partielles des mesures par rapport aux paramètres), L (matrice des seconds membres ou écart entre valeur théorique et valeur mesurée) et P (matrice de poids des mesures).

3. Les deux Etapes du Traitement Elémentaire

La présentation qui vient d'être faite du traitement proposé une amélioration simultanée des deux types de paramètres. En fait, on peut procéder en deux étapes:

– la première étape consiste à améliorer les seuls paramètres intérieurs

* Inconvénient qu'il faut se garder d'exagérer surtout si la période étudiée n'est pas trop longue (moins de 3 jours par exemple).

en figeant les paramètres extérieurs à leur valeur de référence fixées pour l'ensemble des traitements élémentaires.

– la deuxième étape consiste, à partir de ces valeurs ainsi obtenues, à libérer l'ensemble des paramètres intérieurs ou extérieurs pour une application du critère des moindres carrés.

L'accroissement proposé sur les paramètres par cette application sera dit accroissement proposé pour bien marquer qu'il sera accepté ou refusé selon les résultats de certains critères à postériori. Les avantages de la méthode sont les suivants:

a) La linéarisation est toujours faite par rapport aux mêmes valeurs de référence des paramètres extérieurs.

b) Elle évite les divergences ou convergences erronées dues à un défaut de diversité des mesures en regard du nombre total de paramètres libres lors de la deuxième étape. De tels effets sont à craindre lorsque certaines composantes du déplacement proposé sont anormalement grandes.

c) Corollaire du précédent, il est possible de n'avoir nulle contrainte quant aux nombres de paramètres extérieurs considérés, même pour un faible diversité de mesures.

Le résultat d'un traitement élémentaire sera constitué par le déplacement de l'ensemble des paramètres intérieurs et extérieurs et de la matrice de covariance associée. En fait, on peut éliminer (au sens de l'Algèbre linéaire) les paramètres intérieurs et obtenir les déplacements sur les seuls paramètres extérieurs avec la matrice de covariance associée (on désignera par COV^* cette matrice par la suite).

4. Exemple du Laser – les Cas Traites

Au premier semestre 1967, trois stations Laser, situées autour du Bassin Méditerranéen (voir carte), ont suivi les satellites DIADEME I et II (respectivement D1-C et D1-D).

Au Tableau 1 sont données les coordonnées de ces stations dans le système Europe 50 et dans un système (C′6) où on a tenu compte d'une évaluation du vecteur centre de masse de la Terre–Centre d'Europe 50. Ce sont les coordonnées cartésiennes des stations dans C′6 qui seront prises comme valeurs de référence.

La période du 8 au 10 Mai sur D1-D ayant été particulièrement favorable, deux cas (les seuls traités à ce jour) ont été traités (en premier), désignés par A et B, et présentés au Tableau 2.

Le cas A et B ont été définis en retenant les passages comportant un nombre important de points et en choisissant un nombre homogène de mesures pour chaque passage sélectionné.

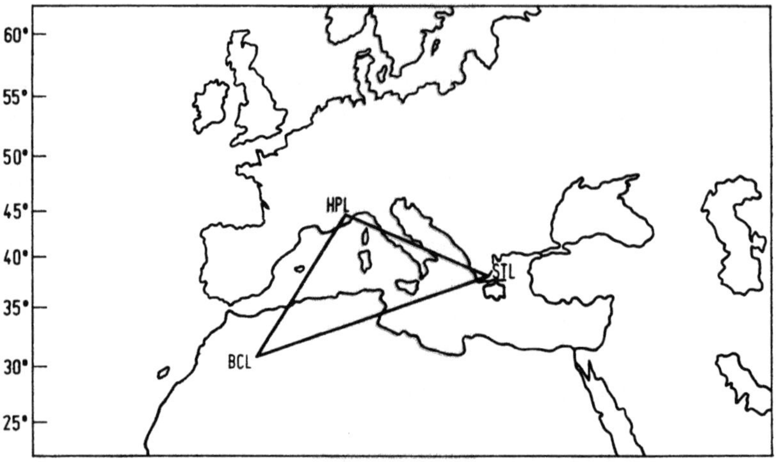

Fig. 1.

Tableau 1. *Systemes de Référence*

		Europe 50	C'6
BCL 82	φ	31,722014	31,720609
	λ	357,581683	357,580241
	h	808,021297	870,660521
	x	5426414,29	5426324,29
	y	−229172,0	−229305,0
	z	3334725,61	3334585,61
STL 85	φ	37,754734	37,753688
	λ	22,828698	22,82770,3
	h	803,116752	810,052553
	x	4654431,9	4654341,9
	y	1959286,8	1959153,8
	z	3884510,7	3884370,7
HPL 89	φ	43,933106	43,932024
	λ	5,713439	5,711903
	h	645,027640	663,239972
	x	4578455,20	4578365,20
	y	458075,23	457942,23
	z	4403267,46	4403127,46

Tableau 2. *Traitement des Mesures Laser effectuées sur D1-D. Cas Traites et Nombre d'Echos*

Cas A / Cas A*	N° de Révolutions 1079	1080	1081	1082	1092	1093	1094	1095	Cas B / Cas B*	N° de Révolutions 1093	1092	1095	1105	1106	1107	1108
BCL 82		7	195/49*	236/50*			186/50*	239/50*	BCL 82		186/51*	239/50*		139/33*	288/55*	245/54*
STL 85	7/0*	42	39		16/0	52	125/60*		STL 85	52	125/52*		45	68/50*	110/50*	
HPL 89		42	92/48*	47		58	68/54*	79/36*	HPL 89	58	68/53*	79/49*		41		

Tableau 3. *Déplacement proposé au Système S.A.O.*

	Cas A	(2σ)	Cas A*	(2σ)	Cas B	(2σ)	Cas B*	(2σ)
BCL 82	60	(12)	54	(18)	26	(4)	30	(7)
	−14	(2)	−14	(2)	−9	(1)	−10	(1)
	−44	(24)	9	(32)	74	(10)	56	(15)
STL 85	3	(9)	3	(10)	−18	(4)	−13	(5)
	4	(4)	6	(4)	−4	(1)	−2	(2)
	−12	(22)	16	(28)	81	(10)	63	(14)
HPL 89	31	(8)	4	(9)	−2	(4)	1	(4)
	24	(21)	17	(3)	13	(1)	14	(2)
	23	(20)	50	(24)	93	(9)	78	(13)
Cohérence Interne en *m*	9		8		5		5	

Tableau 4. *Vecteurs et Sigmas associés*

Cas A*

x_1	x_2	x_3	x_4	x_5	x_6	x_7	x_8	x_9
−0,08	−0,21	0	0,07	0,10	−0,41	0,18	0,81	0,25
0,57	0,01	−0,10	0,81	−0,02	0,01	−0,07	0,02	−0,01
0	−0,42	0,03	0,03	0,12	−0,56	0,16	−0,26	−0,63
−0,08	0,57	0,26	0,15	0,43	−0,22	0,54	−0,2	0,14
0,57	0,14	−0,61	−0,45	0,12	−0,15	0,20	−0,01	−0,06
0	0,53	0	−0,06	0,24	−0,06	−0,49	0,36	−0,53
−0,08	−0,35	−0,15	0,05	0,84	0,23	−0,24	−0,09	0,16
0,57	−0,15	0,72	−0,32	0,11	0,07	−0,05	0,07	0,03
0	−0,10	−0,05	0,02	0,05	0,61	0,55	0,29	−0,46

σ_1	σ_2	σ_3	σ_4	σ_5	σ_6	σ_7	σ_8	σ_9
0,02	0,70	1,20	1,34	2,38	3,46	6,14	7,47	24,83

Cas B*

x_1	x_2	x_3	x_4	x_5	x_6	x_7	x_8	x_9
−0,08	−0,21	0,05	−0,11	0,03	0,54	0,44	0,63	0,21
0,57	0,01	0,59	−0,56	−0,04	−0,05	0	−0,02	−0,01
0	−0,39	−0,03	−0,07	0,03	0,58	−0,13	−0,34	−0,61
−0,08	0,61	0,23	0,09	0,37	0,47	−0,43	0,01	0,14
0,57	0,14	−0,74	−0,22	0,16	0,11	−0,06	0,07	0,06
0	0,5	0,01	0,05	0,23	−0,11	0,61	0,01	−0,55
−0,08	−0,37	0,05	−0,08	0,87	−0,20	0,08	−0,13	0,15
0,57	−0,15	0,20	0,77	0,06	0,06	0,07	0,03	0,03
0	−0,10	0,02	0,02	0,12	−0,28	−0,46	0,67	−0,47

σ_1	σ_2	σ_3	σ_4	σ_5	σ_6	σ_7	σ_8	σ_9
0,01	0,39	0,62	0,83	1,21	1,77	2,89	3,47	12,18

Le déplacement proposé des stations est donné au Tableau 3. La matrice de $COV\alpha$ (sur les seuls paramètres extérieurs à savoir les déplacements des stations) a été diagonalisé et a permis de fournir des combinaisons linéaires des accroissements des coordonnées cartésiennes des stations, statistiquement indépendantes.

Elles sont données au Tableau 4.

D'autres combinaisons linéaires de ces mêmes accroissements ont été définis à cause de leur signification physique et désignés sous le nom de «formes physiquement caractéristiques». Ce sont:

– les trois longueurs des cordes (ce qui caractérise le triangle en forme).

– la distance centre de masse de la Terre au barycentre des stations, la cote «Z» de ce même barycentre et sa longitude, d'ailleurs figée (ce qui caractérise le triangle en position).

– enfin, le déplacement selon les directions tangentes respectivement au méridien et au parallèle d'un vecteur normal au plan des stations (sa longueur a été choisie égale à la moyenne des cordes) et la rotation totale* des stations dans leur plan (ce qui caractérise le triangle en orientation). Les valeurs de ces formes sont données dans le Tableau 5.

Tableau 5. *Formes physiquement privilégiées*

	Cas A*	(σ)	Cas B*	(σ)
Acroissement de la corde:				
Béchar-Stéphanion	36	(5)	23	(2)
Stéphanion-H^{te} Provence	1	(4)	−11	(2)
H^{te} Provence-Béchar	70	(9)	42	(4)
Acroissement de la distance centre de masse barycentre des stations	32	(5)	45	(3)
Z du barycentre	25	(13)	66	(7)
Longitude du barycentre	0		0	
Déplacement en latitude de la normale* au plan des stations	−27	(3)	−25	(1)
Déplacement en longitude de la normale* au plan des stations	1	(8)	−4	(4)
Rotation totale des stations dans leur plan	−9	(8)	1	(4)

* Il s'agit d'un vecteur ayant comme longueur, la longueur moyenne des 3cordes.

* Pour préciser cette définition, le mieux est encore de donner sa définition mathématique. Si G est le barycentre des stations S_i

$$S_i\left(\boldsymbol{O}\,\boldsymbol{G} = \frac{1}{3}\sum_i \boldsymbol{O}\,\boldsymbol{S}_i\right)$$

et $\boldsymbol{n}$ un vecteur unité normal au plan du triangle des stations, la quantité en question est égale à:

$$\sum_{i=1}^{3} \frac{(\boldsymbol{n} \wedge \boldsymbol{G}\,\boldsymbol{S}_i)\cdot \delta\,\boldsymbol{S}_i}{|\boldsymbol{G}\,\boldsymbol{S}_i|}.$$

Commentaires sur des Resultats

Vu le nombre faible de cas traités, les conclusions sont difficiles à dégager. Le mieux est de les comparer aux résultats analogues obtenus par J. P. CHASSAING sur des mesures Doppler. La différence essentielle vient de ce que, dans le cas Laser, les σ évalués sont plus faibles. Ceci s'explique par la meilleure qualité des mesures. Il ne s'ensuit pas forcément que les résultats soient meilleurs car la limite vient, dans deux cas, essentiellement de limitations de diversité de mesures et de qualité de la théorie.

Short Arc Optical Survey of the GEOS North American Network

By

J. J. Lynn

DBA Systems, Inc., Melbourne, FL, USA

Abstract. Optical observations of GEOS-I over North America were employed in a short arc reduction to improve the survey coordinates of 13 tracking stations. The dependence of the final adjusted surveys on the adopted value of the gravitational parameter and the coordinates of the earth's center of mass were analyzed. The results indicate that the accuracy of dynamically derived scale is dependent on the coordinates of the center of mass relative to the tracking network. Center of mass uncertainties of 20 m may result in errors of 5 ppm in the derived scale for the network.

Introduction

The potential application of short arc dynamic techniques for the recovery of unknown station surveys has been discussed by Brown [1]. In particular, he shows that "scale" can theoretically be extracted from short arc optical observational data, and therefore, that baselines of known length are not essential to recovery of the unknown stations' coordinates. It is, however, essential that, for a solution to be determinate, at least one of the observing stations is held totally fixed.

In order for this technique to be of practical value, any absolute errors in the coordinates of this fixed station relative to the center of mass should have a minimal effect on the final recovered coordinates relative to the fixed station. Since "scale" is obtained strictly from orbital dynamics, it follows that the absolute magnitude of the recovered intrastation distances depends directly on the value chosen for the gravity parameter.

In this report, we shall be primarily concerned with studying the effects of 1. errors in the gravity parameter, μ; and 2. errors in the assumed position of the fixed station, on the precise determination of the positions of thirteen Minitrack Optical Tracking Stations located on the North American continent.

Three estimates for the coordinates of these stations are derived from the observational data set. Each estimate is based on a different choice for the position of the fixed station. The positions chosen are those associated with the following Datums: NAD-27, SAO-C5, and MERCURY. Final estimates are given as geocentric cartesian coordinates with positive X directed toward the Greenwich meridian, positive Z along the polar axis toward the north pole, and Y in a direction which makes this a right-handed system. All references throughout this report to a cartesian coordinate system, or to X, Y, Z coordinates follow the above definition.

The Optical Tracking Network

The permanent network includes thirteen North American Minitrack Optical Tracking Stations belonging to NASA. In all our short arc dynamic reductions, the location of the optical tracker at Columbia, Missouri,

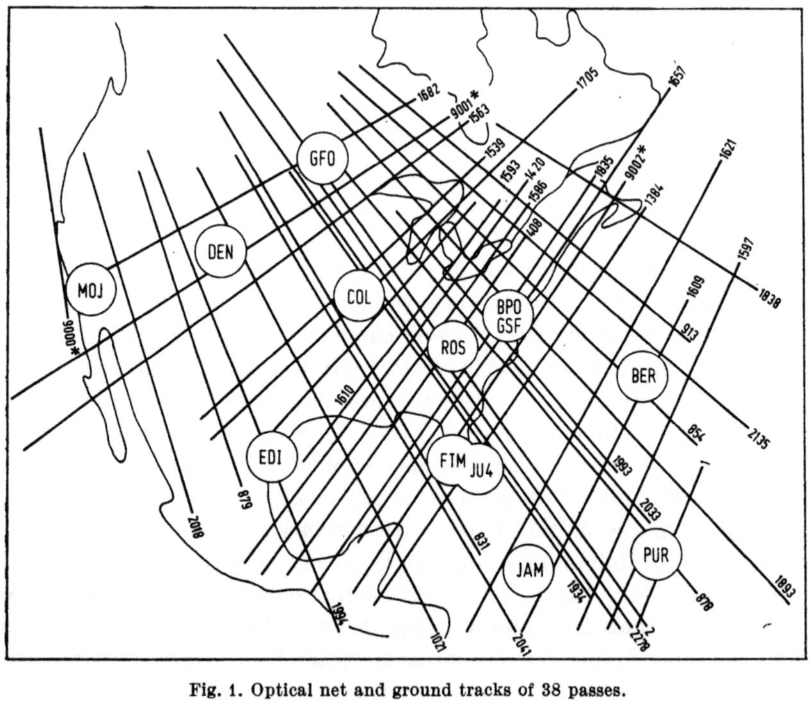

Fig. 1. Optical net and ground tracks of 38 passes.

was chosen as the fixed origin. All other stations were completely free to adjust their positions simultaneously with the six initial conditions of 38 oberved passes of the GEOS-I satellite. The ground tracks of these selected passes and their relationship to the tracking stations are illustrated in Fig. 1. As many as seven well spaced groups of flashes were observed on some passes, but the typical number is four or five. Reorientation of the cameras resulted in as many as four separate groups of flashes from a single camera on a pass. In general, an attempt was made to exercise each camera to the maximum practical extent on each possible satellite pass.

Data Reduction

The actual observational material was obtained from the NASA Data Bank in the form of cards in the GEOS-I format. The given UTC observation times were transformed to UT-I and corrections were applied to the given right ascension and declination values to allow for polar motion relative to the adopted IPMS mean pole of 1900–1905. Simultaneous multipass short arc reductions were performed by means of the Geodetic Data Adjustment Program (GDAP). A description of this program has been given by LYNN [2]. Briefly, this is a rigorous generalized least squares or minimum variance program, capable of performing the simultaneous recovery of an unlimited number of independent orbits and other dependent tracking error models, or survey parameters. A measure of the "goodness" of the data is afforded by the computed quadratic loss function which is, of course, the object function to be minimized by the program algorithm, and represents the sum of squares of the weighted measurements residuals. An estimate of the unit variance is given by dividing this quadratic loss function by the total number of degrees of freedom or the number of observations less estimated parameters. As mentioned previously by BROWN [1] in his exposition on the merits of the short arc approach, errors in the adopted geopotential functions are almost totally absorbed in the recovered epoch initial conditions. We have thus chosen a geopotential model which includes only the zonal harmonics, J_2 through J_7, as sufficient for these reductions. All orbit integrations are performed by numerical power series expansions developed by HARTWELL [4].

Results

Table 1 illustrates the conditions under which eight separate reductions, utilizing the same observational material, were performed. Cases 1, 2, and 3 differ only in the chosen coordinates of the fixed station (Columbia, Missouri) relative to the center of mass. For case 1, the coordinates are derived from the published NAD-27 Datum's latitude, longitude, and height (reference [3]) utilizing the listed (CLARKE, 1866) semimajor axis and flattening. For cases 2 and 3, the coordinates are similarly derived from the given SAO-C5 and MERCURY Datums respectively [3]. From the listed values of the quadratic loss functions, it would

Table 1.

Case No.	Geocentric Cartesian Coordinates of Fixed Station in m	Gravity Parameter μ in m^3/sec^2	Quadratic Loss Function	Mean Error in Arc Seconds
1	$X = -191\,260.6$ $Y = -4967\,427.4$ $Z = 3983\,083.7$	3.986013 E14* (Based on PJL Value)	4437	1.210
2	$X = -191\,285.6$ $Y = -4967\,284.5$ $Z = 3983\,257.1$	3.986013 E14*	4287	1.190
3	$X = -191\,285.6$ $Y = -4967\,314.7$ $Z = 3983\,307.3$	3.986013 E14*	4276	1.188
4	Same as in Case 3	3.986032 E14	4276	1.188
5	$X = -191\,307.5$ $Y = -4967\,314.7$ $Z = 3983\,307.3$	3.986032 E14	4286	1.190
6	$X =$ Same as Case 3 $Y = 4967\,364.7$ $Z =$ Same as Case 3	3.986032 E14	4284	1.189
7	$X =$ Same as Case 3 $Y = -4967\,264.7$ $Z =$ Same as Case 3	3.986032 E14	4276	1.188
8	$X =$ Same as Case 3 $Y =$ Same as Case 3 $Z = 3983\,257.3$	3.980632 E14	4298	1.190

appear that a slightly better adjustment is afforded when the listed case 3 conditions are utilized. The only difference between case 4 and case 3 is in the adopted value of the gravity parameter, μ. As would be expected, the effect of this change in μ results in a scale change as we shall show later on. The last four cases, namely 5, 6, 7, and 8, represent our attempt to ascertain the effect of a change in each of the coordinates of the fixed station on (a) the quadratic loss function and (b) the final adjusted station coordinates relative to the fixed station. As can be seen from the table, the value of the quadratic loss function (and the mean error) remains essentially constant. This indicates that it is rather intensitive to small (on the order of 100 m) changes in the position of the fixed station relative to the center of mass (assumed origin of the cartesian system). Consequently, it would appear that in the short arc context an attempt to recover, or estimate, the positions of such a fixed station with reasonable accuracy is just not feasible.

After obtaining the best station position estimates for each of the eight cases, the chord distances from the fixed origin to each station were computed. Differences in these distances, normalized by the actual distance represent proportional differences in chord distances among the eight cases. Note that since the actual stations can be considered as fixed points on a rigid body, all interstation distances should remain invariant. Any variations can thus be directly attributed to the different case conditions.

Table 2. *Average of Proportional Differences in Chord Distances to Fixed Origin in Parts per Million*

Case 1–2	Case 1–3	Case* 2–3	Case 1–4	Case 2–4	Case 3–4	Case 5–4	Case* 6–4	Case* 7–4	Case 8–4
−3.3	2.7	6.0	0.5	3.8	−2.2	−0.4	−13.3	12.9	3.7

* Close agreement between average and individual station values.

To investigate the effect of an error in the adopted value of the gravity parameter μ, on the final results, let us look at the column listing the differences between cases 3 and 4. The listed mean value for this column is -2.2 ppm. This is, of course, not surprising since, as Kaula [5] has shown, the proportional error in scale $\Delta S/S$ attributable to an error $\Delta\mu$ in μ is given by

$$\Delta S/S = \frac{1}{3}(\Delta\mu/\mu)\,.$$

From Table 1, we see that the difference between the adopted values for μ for cases 3 and 4 is -4.75 ppm. This should then translate into a proportional scale difference of -1.58 ppm. It thus appears that, of the listed $-2.2.$ ppm proportional scale change, about $-.6$ ppm remain unaccounted for by the above.

Now, let us take a look at the effect of the adopted position of our fixed station on the adjusted interstation chord distances. Examination of Table 2 reveals that, while in some cases this results in a systematic scale change, in others it is more random. For example, we know (Table 1) that the only differences between cases 1, 2, and 3 are in the adopted coordinates of our fixed station relative to the center of mass. From Table 2, it is apparent that a shift in the coordinates of our fixed station from those adopted in case 2 to those in case 3, results in a systematic scale change of 6 ppm. This make all final interstation chord distances for case 3 to be shorter than those of case 2 by this constant proportionality factor. On the other hand, no such systematic scale effect is in evidence when we shift from case 1 to case 2 or case 3.

A better understanding is afforded when we compare the results for cases 5, 6, 7, and 8 with those of case 4. Case 5, for example, differs from case 4 (see Table 1) only in the adopted value of the X coordinate of our fixed station. From Table 2, we see that this change in the position of our fixed station did not cause a systematic proportional change of scale. The situation is quite different for cases 6 and 7 where the Y coordinate of the fixed station was perturbed by -50 and $+50$ m respectively. Here we evidence a very pronounced systematic proportional scale change as a result of the perturbation in the position of our fixed station. For case 8, where the Z coordinate of the fixed station was perturbed by -50 m, there is some effect on scale, but not quite as pronounced.

We feel that these variations are a function of the relative tracking geometry. Whether or not a change in the adopted coordinates of our fixed station affects the final interstation chord distances depends on the ability of the simultaneously recovered orbital elements to absorb such station shifts. This turn depends on the direction of such a shift relative to orbital geometry and station location.

The final adjusted geocentric cartesian coordinates (case 3) are listed in Table 3 together with the average difference in these coordinates for cases 1 through 4 (Table 4).

Table 3. *Final Cartesian Coordinates for the 13 Station Optical Net*

Station	X [m]	Y [m]	Z [m]
GSF	1130738.7	−4831365.9	3994184.0
BPO	1118066.0	−4876372.6	3943012.1
JU4	976298.5	−5601427.7	2880287.9
FTM	807886.0	−5652019.1	2833552.0
BER	2308248.8	−4873632.3	3394611.7
COL	−191261.4	−4967320.1	3983305.3
DEN	−1240448.6	−4760270.9	4049033.5
EDI	−828467.9	−5657497.9	2816866.4
GFO	−521682.4	−4242087.6	4718772.2
JAM	1384193.5	−5905696.4	1966588.6
MOJ	−2357218.2	−4646367.3	3668362.6
PUR	2465083.4	−5534952.1	1985560.5
ROS	647539.8	−5177964.1	3656754.7

Table 4. *Average Shift in Adjusted X, Y, Z, Station Coordinates in* m

Case 1–2			Case 1–3			Case 2–3			Case 3–4		
X	Y	Z	X	Y	Z	X	Y	Z	X	Y	Z
23	−140	177	−2	−113	−223	−26	27	−47	0	0	0

Conclusions

Optical short arc dynamic approach to satellite geodesy may provide an effective means for the establishments of intracontinental nets. The ultimate limit on the precision and, consequently, the usefulness of the adjustment depends on the uncertainty in the adopted value of the gravitational parameter, and on the error in the position of the center of mass relative to a fixed observing station. Present uncertainty in the value of μ may cause a 1 to 2 ppm proportional error in scale. Errors of 20 m in the center of mass may result in a possible proportional scaling error of 5 ppm depending on the relative tracking geometry. The apparent lack of sensitivity in the mean error to changes in the position of the fixed observing station precludes the simultaneous recovery of the center of mass, unless the observed data set is significantly extended.

For the GEOS North American Optical Network the best adjustment, in the sense of smallest mean error, was obtained when the coordinates of the station at Columbia were fixed on either MERCURY, or SAO-C5 Datums. The final results for these two cases differ by a constant scaling factor of 6 ppm. The uncertainties in the final surveys, which are due to the noise associated with the tracking data, are on the order of three six meters. These could, of course, be reduced by expanding the present observational data set.

References

1. Brown, D. C.: Short Arc Optical Survey of the GEOS North American Tracking Network. NASA X-550-68-439, 1968.
2. Lynn, J. J.: Geodetic Data Adjustment Program (GDAP). Proceeding of the GEOS Program Review Meeting, 12–14 December 1967.
3. Wolf Research & Development Corp.: GEOS-I Satellite Tracking Station Positions on the SAO Standard Earth (C-5). NASA X-552-68-70, 1967.
4. Hartwell, J. G.: Integration of Close Satellite Orbits and Solution of First Variational Equations by Recursive Analytical Continuation. Presented at Sixth Western National Meeting 1966, American Geophysical Union.
5. Kaula, W. M.: Comparison and Combination of Satellite with Other Results for Geodetic Parameters. Presented at Second International COSPAR Symposium, Athens 1965.

Traitement des Mesures Doppler en Géodésie Semi-Dynamique

By

J.-P. Chassaing

Centre National d'Etudes Spatiales, Brédigny-sur-Orge, France

Abstract. The purpose of this paper is to show the results obtained, in the field of semi-dynamical geodesy, by processing of Doppler data collected in three mediterranean ground stations:

Stephanion (STD) (Greece), Colomb-Béchar (BCD) (Algeria), Haute-Provence (HPD) (France).

The orbit parameters, the position of the stations and one frequency parameter for each station, are simultaneously computed with a differential correction calculation using the least squares method.

The program of differential correction is based on a numerical integration of the equations of motion (Cowell's method to the 8th order).

All the disturbances that have a known effect can be taken into account in the satellite motion theory.

During the periods of treatment that will be considered in this study, the values of the Earth's potential will be those of "Standard Earth" (or NWL.5E.6); Jacchia's model provides us with densities for the atmospheric drag.

Displacement of the rectangular coordinates of the stations are given with reference to the C6 system. The following results have been obtained:

As far as D1.D. is concerned, the most interesting period to be treated is the period including 2 series of observations in succession (1.25 day) but good results can also be obtained with 3 orbits in succession.

The reliability of the results has been estimated with regards to their scattering and their confidence intervals:

In making the estimation of the quality of the result, the following items have been considered:

the scattering,

confidence intervals,

quality of the representation of the observations (interval consistency) we have also pointed out the special importance of the resonance harmonics for the interval confidence of the computations concerning the "TRANSIT satellite 66.76.A".

The study of the covariance matrix shows that the highest eigen value is associated to an eigen vector with 3 leading vector components Z_1, Z_2, Z_3.

The reduction of the covariance matrix to a sub-matrix including only the external parameters (positions of the stations) makes possible the definition of 9 preferential physical forms concerning the dimensions and the orientation of the triangle as well as the position of the gravity center of the triangle.

The annexed tabulated values show the whole processing corresponding to 7 basic cases of 1.25 day.

One can see that these values square well with those formerly obtained by different methods (Mr. Lefebvre's paper).

These results lead us to conclude as follows:

in spite of the following difficulties:

the inherent difficulties of the semi-dynamical geodesy

the area included in the triangle is too restricted and the range of measurements not broad enough,

there are some loop-holes in the theory (resonance harmonics are not so well-known when D1.D is concerned).

We may assume the following points:

it is possible to get a good representation of the results and an accuracy ranging about the accuracy of the measurement.

2 (out of 3) coordinates (X, Y) of each station can be obtained, the scattering being rather small and ranging about the confidence intervals; Z is not so accurately determined.

Nevertheless, it is possible to deduce the dimensions of the triangle with an accuracy of 10 m as well as the orientational parameters.

the center of masses can be located with regards to the center of a reference ellipsoid which is set by means of classical geodesy.

I. Introduction

On donne les résultats obtenues en géodésie semi-dynamique par traitement des mesures Doppler effectuées en trois stations méditerranéennes [Stéphanion (STD), Colomb-Béchar (BCD), Haute-Provence (HPD)].

Par correction différentielle, utilisant la méthode des moindres carrés, on détermine à la fois les paramètres d'orbite, les positions des stations, un paramètre de fréquence par station. Le programme de correction différentielle est basé sur une intégration numérique des équations du mouvement (méthode de Cowell d'ordre 8), la théorie des satellites permet de tenir compte de toutes les perturbations connues; pour les périodes de temps envisagées, nous prenons le potentiel terrestre (Standard Earth ou NWL.5E.6) et le frottement atmosphérique (Modèle de Jacchia).

Les déplacements proposés aux coordonnées rectangulaires des stations sont données par rapport au système C 6*.

On obtient les résultats suivants:

1. Pour le satellite D1.D, la période de traitement la plus interessante est celle qui groupe 2 séries d'observations consécutives (soit, 1,25j), bien que le traitement de 3 orbites consécutives permette d'obtenir de bons résultats. On a examiné la validité des résultats relativement à leur dispersion et aux intervalles de confiance affichés, et relativement à la qualité de la représentation des observations, (cohérence interne). On montre également l'importance des harmoniques de résonnance, quant à la cohérence interne sur le satellite TRANSIT 6676.A.

---

* Il n'a pas été tenu compte du mouvement du pôle dans les déplacements.

2. L'étude des matrices de covariance montre qu'à la plus grande valeur propre est associé un vecteur propre ayant 3 composantes prépondérantes suivant Z_1, Z_2, Z_3.

La réduction de la matrice de covariance à une sous-matrice ne contenant que les paramètres externes (position des stations) permet de définir, 9 formes physiquement privilégiées concernant les dimensions et l'orientation du triangle ainsi que la position du centre de gravité du triangle.

On donne sous forme de tableau le traitement global correspondant à 7 cas élémentaires de 1,25 j.

On constate que l'accord est excellent sur les cordes comparativement à des résultats obtenus par des méthodes différentes (géodésie géométrique).

3. La comparaison des coordonnées obtenues (ramenées au pôle 1900–1905) avec les coordonnées du système Europe 50 permet de calculer les composantes du vecteur «Centre de Masses – Centre de l'ellipsoïde Europe 50». L'accord est également excellent avec les valeurs précédemment publiées compte-tenu de la position minime occupée par le triangle méditerranéeen sur le globe terrestre.

A. Influence de la période de traitement

Tableau 1. *Satellite D1.D Théorie: Standard Earth et Frottement Atmospherique. Etude de la Période de Traitement Doppler Déplacement proposé au Systéme S.A.O.*

Période de Traitement		Déplacements et intervalles de confiance moyens [m]		Dispersion	Cohérence interne en cycles
0,2 jour (3 Orbites)	BCD	120	(90)	30	
		14	(20)	15	
		−174	(160)	160	
	STD	101	(95)	75	
		30	(15)	15	0,43
		−130	(130)	115	
	HPD	148	(100)	40	
		16	(15)	15	
		−94	(160)	110	
1,25 jour	BCD	46	(10)	25	
		−1	(10)	10	
		−28	(80)	120	
	STD	2	(15)	25	
		16	(10)	10	0,63
		16	(70)	110	
	HPD	45	(10)	15	
		2	(10)	7	
		0	(70)	85	

Tableau 1. (Suite)

Période de Traitement		Déplacements et intervalles de confiance moyens [m]		Dispersion	Cohérence interne en cycles
3 jours	BCD	42	(25)	10	
		−6	(15)	8	
		90	(140)	250	
	STD	18	(25)	70	
		22	(20)	30	1,8
		85	(125)	220	
	HPD	40	(20)	30	
		3	(15)	8	
		71	(120)	170	

B. Influence du modèle

a) Satellite D1.D – Période de 3 jours (Déplacement en m)

Théorie	Standard Earth	NWL.5E.6	NWL.5E.6 (tronqué à 7,7)
BCD	42	32	34
	−6	−5	−5
	90	63	41
STD	18	−2	1
	22	21	24
	85	85	59
HPD	40	22	25
	3	−3	−4
	71	70	62
Cohérence interne moyenne	1,8 cycles	1,8 cycles	1,65 cycles

b) Satellite TRANSIT 6676A

Le fait interessant est de montrer l'importance des harmoniques de résonnance.

Sur les 3 cas traités, on obtient:

	TA	TB	TC
Standard Earth	0,94 cycles	0,87	0,54
Anderlé (7,7)	1,03	1,65	3
Anderlé	0,84	0,80	0,87

On remarque, que les différences ici constatées, n'apparaissent pas sur D1.D. Il semble donc, que les termes 13,13; 15,13; 15,14; du modèle NWL.5E.6 seraient uniquement valables pour des orbites du type TRANSIT.

II. Interprétation

La décomposition en valeurs propres et vecteurs propres de la matrice de covariance, montre que le vecteur propre correspondant à la plus grande valeur propre a 3 composantes prépondérantes suivant z_1, z_2, z_3.

La réduction de la matrice de covariance a une sous-matrice ne contenant que les paramètres externes; nous donne 3 groupes de valeurs propres.

Tableau 2. *Vecteurs Propres et Sigmas Associés*

X_1	X_2	X_3	X_4	X_5	X_6	X_7	X_8	X_9
−0,08	−0,30	−0,35	0,32	−0,46	0,39	−0,42	0,36	0,01
0,57	−0,02	0,14	−0,08	0,37	0,71	−0,01	0,03	0
0	−0,22	−0,32	0	−0,21	0,2	0,6	−0,29	0,57
−0,08	0,61	0,12	−0,45	−0,49	0,27	0,14	0,25	0,01
0,57	0,45	−0,56	0,24	0,01	−0,3	0	0,08	0,01
0	−0,09	0,11	−0,11	0,29	−0,21	0,04	0,74	0,53
−0,08	0,25	0,41	−0,78	−0,05	0,11	0,35	0,13	−0,03
0,57	−0,34	0,44	−0,06	−0,52	−0,29	0,02	0	0
0	0,29	0,28	0,12	−0,03	0,01	0,56	−0,38	0,63
σ_1 [m]	σ_2 [m]	σ_3 [m]	σ_4 [m]	σ_5 [m]	σ_6 [m]	σ_7 [m]	σ_8 [m]	σ_9 [m]
0,0006	1,66	1,86	2,01	2,08	2,26	2,93	3,26	21,67

1. Une valeur propre correspondant à la condition que nous avons imposée: fixation de la longitude moyenne du centre de gravité du triangle.

2. Sept valeurs correspondant à des relations linéaires entre les coordonnées des stations.

3. Une valeur très forte correspondant au Z moyen mal déterminé.

On peut examiner les résultats, cas par cas, pour une période de traitement élémentaire, on peut faire la moyenne des valeurs ou bien procéder à un traitement global de matrices.

Nous avons fait apparaître, ci-dessous, 9 formes statistiquement indépendantes, et il est interessant de faire apparaître 9 formes physiquement indépendantes relatives à la dimension et l'orientation du triangle ainsi qu'à la position du centre de gravité de ce triangle.

Tableau 3. *Formes Physiquement Privilegiées*

	Doppler		Laser			
	Traitement global de 7 cas élémentaires de 1,25 jours	(σ)	Cas A*	(σ)	Cas B*	(σ)
Acroissement de la corde Béchar-Stéphanion	36	(3)	36	(5)	23	(2)
Acroissement de la corde Stéphanion-H^{te}Provence	4	(3)	1	(4)	−11	(2)
Acroissement de la corde H^{te}Provence-Béchar	17	(4)	70	(9)	42	(4)
Acroissement de la distance centre de masse barycentre des stations	18	(8)	32	(5)	45	(3)
Z du barycentre	−10	(12)	25	(13)	66	(7)
Longitude du barycentre	0		0		0	
Déplacement en latitude de la normale* au plan des stations	−19	(3)	−27	(3)	−25	(1)
Déplacement en longitude de la normale* au plan des stations	−26	(5)	1	(8)	−4	(4)
Rotation totale des stations dans leur plan	52	(5)	−9	(8)	1	(4)

* Il s'agit d'un vecteur avant comme la longueur, moyenne des 3 cordes.

On trouve ces valeurs sur le Tableau 3 ainsi que les valeurs analogues obtenues sur 2 cas Laser.

Nous donnons également, ci-dessous, les valeurs moyennes de l'accroissement des cordes obtenues pour les différentes périodes de traitement.

	0,2j	1,25j	3j
BCD-STD	35(9) m	40(10)	32(17)
STD-HPD	19(14)	6(10)	17(15)
HPD-BCD	24(20)	21(23)	8(23)

Les formes physiquement privilégiées (traitement global):

BCD-STD	36(3)
STD-HPD	4(3)
HPD-BCD	17(4)

Ces valeurs sont en bon accord avec d'autres résultats, notamment ceux obtenus par géodésie géométrique (voir exposé de M. LEFEBVRE).

III. Position du Centre des Masses par Rapport a l'Ellipsoide Europe 50

Le traitement global des cas de 1,25j donne pour les composantes du vecteur «Centre des Masses – Centre de l'ellipsoïde Europe 50»*:

	ΔX	ΔY	ΔZ
BCD	−56	138	−142
STD	−88	−122	−119
HPD	−56	−123	−145

Si l'on compare les valeurs moyennes avec des valeurs précedemment trouvées, on a:

	Semi-dynamique [m]	Veis [m]	Dynamique [m]	Levallois-Dufour [m]
ΔX	−67	−90	−68	−30
ΔY	−128	−133	−126	−100
ΔZ	−135	−140	−124	−150

Conclusions

Ces résultats montrent:

que malgré les difficultés inhérentes à la semi-dynamique,

que malgré la position minime occupée par le triangle et la faible diversité des mesures,

malgré une théorie incomplète (harmoniques de résonnance mal connus pour D1.D, etc.).

On peut:

Avoir une bonne réprésentation des observations et obtenir une précision voisine de la mesure,

obtenir 2 des coordonnées de chaque station (X, Y) avec une faible dispersion, cette dernière étant voisine des intervalles de confiance, la composante Z étant plus mal déterminée,

néanmoins, retrouver les dimensions du triangle à une dizaine de mètres près ainsi que des paramètres d'orientation,

positionner le centre des masses par rapport au centre d'un ellipsoïde de référence obtenu par géodésie classique.

Nous remercions Monsieur Roumiguier qui a réalisé la programmation correspondant au paragraphe II.

* Correction faite du mouvement du pôle (Pôle moyen de la date – Pôle 1900 à 1905).

Références

CHASSAING, J.-P.: Integration Numerique des Equations du Mouvement d'un Satellite Artificiel par la Methode de Cowell. Exposés sur différents problèmes de calcul d'orbites et de géodésie spatiale. Févr. 1968.

Standard Earth, Volume I. Smithsonian Astrophysical Observatory, Spec. Rep. 200.

JACCHIA, L. G.: Static Diffusion Models of the Upper Atmosphere with Empirical Temperature Profiles. Research in Space Science, S.A.O. Spec. Rep. n° 170.

LEVALLOIS, M.: Compte-Rendu de l'Académie des Sciences de Paris 1967.

VEIS, G.: The Determination of the Radius of the Earth and Other Geodetic Parameters as Derived from Optical Satellite Data. 14ème Ass. Gen. de l'U.G.G.I., octobre 1967, Lucerne.

GAPOSCHKIN, E. M., WRIGHT, J. P.: General Relativity in Satellite Orbits. S.A.O. Spec. Rep. 283.

CHASSAING, J.-P., LEFEBVRE, M.: Exposés sur différents problèmes de calcul d'orbites et de géodésie spatiale: Resultats Preliminaires de la Campagne Diadème en Géodésie Semi-Dynamique.

ANDERLE, R. J.: Geodetic Parameter set NWL 5 E-6 – Based on Doppler Satellite Observations. NWL Report 1978.

Comparative Analysis of Different Results on the European Datum from Spatial Geodesy

By

M. Lefebvre

Centre National d'Etudes Spatiales, Brétigny-sur-Orge, France

Abstract. We analyse results obtained in dynamical and geometrical geodesy on the European net using Doppler, Laser and photographic measurements during DIADEME and RCP 133 campaigns. We compare the different measures and study the scale and orientation of the Europe 50 system.

The aim of the paper is to compare and analyse different results obtained by spatial geodesy, over these last years from stations in the Europe 50 datum.

Several experiments have been made, most of them in cooperative program:

1965 – triangulation between France and Algeria with simultaneous photographical observations of Echo satellite by IGN.

1966 – simultaneous photographical observations of the flashes of GEOS A.

Simultaneous Doppler observations from Nice and Beyrouth using DIAPASON satellite.

1967 – Doppler observations from three stations located in Stéphanion (Greece), Colomb-Béchar (Algieria) and Haute-Provence (France), using the DIADEME satellites.

Laser observations (from the same stations) and photographical observations (from the SAO network) of all the satellites provided with retroreflectors = D1.C, D1.D, BEB, BEC and GEOS A.

1968 – simultaneous Laser and photographical observations from San-Fernando (SAO Baker-Nunn and French Laser station) Haute-Provence (Schmidt telescope, French Laser stations and Antares Camera in Nice), Dyonisios (SAO Baker-Nunn and Laser stations) using the same satellites and GEOS B.

All the data were reduced and used in different ways.

I. Methods

1. Geometrical Method

Each simultaneous observations from n stations with p measurements give $(p-3)$ equations of condition. If we have l equations of condition with $l \geq n$, it is possible to compute all the relative positions of the stations.

Practically $l \gg n$ and the best solution is given by a least squares adjustement.

The final accuracy is related to the precision of the measurements and to the strength of the geometrical configuration.

During DIADEME experiment, R. Bivas used simultaneous observations from San Fernando Baker-Nunn and the three Laser stations. The geometrical configuration was sufficent but not optimal. The results issued from the thesis of R. Bivas are in the first column Table 1.

As to the RCP 133 experiment about four hundred simultaneous observations were successfully performed. The repartition of the satellite positions is very satisfactory and we think that the final accuracy will be about one meter. Only a few photographical observations were reduced and preliminary results of the station-station vector Haute-Provence San Fernando are in column 2.

2. Orbital Method

The orbit of the satellite is computed with the Standard Earth model, the C 6 system for the coordinates of the stations and the Baker-Nunn observations. – The accuracy of the satellite positions is about 20 to 40 m.

The measurements from the unknown stations were used to relocate these stations.

This method was used by SAO to compute the positions of Harvestua Baker-Nunn, Zimmerwald, Haute-Provence, and Malvern with the photographical observations of the GEOS A flashes. The scale is connected with the adopted value of $GM = 398\,601.3$ km^3/sec^2.

In the same way, the orbit of SAO computed by SAO and the French Laser observations were used by CNES to relocate the stations. In this case the scale depends on the adopted value for GM and the light velocity; results are given in Tables 1 and 2.

3. Semi-dynamical Method

In this method, we determine orbital parameters and coordinates of the stations, but the earth potential and other forces are fixed to the best values (Standard Earth for the potential).

Table 1.

	C 6	Méthode Orbitale		Méthode Semi-Dynamique		Géometrie		Moyenne
		Flashes GEOS A	Laser GEOS A	Doppler D1	Laser	+ San Fernando á titre indicatif		
San Fernando	−88							−88
	−129							−129
	−126							−126
Harvestua	−95	−104						−100
	−120	−128						−124
	−137	−130						−133
Shiraz	−91							−91
	−145							−145
	−160							−160
Colomb-Béchar			−56	−44	−53		−82	−51
			−150	−134	−145		−135	−143
			−132	−168	−108		−153	−136
Stéphanion			−84	−87	−95		−101	−89
			−97	−116	−131		−105	−114
			−134	−124	−100		−152	−120
Haute-Provence		−88	−65	−46	−88	−104	−83	−72
		−132	−130	−131	−138	−114	−116	−133
		−82	−105	−140	−76	−139	−120	−101

Table 1. (Continued)

	C 6	Méthode Orbitale		Méthode Semi-Dynamique		Géométrie	Moyenne
		Flashes GEOS A	Laser GEOS A	Doppler D 1	Laser	+ San Fernando á titre indicatif	
Malvern		−57 −135 −104					−57 −135 −104
Zimmerwald		−69 −147 −97					−69 −147 −97

Centre de Gravité – Centre Ellipsoide Europe 50

	C 6	Orbitale		Semi-Dynamique		Levallois-Dufour	Moyenne générale
		Flashes GEOS A	Laser	Doppler	Laser		
Δx	−90	−80	−68	−59	−79	−30	−75
Δy	−133	−136	−125	−127	−138	−100	−132
Δz	−140	−104	−124	−144	−95	σ−152	−121

Table 2.

Vecteurs		Géodésie Géométrique			
		DIADEME	RCP	GEOS A	$\overline{m}$
SFE-HPL	$-0.400\ dx$	-16	$+6(2)$		
	$+0.775\ dy$	$+15$	$+13(2)$		
	$+0.484\ dz$	-13	$+6(2)$		
1301	dp	$+12$	$+11$		
SFE-BCL	$+0.507\ dx$	$+6$			
	$+0.515\ dy$	-6			
	$-0.689\ dp$	-27			
632	dp	$+18$			
SFE-STL	$-0.175\ dx$	-13			
	$+0.983\ dy$	$+24$			
	$+0.044\ dz$	-26			
2557	dp	$+24$			
HPL-BCL	$+0.555\ dx$	$+22$			
	$-0.450\ dy$	-22			
	$-0.699\ dz$	-13			
1527	dp	$+31$			
HPL-STL	$+0.048\ dx$	$+3$			
	$+0.944\ dy$	$+9$			
	$-0.326\ dz$	-13			
1590	dp	$+12$			
BCL-STL	$-0.323\ dx$	-18			
	$+0.917\ dy$	$+30$			
	$+0.230\ dz$	0			
2384	dp	$+33$			

The method is analysed in the papers from B. Lago and J. P. Chassaing in this symposium. They apply to the Laser and Doppler data respectively and the results are given in Tables 1 and 2.

An other original method for the treatment of Doppler Data using Kalman filters and due to Monsieur Pollak (SEREB) give others results in Table 2.

II. Comparison and Analysis

1. Geocentric Coordinates of the Stations

All the results are in Table 1.

Semi-dynamical methods give the datum shifts with a good accuracy in spite of the smallness of the triangle.

The results of different stations for the same stations are in good agreement.

Tableau Resultats

Géodésie orbitale		Semi-dynamique						
Laser	GEOS	DOP. TRA.	DOP. DID	m	KALMAN	Laser		$\overline{m}$ générale
+25	0							+4
+1	−3							+7
+20	+44							+14
+2	+24							+12(11)
+32								+19
−21								−14
−6								−17
+10								+14
+4								−5
+32								+28
−8								−17
+30								+27
+9		−11	+3	−4	−12	+39	+6	+8
−20		−8	−3	−6	−40	−28	−20	−20
−27		−24	−28	−26	−11	−15	−21	−20
+32		+16	+32	+19	+20	+56	+29	+30(14)
−19		−18	−41	−29	−9	−8	−19	−15
+33		+7	+15	+11	−6	−13	+7	+8
−22		−12	+16	+2	−15	−24	−11	−12
+39		+11	+6	+8	0	−5	+10	+11(15)
−28		−7	−44	−25	+7	−46	−24	−23
+52		+15	+18	+17	+34	+14	+27	+27
0		+12	+44	+28	−4	+7	+11	+10
+51		+18	+40	+29	+28	+30	+33	+33(11)

The mean values for the components of the vector center of gravity – Center of Europe 50 datum are in good agreement with the vector deduced by VEIS from the Standard Earth. Neverless the difference in the x component seems significative. (The dx for Malvern, Haute-Provence, Colomb-Béchar, Zimmerwald are smaller). The dz from Haute-Provence and Zimmerwald are significatively smaller.

The adopted value of *GM* is the same in all the determinations; for the Laser and Doppler data, the light velocity is also used to define the scale.

2. Station-Station-Vectors

The Table 2 summarises all the results:

The agreement between the different determinations is good and quite satisfactory on the length of chords. In all the cases the form of the triangle for Colomb-Béchar, Stéphanion, Haute-Provence is invariant when the orientation is not the same. The general mean values are very near the geometrical geodesy values that we consider as the best.

This good agreement was used to study the scaling factor of the Europe 50 datum; lengths of the chords obtained by spatial geodesy were compared to the survey values.

We find for the stations of the DIADEME experiment a scaling factor of $+8 \times 10^{-6}$ with a residual shift of 17 m for the Colomb-Béchar station.

The results for Europe 50 are:

	scaling factors	scale	length
San-Fernando – Harvestua (Standard Earth)	$+3.6 \times 10^{-6}$	*GM*	2900
GEOS A – Flashes	$+4.0 \times 10^{-6}$	*GM*	7300
DIADEME and RCP Experiments (Geometrical Methods)	$+8.7 \times 10^{-6}$	*C*	5500
DIADEME (Semi-dynamical)	$+4.0 \times 10^{-6}$	*GM* and *C*	1600
Mean Weighted Value	$+5.5 \times 10^{-6}$	(± 3)	17300 km

For North American Datum:

Standard Earth	$+5.3 \times 10^{-6}$	*GM*	12.700
9001 – 9010 – (Geometrical Method with Laser)	$+4.6 \times 10^{-6}$	*C*	1500
Mean Weighted Value	$+5.3 \times 10^{-6}$	(± 3.5)	14.200

Perhaps, it is a little surprising to find a mean scaling factor. But with LAMBECK, one notes that the definitions of the scales are not the same: in the geometrical method the basis is the light velocity, in the dynamical method the value of *GM*, in the classical geodesy, the values of several base lines.

The propagation of the errors in the network is unknown and it is possible that these appears a mean error, (uncertainty on the heights).

The errors on the light velocity is less than 0.5 km/sec equivalent to 1.5×10^{-6} or the relative scaling factor.

N.B. – More recently, LAMBECK published global determination for the stations deduced from geometrical method and rescaled with Laser data. The comparison on the station-station vector Haute-Provence – San Fernando shows a very good agreement on the length of the chords (+ 1 m), the differences with the RCP results for the orientation are respectively: 0.2, 4, and 9×10^{-6}.

Conclusion

The accuracy and the great number of results show the ability to compute a new geodetical European network with an accuracy of 10^{-6}. The geometrical methods seem the best way for the relative coordinates especially with colocated Laser and photographical stations. The geocentric coordinates can be computed with 10 m accuracy with the semi-dynamical method.

Nevertheless, the comparison requires an accurate analysis of each method (definition of reference system and scale) before mixing different determinations. Conclusions on orientation, heights, scaling factors, values of the constants can be derived from this kind of treatment.

References

Standard Earth, Vol. 2 et 3. Smithsonian Astrophysical Observatory, Spec. Rep. 200.

Bivas, R.: La Télémètrie Spatiale par Laser, Application Géodésique. Thèse présentée à la Faculté des Sciences de l'Université de Paris (19 Juillet 1968).

Forterre, M.: Communication privée.

Levallois, M.: Compte-rendu à l'Académie des Sciences de Paris 1967.

Veis, G.: The Determination of the Radius of the Earth and Other Geodetic Parameters as Derived from Optical Satellite Data. 14ème Ass. Gen. de l'U.G.G.I. octobre 1967, Lucerne.

Lambeck, M.: Scaling a Satellite Triangulation Net with Laser Range Measurements. Prague 1968.

Smithsonian Astrophysical Observatory: Geodetic Satellite Results During 1967. Spec. Rep. 264.

Comparisons and Combinations of Geodetic Parameters Estimated from Dynamic and Geometric Satellite Solutions and from Mariner Flights*

By

K. Lambeck

Smithsonian Astrophysical Observatory, Cambridge, MA, USA

Abstract. The use of data from different sources for estimating geodetic parameters serves dual purposes – one of comparison, to obtain realistic accuracy estimates for the individual solutions and to resolve any inherent biases in these results, and one of combination, to obtain the most complete and reliable set of parameters. In this solution three sources of data are combined: the recent SAO dynamic and geometric solutions and the recent Jet Propulsion Laboratory solution from Mariner IV and V missions.

The success of the dynamic solution depends in part on the ability to separate the residuals between observed satellite positions and positions computed from orbital theory into components due to station position errors and due to neglected or incorrect non-zonal harmonic coefficients. The combination of the first two data sets is obviously important for this purpose. The comparison and combination of this satellite solution with the JPL solution is important as the two are essentially independent. Comparison of the individual solutions against the combination indicates that the SAO Baker-Nunn stations coordinates are determined with an average accuracy of 10 m or better relative to the earth's mass center.

1. Introduction

Data from various sources can be utilized to determine geodetic parameters, but characteristic of most of these methods are that they are incomplete in one way or another and that inadequacies in the mathematical models used often lead to unrealistic accuracy estimates. Consequently, the use of data from different sources serves two purposes: one of comparison, to obtain realistic accuracy estimates and to resolve any biases in the results; and one of combination, to obtain the most complete and reliable set of parameters.

The two satellite methods both provide information on station locations but use essentially independent techniques. The dynamic or or-

* This work was supported in part by grant NGR 09-015-002 from the National Aeronautics and Space Administration.

bital method provides coordinates that are relative to the earth's mass center and that are characterized by small propagation of variances from one station to the next. These coordinates are determined, together with the nonzonal harmonics of the earth's potential, by analyzing the residuals between the observed satellite positions and the positions computed from orbital theory. The success of the solution depends in part on the ability to separate the two types of parameters. The accuracy of the coordinates determined by the geometric solution is very dependent on the station configurations, and the propagation of variances rapidly becomes unfavorable. But the geometric method does provide strong solutions for the directions of station-to-station vectors.

Comparisons of the directions between stations computed by the two methods provide a means of establishing their relative accuracies, and the combination gives improved station coordinates, and possibly improved tesseral harmonics by assisting in the separation of the orbital residuals into the two principal components mentioned above. Under certain circumstances the dynamic solution could yield poor results for longitude – for example, when only range or range-rate observations are used – owing to correlations between this coordinate, time, and the right ascension of the nodes. The geometric solution could now be used for orienting the dynamic solution.

The Jet Propulsion Laboratory (JPL) has used data from their deep space probes to solve for – among other parameters – station longitudes and the distances to the earth's spin axis. Most of their radars are connected to nearby Baker-Nunn sites by ground survey, so that a valuable and completely independent comparison of results from the two systems is possible. The JPL data are particularly important in two instances where the geometric solution is either very poor (9002, South Africa) or nonexistent (9003, Australia). In return, any systematic differences between the JPL and Smithsonian Astrophysical Observatory (SAO) longitude definitions, as well as the latitude of the former stations, can be established.

2. Results

In the solution presented here, the following three sources of data are used:

A. The new SAO dynamic satellite solution obtained by Gaposchkin [1],

B. The new SAO geometric satellite solution obtained by Lambeck [2] and

C. The recent JPL results from Mariner IV and V encounters obtained by Mottinger [3].

Observations from 34 stations have been used in Gaposchkin's solution. These include 17 SAO Baker-Nunns, 4 USAF Baker-Nunns, 4 Goddard Range and Range-Rate sites (GRARR), 5 laser sites, and 3 European optical sites. Some of these stations have been treated as colocated (Fig. 1). GAPOSCHKIN also solves for tesseral harmonics up to degree and order 12 plus 50 resonant terms. Approximately 63000 observations have been used in this solution.

Lambeck's geometric solution is for the coordinates of 38 stations*, of which 27 are in common with the stations used in the dynamic solution. Observations from several Minitrack Optical Tracking System (Mots) cameras and from six European optical sites have been used in addition to the Baker-Nunn data (Fig. 1). Approximately 50000 observations have been used in this solution – a number that is comparable to the data used in the dynamic solution. To obtain estimates of the agreement between the two solutions, the directions between stations are computed for the two solutions and compared. This test gives a better indication of relative accuracies than does a direct comparison of the two sets of coordinates, because the geometric solution is characterized by strong correlations between station coordinates. Fig. 2 illustrates some results of the comparisons. For the fundamental Baker-Nunn network (stations numbered 9001 to 9012), the variance of the differences in the directions computed from the two solutions is $(1.1'')^2$. For the new SAO Baker-Nunn stations, this variance is larger $(1.9'')^2$, as not so many observations are available for these stations as for the previous group. The average variances of the same directions have been computed for the geometric solution [2]; they can also be computed for the dynamic solution by using the covariance matrix obtained by GAPOSCHKIN. For the fundamental Baker-Nunn net, the results are $(0.5'')^2$ for the geometric solution and $(0.4'')^2$ for the dynamic solution. For the second group of stations, the corresponding results are $(0.9'')^2$ and $(0.7'')^2$. Now, because of its complexity the dynamic solution gives a covariance matrix that is characteristically overoptimistic, although the correlations between the parameters appear to be realistic. Typically, the dynamic solution gives accuracy estimates of a few meters for the station positions. The geometric solution, on the other hand, does appear to give realistic accuracy estimates, so that the "scaling factor" of the covariance matrix of the dynamical solution can be estimated from the above variances. For the fundamental group of stations, the factor is $k^2 = [(1.1)^2 - (0.5)^2]/(0.4)^2 \approx 5$; for the second group of stations, the factor is $k^2 = [(1.9)^2 - (0.9)^2]/(0.7)^2 \approx 5$.

---

* The 1969 solution has been modified to include recently acquired data from 8019 and 9077.

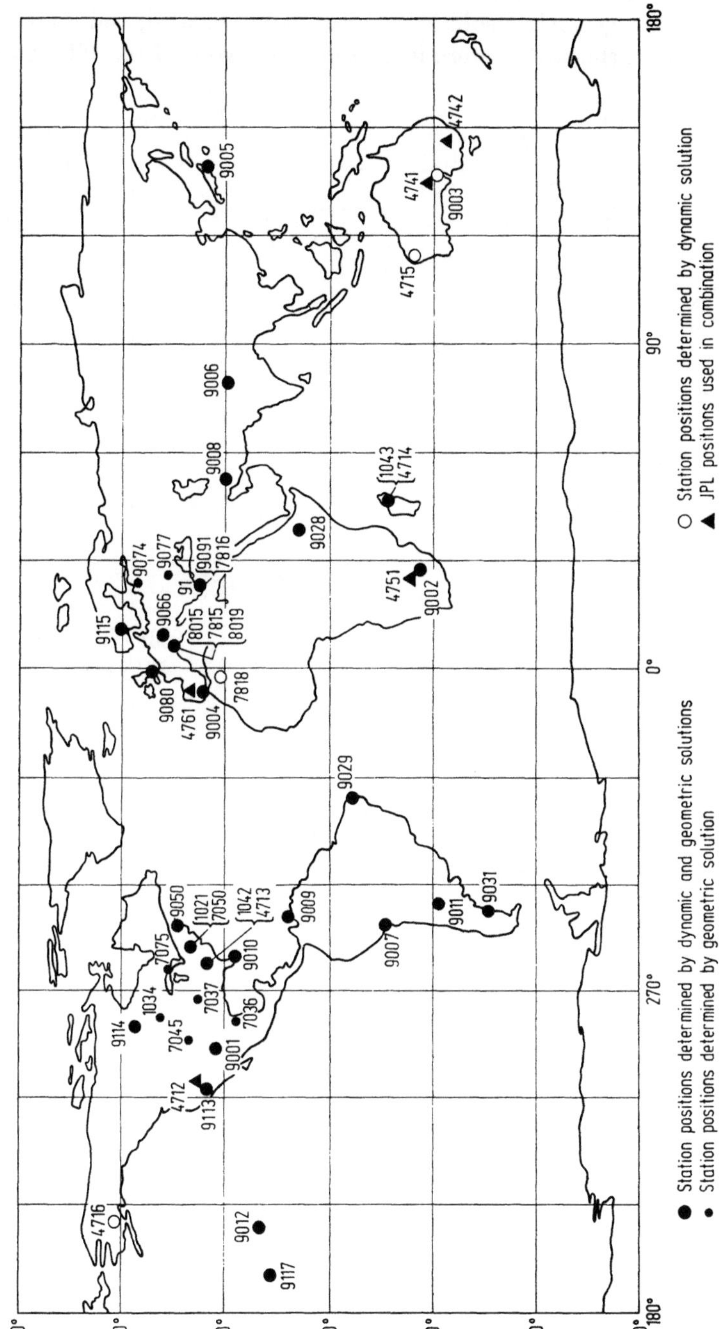

Fig. 1. Location of stations used in the combination solutions. Collocated stations are bracketed together.

In the combination of the two satellite solutions, it must be remembered that the geometric solution is essentially unscaled, and its origin is more or less arbitrarily determined (the coordinates of the 1966 Stan-

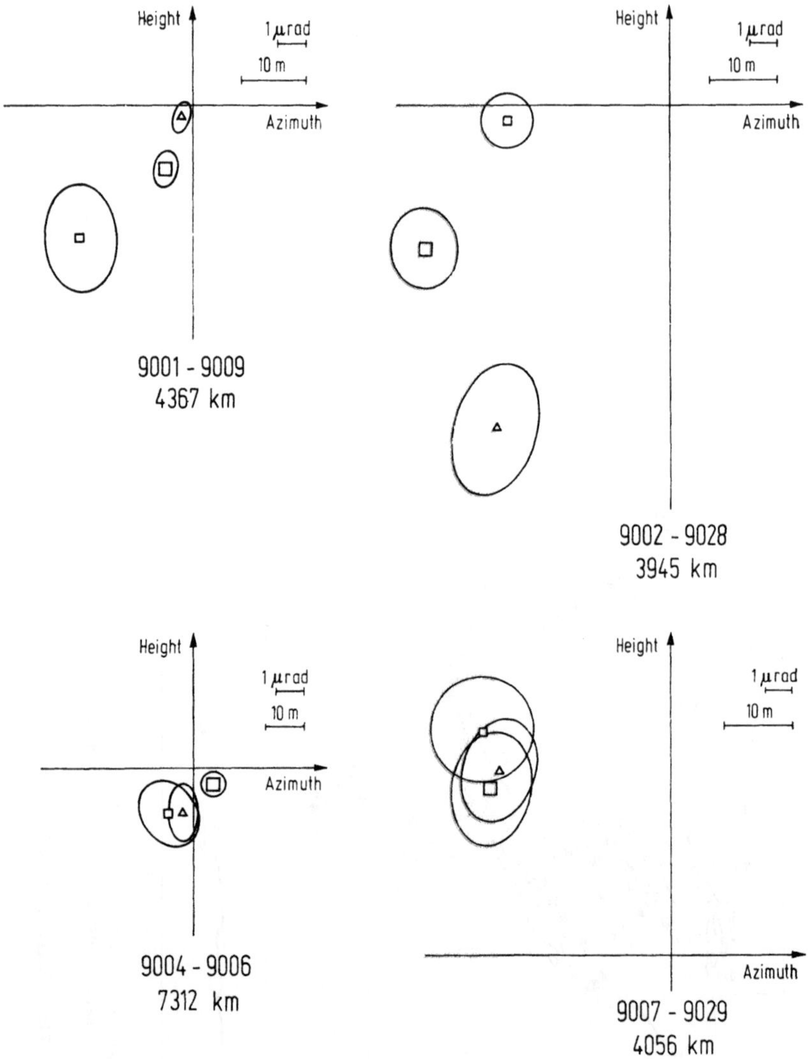

Fig. 2. Representation of some comparisons for directions between stations derived from the geometric solution △, the dynamic solution ▫, and the combination solution □.

dard Earth solution for station 9010 are used as the origin, and the distance computed from the coordinates of stations 9001 and 9010 is used for the scale). Thus, in the transformation linking the two solutions, one scale and three translation parameters must be introduced. A rotation term is also introduced into the transformation to determine whether a systematic longitude discrepancy exists between the two solutions. In the combination, the covariance matrices derived from the two individual solutions are used with the dynamic solution matrix scaled by the k^2 factor.

The results of Mottinger's solution for the JPL stations can be compared with the SAO combined satellite solution through the use of terrestrial survey data (see Fig. 1). In the case of stations 4712, 4741, and 4751, the distances to the nearest Baker-Nunn sites are small, and the survey inaccuracies should not have a significant influence on the comparison of results. For station 4761 the distance to the nearest Baker-Nunn is about 320 km, and for station 4742 the distance is of the order of 1160 km. For the latter station it has been possible to compute a "correction" to the difference in survey coordinates between the JPL radar and the Baker-Nunn at 9003. A recent solution by ANDERLE [4] has determined the coordinates of 8 stations tied directly to the Australian Geodetic Datum (AGD). A comparison of the satellite solution with the AGD shows significant distortions in the ground-survey net (LAMBECK, unpublished), distortions that affect the differences in the AGD x and y coordinates of 4742 and 9003 by about 10 m. The Anderle solution can be related to the SAO 1966 reference system by using the parameters determined by GAPOSCHKIN [5]. Thus, the AGD system can be related to the SAO system, and the differences between the survey coordinates of 4742 and 9003 in this modified system are used to relate the JPL station to the Baker-Nunn site.

The reduced JPL coordinates for the Baker-Nunn sites show a significant longitude difference when compared with the combined satellite solution. Thus, the JPL data are related to the combined satellite solution by introducing a second longitude rotation term as well as a second scale parameter. When data from different flights are used, it may be necessary to introduce additional scale and rotation terms for each mission. Accuracy estimates for the JPL solution have been given by MOTTINGER, and the survey coordinates are treated as stochastic quantities.

Consequently, the final combined set of data consists of a set of equations relating the unknown station coordinates, tesseral harmonics, and a variety of transformation parameters to functions of observed quantities.

Table 1 gives the coordinates of the final combination solution. Referred to as the C.B. system, they replace the 1966 C.6. set. The C.B. reference defines a coordinate system scaled by $GM = 398\,603.2 \times 10^9\,\mathrm{m^3\,sec^{-2}}$, an orientation defined by the 1903 mean pole and the 1968 mean observatory of the Bureau International de l'Heure. The earth's mass center is supposedly at the point 0, 0, 0 ± 3 m.

Table 1. *Final Coordinates of Station Positions derived from Combination Solution**

Station	X	Y	Z	Code	σ	Comments
		SAO Baker-Nunn and K-50 data				Collocated with
9001	−1.535751	−5.166997	3.401045	c	6	7901
9002	5.056130	2.716504	−2.775800	b	7	
9003	−3.983767	3.743101	−3.275596	a	6	9023
9004	5.105590	−0.555221	3.769681	c	7	
9005	−3.946702	3.366293	3.698832	b	10	
9006	1.018193	5.471111	3.109621	c	7	
9007	1.942775	−5.804080	−1.796931	c	6	
9008	3.376881	4.403982	3.136235	c	9	
9009	2.251829	−5.816911	1.327158	c	6	
9010	0.976291	−5.601393	2.880242	c	6	
9011	2.280590	−4.914565	−3.355430	c	7	
9012	−5.466040	−2.404285	2.242166	c	7	
9023	−3.977757	3.725116	−3.303064	a	6	9003
9028	4.903733	3.965199	0.963860	c	12	
9029	5.186475	−3.653849	−0.654329	d	17	
9031	1.693819	−4.112335	−4.556664	c	13	
9050	1.489749	−4.467474	4.287294	d	12	
9091	4.595157	2.039434	3.912674	d	10	
		USAF Baker-Nunn				
9113	−2.450003	−4.624422	3.635032	d	8	
9114	−1.264836	−3.466892	5.185463	d	15	
9115	3.121277	0.592630	5.512687	d	17	
9117	−6.007391	−1.111843	1.825719	d	18	
		European Optical Sites				
8015	4.578326	0.457970	4.403201	c	8	7815, 8019
8019	4.579467	0.586608	4.386424	e	8	7815, 8015
9066	4.331319	0.567511	4.633103	d	8	
9074	3.183899	1.421456	5.322802	e	13	
9077	3.907423	1.602406	4.763921	e	16	
9080	3.920181	−0.134731	5.012709	d	10	

A code is given in Table 1 to indicate which of the two satellite solutions has the dominant influence on the combination results. Accuracy estimates for the coordinates are also given in this table. One-figure accuracy estimates suffice, since the correlation between coordinates of different stations is small and since for most stations the accuracy at any one point is practically equal in all directions. These estimates have been obtained from the covariance matrix of the combination solution and, because of the k^2 factor introduced, should be fairly realistic. The directions between stations have also been computed for the combination solution and compared with the two individual satellite solutions (Fig 2). These comparisons are in general agreement with the estimates in Table 1.

Table 1 (Continued)

Station	X	Y	Z	Code	σ	Comments
			Mots Cameras			Collocated with
1021	1.118024	−4.876306	3.942972	e	8	7050
1034	−0.521699	−4.242049	4.718724	e	10	
1042	0.647516	−5.177924	3.656703	e	8	4713
1043	4.091894	4.434272	−2.064692	e	20	4714
7036	−0.828490	−5.657454	2.816812	e	10	
7037	−0.191283	−4.967278	3.983258	e	10	
7039	2.308231	−4.873591	3.394578	e	10	
7040	2.465060	−5.534916	1.985515	e	10	
7045	−1.240474	−4.760229	4.048991	e	10	
7075	0.692626	−4.347057	4.600477	e	10	
7076	1.384171	−5.905677	1.966538	e	10	
			Lasers			
7050	1.130669	−4.831358	3.994099	a	8	1021
7815	4.578368	0.457954	4.403156	a	8	8015, 8019
7816	4.654337	1.959148	3.884390	a	10	9091
7818	5.426335	−0.229363	3.334598	a	10	
7901	−1.535752	−5.166998	3.401046	a	6	9001
			GRARR			
4713	0.647190	−5.178327	3.656132	a	8	1042
4714	4.091403	4.434128	−2.066044	a	20	1043
4715	−2.328267	5.299698	−2.669332	a	15	
4716	−2.282509	−1.453427	5.756748	a	20	

* The explanation for the code is as follows:
a Dynamic solution only.
b Dynamic solution dominant.
c Dynamic and geometric solution about equal influence.
d Geometric data dominant.
e Geometric data only.

Summing up: the coordinates of the fundamental Baker-Nunn sites, stations 9001 to 9012 and 9023, have standard deviations of about 10 m or less, while the coordinates of the other Baker-Nunn cameras, including the USAF instruments, are determined with standard deviations in the 15- to 20-m range. The Mots, GRARR, and Laser sites in the United States are also determined with standard deviations of about 10 m, this high accuracy being due to the collocation of instruments at several sites and to the strong geometric solution in this area. The stations in Europe are also well determined for the same reasons.

The corrections, resulting from the combinations solution, to Gaposchkin's 1969 tesseral harmonics are small, less than 2×10^{-8}. The correlation coefficients between station coordinates and harmonic coefficients are also small; the maximum is 16%, and the average is about 4%. These results therefore suggest that the effects of uncertainties in the station coordinates and in the tesseral harmonics on the observed orbit residuals have been effectively separated. Thus, in future iterations it will be possible to keep the coordinates of the well-determined stations fixed (using the combination-solution coordinates) and solve for a higher order gravity field [1].

The combination solution indicates no significant difference in the longitudes of the two satellite solutions, the result being -1.4 ± 0.7 μrad.

On the other hand, the longitude difference between the satellite solution and the JPL solution is significant: $\Delta\lambda = -5.6 \pm 0.7$, the JPL longitudes being to the east of the satellite solution. Only a small part of this, -0.8 μrad, can be attributed to the differences in the UT1 time systems used by the two agencies. The scale factor for this solution is negligible, $0.5 \pm 0.6 \times 10^{-6}$, as would be expected since both SAO and JPL have used the same value for GM. Table 2 gives the residuals between the satellite solution and the JPL solution after the systematic rotation difference has been removed. The coordinates of the JPL sites in the C.B. system are given in Table 3.

Table 2. *Comparison of JPL and SAO Results: $d\lambda_i$ is the Longitude Difference Between the two Solutions for each Comparison; $\overline{d\lambda}$ is the Mean Longitude Difference; and r is the Distance to the Spin Axis*

Station JPL–SAO	$\lambda_{SAO} - \lambda_{JPL} = d\lambda_i$ [μrad]	$\overline{d\lambda} - d\lambda_i$ [μrad]	[m]	$r_{SAO} - r_{JPL}$
4751–9002	−5.65	+0.09	+0.5	+6.9
4741–9003	−7.33	+1.77	+9.7	−2.2
4742–9003	−6.54	+0.98	+5.1	−6.3
4761–9004	−4.60	−0.96	−4.7	+1.2
4712–9113	−4.35	−1.21	−6.3	+2.4

The results in Table 2 support the accuracy estimates obtained for the Baker-Nunn sites at 9002, 9003, 9004, and 9113, particularly when it

Table 3. *Coordinates of JPL Stations in the C.B. System*

Station	X	Y	Z
4712	−2.350459	−4.651970	3.665628
4741	−3.978705	3.724862	−3.302243
4742	−4.460971	2.682430	−3.674646
4751	5.085457	2.668245	−2.768744
4761	4.849240	−0.360295	4.114885

is recalled that these residuals are the combined effect of uncertainties in the satellite solution, in the JPL solution, and in the surface-survey data connecting the two types of instrumentation.

References

1. Gaposchkin, E. M.: Improved Values for the Tesseral Harmonics of the Geopotential and Station Coordinates. To be presented at XII Meeting of COSPAR, Prague, May 1969.
2. Lambeck, K.: A Spatial Triangulation Solution for a Global Network and the Position of the North American Datum within it. Presented at the Annual Meeting AGU, April 21–25, 1969, Washington, D.C.
3. Mottinger, N. A.: Breaking the 10 Meter Level in Obtaining Consistent Station Location Solutions from the Reduction of deep Space Probe Data. Presented at the Annual Meeting AGU, April 21–25, 1969, Washington, D.C.
4. Anderle, R. J.: Geocentric Positions of Sites based on Doppler Satellite Observations made under the Auspices of the Commonwealth of Australia. NWL Report No. 2126 (1967) 14 pp.
5. Gaposchkin, E. M., in: Dynamical Determination of Station Locations using GEOS 1 Data. Smithsonian Astrophys. Obs. Spec. Rep. No. 264 (1964) 25.

Tidal Potential with Location-Dependent Amplitude and Phase Angle

By

W. M. Kaula

University of California, Los Angeles, CA, USA

Abstract. The tidal disturbing function has been developed in a form such that the Love number k and lag angle δ are expressed as a sum of spherical harmonics which are functions of latitude and longitude of the Earth. Solutions are made for the zonal harmonics to fit results obtained from artificial satellites. However, it does not appear that the time scale problem of the lunar orbit evolution can be solved without an appreciably smaller overall lag in the Earth in the past.

Unabridged paper is published in the Astronomical Journal, 74 (1969) 1108 to 1114.

On the Determination of Field Parameters Using a Generalized Fourier-Analysis*

By

M. Schneider and **Ch. Reigber**

Institut für Astronomische und Physikalische Geodäsie,
Technische Hochschule, München, FRG

Abstract. The deviations of the satellite motion from an arbitrarily chosen reference orbit, which can be generated either analytically or numerically, are analyzed. This is done using a generalized Fourier-analysis. The data thus obtained are introduced into a systen of determining equations for field parameters, e.g. potential coefficients.

This method has extensively been tested on the basis of simulated satellite positions. Applicability conditions, the influence of (normally distributed) observational errors, linear correlation between field parameters and truncation error have been studied. Preliminary results of these tests will be given.

1. Outline of the Method

The motion of a dynamical system may be represented by the n-dimensional vector $\boldsymbol{a}(t)$, where n denotes the number of degrees of freedom and t the time. When $\boldsymbol{a}(t)$ is the solution of the boundary-value problem defined through

$$\frac{d^2 \boldsymbol{a}}{d t^2} = \boldsymbol{f}(t; \boldsymbol{a}, \dot{\boldsymbol{a}})$$

$$\boldsymbol{a}(t_A) = \boldsymbol{a}_A; \quad \boldsymbol{a}(t_B) = \boldsymbol{a}_B$$

then it can be expanded as

$$\boldsymbol{a}(t) = \overline{\boldsymbol{a}}(t) + \sum_{\nu=1}^{\infty} \boldsymbol{a}_\nu \sin\left(\nu \pi \frac{t - t_A}{t_B - t_A}\right) \tag{1}$$

where

$$\overline{\boldsymbol{a}}(t) = \frac{t_B - t}{t_B - t_A} \boldsymbol{a}_A + \frac{t - t_A}{t_B - t_A} \boldsymbol{a}_B .$$

* Contribution No. 60 of the Institute of Astronomical and Physical Geodesy, Technical University Munich.

The coefficients $\boldsymbol{a}_\nu$ obey the infinite system of equations

$$\boldsymbol{a}_\nu = -\frac{2}{\nu^2 \pi^2} \int_{t_A}^{t_B} \sin\left(\nu \pi \frac{t' - t_A}{t_B - t_A}\right) \boldsymbol{f}(t';\ \boldsymbol{a}, \dot{\boldsymbol{a}})\, d t' \tag{2}$$

$$(\nu = 1, 2, \ldots \infty).$$

The n-dimensional vector $\boldsymbol{f}$ generally contains a number of parameters c_n. When this occurs in a linear form, i.e.

$$\boldsymbol{f} = \sum_{n=0}^{N} c_n \boldsymbol{k}_n + \boldsymbol{z}$$

$\boldsymbol{k}_n$ functions depending only on t, $\boldsymbol{a}$, $\dot{\boldsymbol{a}}$,

$\boldsymbol{z}$ completely known remainder of $\boldsymbol{f}$,

the Eq. (2) read

$$\sum_{n=0}^{N} c_n \boldsymbol{C}_{n,\nu} = \boldsymbol{U}_\nu \qquad (\nu = 1, 2, \ldots \infty) \tag{3}$$

where

$$\boldsymbol{C}_{n,\nu} = \int_{t_A}^{t_B} \sin\left(\nu \pi \frac{t' - t_A}{t_B - t_A}\right) \boldsymbol{k}_n(t';\ \boldsymbol{a}, \dot{\boldsymbol{a}})\, d t';$$

$$\boldsymbol{U}_\nu = -\frac{\nu^2 \pi^2}{2} \boldsymbol{a}_\nu - \int_{t_A}^{t_B} \sin\left(\nu \pi \frac{t' - t_A}{t_B - t_A}\right) \boldsymbol{z}(t';\ \boldsymbol{a}, \dot{\boldsymbol{a}})\, d t'.$$

When the solution (1) has been determined from observations, then the vectors $\boldsymbol{C}_{n,\nu}$ and $\boldsymbol{U}_\nu$ are completely known. The equations (3) can then be used for determining the parameters c_n.

There is a great variety of possible applications of the above procedure. It can be used for example in determining masses of celestial bodies from lunar and interplanetary orbits; furthermore one can derive station coordinates of observing stations avoiding the knowledge of the earthcentered motion.

In this paper we present some investigations concerned with the determination of the earth's gravitational field. This research is part of a cooperative project between the National Institute for Astronomy and Geophysics of France and the Institute for Astronomical and Physical Geodesy in Munich.

2. Application to Dynamical Satellite Geodesy

Comparing the actual motion $\boldsymbol{r}(t)$ of a satellite with an arbitrarily chosen reference orbit $\boldsymbol{r}_n(t)$, one obtains for the deviations $\boldsymbol{d}(t) = \boldsymbol{r}(t) - \boldsymbol{r}_n(t)$ the differential equation

$$m\ddot{\boldsymbol{d}}(t) = \boldsymbol{F}_{\text{eff}}(t; \boldsymbol{d}, \dot{\boldsymbol{d}}) = \boldsymbol{F}(t; \boldsymbol{r}_n + \boldsymbol{d}, \dot{\boldsymbol{r}}_n + \dot{\boldsymbol{d}}) - \boldsymbol{F}_n(t; \boldsymbol{r}_n, \dot{\boldsymbol{r}}_n),$$

where $\boldsymbol{F}$ denotes the actual force field and $\boldsymbol{F}_n$ the reference field. The reference orbit can be generated either analytically or by numerical integration.

When one looks for the parameters of the earth's gravitational field, one can represent this field for example by means of the spherical harmonics expansion of its potential. The determining equations for the potential coefficients C_{nm} and S_{nm} then read

$$\sum_{n=0}^{\infty} \sum_{m=0}^{n} (C_{nm} \boldsymbol{A}_{nm,\nu} + S_{nm} \boldsymbol{B}_{nm,\nu}) = \frac{\nu^2 \pi^2}{2} \boldsymbol{d}_\nu + \boldsymbol{M}_\nu + \boldsymbol{N} \quad ,$$

where

$$\boldsymbol{M}_\nu = \frac{1}{m} \int_{t_A}^{t_B} \sin\left(\nu \pi \frac{t' - t_A}{t_B - t_A}\right) \boldsymbol{Z}(t'; \boldsymbol{r}, \dot{\boldsymbol{r}})\, dt';$$

$$\boldsymbol{N}_\nu = -\frac{1}{m} \int_{t_A}^{t_B} \sin\left(\nu \pi \frac{t' - t_A}{t_B - t_A}\right) \boldsymbol{F}_n(t'; \boldsymbol{r}_n, \dot{\boldsymbol{r}}_n)\, dt';$$

$$\boldsymbol{A}_{nm,\nu} = -k^2 M R^n \int_{t_A}^{t_B} \sin\left(\nu \pi \frac{t' - t_A}{t_B - t_A}\right) \nabla \left(\frac{\cos m\lambda \cdot P_n^m}{r^{n+1}}\right) dt';$$

$$\boldsymbol{B}_{nm,\nu} = -k^2 M R^n \int_{t_A}^{t_B} \sin\left(\nu \pi \frac{t' - t_A}{t_B - t_A}\right) \nabla \left(\frac{\sin m\lambda \cdot P_n^m}{r^{n+1}}\right) dt',$$

and k^2 denotes the gravitational constant, M and R respectively the earth's mass and equatorial radius.

Every available orbital arc supplies us with one system of determining equations. From these one can collect the final system for the determination of the parameters by means of a least squares adjustment.

3. Results from Trial Computations

The method has been tested for the case where only zonal terms are present in the gravitational potential and the remainder $\boldsymbol{Z}$ vanishes.

When using a reference orbit $\boldsymbol{r}_n(t)$ according to

$$\boldsymbol{F}_n = \nabla \left\{ \frac{k^2 M}{r_n} \left[1 + \sum_{n=2}^{\varkappa} \left(\frac{R}{r_n} \right)^n C'_{no} P_n \right] \right\}$$

where C'_{no} denotes an approximate value for C_{no}, the determining equations for the potential coefficients C_{no} read

$$\sum_{n=2}^{\varkappa} \Delta C_{no} \boldsymbol{A}_{no,\nu} + \sum_{n=\varkappa+1}^{\infty} C_{no} \boldsymbol{A}_{no,\nu} = \frac{\nu^2 \pi^2}{2} \boldsymbol{d}_\nu + \sum_{n=2}^{\varkappa} C'_{no} (\boldsymbol{A}'_{no,\nu} - \boldsymbol{A}_{no,\nu})$$

$$+ k^2 M \int_{t_A}^{t_B} \sin \left(\nu \pi \frac{t' - t_A}{t_B - t_A} \right) \left(\frac{\boldsymbol{r}_n}{r_n^3} - \frac{\boldsymbol{r}}{r^3} \right) d t' ,$$

where $\Delta C_{no} = C_{no} - C'_{no}$ and $\boldsymbol{A}'_{no,\nu}$ $\boldsymbol{A}_{no,\nu}$ are the aforementioned integrals computed with $\boldsymbol{r}_n$ and $\boldsymbol{r}$ respectively.

The trial computations are based on simulated satellite positions $\boldsymbol{r}(t_i)$, generated by numerical integration using some nominal values C_{no}. One then can evaluate the deviations $\boldsymbol{d}(t_i)$ from the numerically generated corresponding reference positions $\boldsymbol{r}_n(t_i)$. Therefrom the coefficients $\boldsymbol{d}_\nu$ are determined by adjustment. Now the integrals occurring in the determining equations are evaluated by means of a high-precision quadrature.

Table 1.

Satellite	i [deg]	Ω [deg]	ω [deg]	a [km]	e
FR 70 (Fictitious)	5.000	0	0	7530.0	0.0200
Pegasus 3	28.887	0	0	6908.6	0.0017
OSO 3	32.865	0	0	6931.9	0.0022
Explorer 29	59.382	235	130	8073.1	0.0717
Explorer 20	79.901	40	40	7316.4	0.0100
Explorer 22	79.690	40	40	7363.0	0.0130

The satellites for which observations have been simulated are listed in Table 1 together with the initial orbital elements. In the following two sections some results are given for the case, when exact observations $\boldsymbol{r}(t_i)$ are introduced. The following section then contains similar results, when using observations affected by computer produced random errors ($r.m.s. = 2$ m).

3.1. Analysis of Short Arcs

Observations have been generated for one complete revolution of each satellite. The time span between two observations was $\Delta t = 40$ sec, i.e. we used about 120 positions $\boldsymbol{r}(t_i)$ for each satellite. The reference orbit corresponded to $C'_{20} \neq 0$, $C'_{no} = 0$ for $n > 2$. The deviation

$$\boldsymbol{d}(t) = \bar{\boldsymbol{d}}(t) + \sum_{\nu=1}^{\infty} \boldsymbol{d}_\nu \sin\left(\nu\pi \frac{t - t_A}{t_B - t_A}\right)$$

has been determined with an accuracy for each component better than 0.05 m. This has been achieved using 60 coefficients $\boldsymbol{d}_\nu$. The final system of determining equations did not contain all the available equations. In Table 2 and 3 some results for C_{no} together with their standard deviations are given.

Table 2. *Solutions for 14, 11, 8 Coefficients C_{no} Using Three Satellites (P3, E20, E29)*

10^6	Nominal Solution	14 Coefficient Solution ($M = 114$)	11 Coefficient Solution ($M = 114$)	8 Coefficient Solution ($M = 114$)
C_2	−1082.639	−1082.637 ± 0.001	−1082.630 ± 0.006	−1082.665 ± 0.021
C_3	2.565	2.565 ± 0.001	2.554 ± 0.006	2.558 ± 0.021
C_4	1.608	1.609 ± 0.001	1.594 ± 0.006	1.680 ± 0.021
C_5	0.174	0.172 ± 0.002	0.161 ± 0.007	0.102 ± 0.024
C_6	−0.542	−0.545 ± 0.002	−0.569 ± 0.008	−0.578 ± 0.026
C_7	0.419	0.422 ± 0.002	0.454 ± 0.009	0.454 ± 0.034
C_8	0.128	0.130 ± 0.003	0.183 ± 0.010	0.016 ± 0.034
C_9	0.022	0.024 ± 0.003	0.054 ± 0.012	0.189 ± 0.041
C_{10}	0.338	0.339 ± 0.003	0.367 ± 0.013	
C_{11}	−0.176	−0.177 ± 0.003	−0.231 ± 0.014	
C_{12}	−0.053	−0.055 ± 0.004	−0.106 ± 0.016	
C_{13}	0.146	0.139 ± 0.005		
C_{14}	0.174	0.169 ± 0.005		
C_{15}	0.065	0.062 ± 0.006		
N		1.12×10^{-10}	6.08×10^{-10}	2.30×10^{-9}

M = number of equations used, N = euclidian norm of residuals.

3.2. Analysis of Long Arcs

For ten revolutions observations have been generated, the time span of which was $\Delta t = 400$ sec, i.e. about 150 observations for each satellite. A reference orbit has been constructed according to $C'_{no} \neq 0$ for $n = 2, 3, 4$ and $C'_{no} = 0$ for $n > 4$. Fitting has been done to < 1 m, i.e. by means of 80 coefficients $\boldsymbol{d}_\nu$. Results are given in Table 4.

The correlation matrix of this solution is similar to that of Table 3 with the exception of some higher correlations between odd potential coefficients.

Table 3. *Correlation Matrix (14 Coefficients Solution)*

	C_2	C_3	C_4	C_5	C_6	C_7	C_8	C_9	C_{10}	C_{11}	C_{12}	C_{13}	C_{14}	C_{15}
C_2	1.00	−0.16	0.06	0.01	−0.28	0.12	−0.06	0.09	0.27	−0.10	0.13	−0.10	−0.05	−0.04
C_3		1.00	−0.21	0.14	0.08	−0.50	0.23	−0.27	0.05	0.41	−0.21	0.38	−0.19	0.11
C_4			1.00	−0.25	0.11	0.20	−0.54	0.28	−0.26	−0.00	0.49	−0.28	0.48	−0.24
C_5				1.00	−0.12	−0.01	0.27	−0.56	0.21	−0.29	−0.08	0.46	−0.25	0.54
C_6					1.00	−0.13	−0.01	0.04	−0.35	0.11	−0.21	0.01	0.30	−0.09
C_7						1.00	−0.30	0.38	−0.03	−0.44	0.27	−0.56	0.25	−0.13
C_8							1.00	−0.41	0.45	0.02	−0.47	0.37	−0.71	0.34
C_9								1.00	−0.24	0.24	0.19	−0.62	0.33	−0.73
C_{10}									1.00	−0.18	0.22	0.11	−0.30	0.17
C_{11}										1.00	−0.20	0.34	−0.04	−0.15
C_{12}											1.00	−0.31	0.57	−0.18
C_{13}												1.00	−0.37	0.70
C_{14}													1.00	−0.34
C_{15}														1.00

Table 4.

10^6	Nominal Solution	FR 70, OSO 3, E22, E29 ($M = 108$)
C_2	-1082.639	-1082.638 ± 0.001
C_3	2.565	2.569 ± 0.001
C_4	1.608	1.611 ± 0.001
C_5	0.174	0.170 ± 0.002
C_6	-0.542	-0.572 ± 0.002
C_7	0.419	0.422 ± 0.001
C_8	0.128	0.147 ± 0.002
C_9	0.022	0.036 ± 0.006
C_{10}	0.338	0.363 ± 0.005
C_{11}	-0.176	-0.206 ± 0.006
C_{12}	-0.053	-0.072 ± 0.004
C_{13}	0.146	0.137 ± 0.004
C_{14}	0.174	0.195 ± 0.006
C_{15}	0.065	0.102 ± 0.009
N		6.54×10^{-12}

3.3. Analysis of Observations Affected by Random Errors

Random errors have been generated on the computer for the deviations $\boldsymbol{d}(t_i)$, the *r.m.s.* of which was 2 m for each component of $\boldsymbol{d}(t_i)$. During one revolution the higher zonal terms ($n \geqq 7$) produce only de-

Table 5.

10^6	Nominal Solution	P3, E20, E29 (*r.m.s.* = 2 m, $M = 135$)
C_2	-1082.639	-1082.683 ± 0.279
C_3	2.565	2.519 ± 0.259
C_4	1.608	1.745 ± 0.269
C_5	0.174	0.229 ± 0.270
C_6	-0.542	-0.760 ± 0.308
N		3.28×10^{-8}

viations less than 1.5 m. (It should be mentioned that the reference orbit osculated the true orbit at time t_A). Therefore we have determined a truncated solution which is given in Table 5. Correlation was very small.

Trial computations for the long arc-case are not yet finished.

4. Conclusion

The problem of determining the $\boldsymbol{d}(t_i)$ from actual satellite observations is now investigated by our French colleagues under the supervision of Dr. F. Barlier.

All computations have been done on the TR4 computer of the Leibniz Computation Center of the Bavarian Academy of Sciences the assistance of which we gratefully acknowledge.

References

Bulirsch, R., Stoer, J.: Numerical Treatment of Ordinary Differential Equations by Extrapolation Methods. Num. Math. 8 (1966).

Businger, P., Golub, G. H.: Linear Least Squares Solutions by Householder Transformations. Num. Math. 7 (1965).

King-Hele, D. G., Cook, G. E., Scott, D. W.: Evaluation of Odd Zonal Harmonics in the Geopotential, of Degree less than 33, from the Analysis of 22 Satellite Orbits. Roy. Aircraft Est., Techn. Rep. 68202.

Kozai, Y.: Improved Values for Coefficients of Zonal Spherical Harmonics in the Geopotential. Smithsonian Astrophysical Observatory Spec. Rep. 264 (1967).

Krylov, V. L.: Approximate Calculation of Integrals, New York: MacMillan Comp. 1962.

Reigber, Ch.: Zur Bestimmung des Gravitationsfeldes der Erde aus Satellitenbeobachtungen. Veröff. der Deutschen Geodätischen Kommission, Reihe C 137, München 1969.

Schneider, M.: Zur dynamischen Nutzung der Bahnen künstlicher Satelliten. Forschungsbericht des Bundesministeriums für wissenschaftliche Forschung W68-55, München 1968.

Correlation Between the Solar Activity Indices and the Use of these Indices in the Study of Motion of the Earth Satellites

By

A. M. Fominov

Institute of Theoretical Astronomy, Leningrad, USSR

Abstract. Correlation between the 10.7-cm solar flux ($F_{10.7}$), the visible solar spot area (S_d) and the geomagnetic planetary index (A_p) have been investigated for the 27-day intervals over the period 1957–1966. $F_{10.7}$ and S_d are the indices connected with the solar UV-radiation. A_p is the index connected with the corpuscular radiation. S_d does not correlate usually with A_p. S_d correlate strongly with $F_{10.7}$ in the course of the 11-year solar cycle. The correlation is intense for the 27-day intervals also and the variations of the regression are especially great during the minimum of solar activity. This fact implies that S_d and $F_{10.7}$ represent the effects of the solar UV-radiation not quite sufficiently. This fact is possible to account for the difficulties occurred when taking into consideration the air drag perturbations from the solar activity by means of the indices mentioned.

The high-atmospheric density variations connected with the variations of the solar activity have been discovered through the motion of the first earth satellites. At first the 27-day variations and then the variations in the 11-year solar cycle of the air density have been investigated. Sporadic variations of the high-atmospheric density have been explored also. It has been found that the variations of the air density are closely connected with the variations of some indices of the solar acrivity.

The influence of the solar activity on the earth atmosphere is manifested chiefly through the ultraviolet and corpuscular radiations of the sun. These radiations do not reach the Earth. They are absorbed in the earth atmosphere. Many indices of the solar activity are connected with the solar ultraviolet (UV) radiation. These are the solar radio-noise in the decimeter range, the Wolf numbers, the solar spot area etc. Penetration of the corpuscular flux into the earth atmosphere causes the magnetic storms. The intensity of these storms can be characterized by the geomagnetic indices. Thus, it is possible to investigate the formal dependence of the air density from the solar activity indices.

Solar radio-noise flux in the 10.7 cm ($F_{10.7}$) is usually considered to be index of UV. As the indexes of the corpuscular radiation either the geomagnetic index (K_p) or the geomagnetic planetary index (A_p) are used. The data on UV and corpuscular radiation are not practically available at present. It has not yet been decided what indices are better to characterize these fluxes. ANDERSON [1] implies that $F_{10.7}$-index represents UV not quite sufficiently. WARD and SHAPIRO [6] have made the harmonic analysis of ten solar activity indices. They have found that some of these indices are in strong correlation with each other.

This work has been undertaken in order to obtain the numerical estimations of the regression and correlate coefficients for some of the indices of the solar activity. The solar radio-noise flux $F_{10.7}$ [2], the visible solar spot area S_d [5] and the geomagnetic planetary index A_p [3] have been used as the indices of the solar activity.

Correlation between the S_d- and A_p-, the S_d- and $F_{10.7}$-indices have been investigated over the period from October 1957 to December 1966. This period has been subdivided into the 27-day intervals beginning with 4th October 1957. In all 125 equal intervals have been formed. The computations have been carried out for three cases. Indices averaged over the period of one day have been used in the first case, that averaged over three and seven days have been used in the second and third cases respectively. The results obtained don't differ significantly from each other in all three mentioned cases. The numerical results have been given here only for the third case of those mentioned above.

The regression coefficients and the correlation coefficients have been determined for the each 27-day interval. The method used can be found in [4]. The regression equations are

$$\left.\begin{aligned} F_{10.7} - \overline{F}_{10.7} &= \varrho_{FS}\,(S_d - \overline{S}_d)\,, \\ S_d - \overline{S}_d &= \varrho_{SF}\,(F_{10.7} - \overline{F}_{10.7})\,, \end{aligned}\right\} \tag{1}$$

where $F_{10.7}$ is expressed in units of $10^{-22}\,\mathrm{W\,m^{-2}\,cps^{-1}}$, S_d is expressed in units of 10^{-6} solar disk area. $\overline{F}_{10.7}$ and $\overline{S}_d$ are the mean values of $F_{10.7}$ and S_d over the 27-day intervals, ϱ_{FS} and ϱ_{SF} are the regression coefficients.

The similar equations take place for S_d- and A_p-indices.

The correlation between S_d and A_p is practically absent. The results of determination of the correlation coefficients r for these indices are given in Table 1. n_1 is the number of the 27-day intervals with r indicated in the first line of Table 1 and with the positive regression coeffi-

Table 1.

r	$0 \leqslant r \leqslant 0.5$	$0.5 < r \leqslant 0.8$	$0.8 < r \leqslant 1$
n_1	37	14	10
n_2	40	20	4

cients, n_2 being that for the case of the negative regression coefficients. S_d and A_p represent UV and corpuscular solar radiation respectively. This fact confirms once more that both of these fluxes are independent of each other. Therefore, it is possible to investigate the influence of the UV and corpuscular radiation on the high-atmospheric density distribution separately. The difficulties can arise only for those time intervals at which S_d and A_p are strongly correlated in phase or counterphase. The initial moments of these 27-day intervals are indicated in the second column of Table 2. Values of r are given in the third

Table 2.

	t_i	r	Sign of ϱ_{FS}
1	20. 1. 58	0.83	+
2	24. 4. 59	0.92	+
3	26. 11. 59	0.93	+
4	19. 1. 60	0.86	−
5	13. 3. 60	0.85	+
6	26. 7. 60	0.84	+
7	11. 11. 60	0.90	+
8	19. 5. 61	0.83	−
9	12. 7. 61	0.87	+
10	24. 11. 61	0.86	+
11	17. 1. 62	0.89	−
12	3. 1. 63	0.96	+
13	10. 2. 66	0.92	+
14	7. 11. 66	0.83	−

column. The signs of regression coefficients are given in the fourth column. For these intervals it is difficult to separate the effects in the air density distribution due to the UV and corpuscular radiation.

Correlation between $F_{10.7}$ and S_d is strong enough (Table 3). N_1 is the number of 27-day intervals with r indicated in the first line of Table 3 and with the positive regression coefficients, N_2 being that for the negative regression coefficients.

Table 3.

r	$0 \leqslant r \leqslant 0.5$	$0.5 < r \leqslant 0.8$	$0.8 < r \leqslant 1$
N_1	10	16	92
N_2	6	1	—

Correlation is practically absent only for 16 cases. The initial moments of these 27-day intervals are indicated in the second column of Table 4. The third column gives the values of r for the corresponding intervals. The signs of the regression coefficients are given in the fourth column of Table 4. For the intervals mentioned in Table 4 the choice of the index can influence the results of the investigation of the air density dependence on the solar activity.

It is seen from Table 3 that the regression coefficients are negative in the case of moderate correlation only for one of the 27-day intervals. This interval is given in the last line of Table 4. $F_{10.7}$- and S_d-indices vary in phase with each other for the remaining cases of the moderate and strong correlation.

Table 4.

	t_i	r	Sign of ϱ_{FS}
1	6. 5. 60	0.34	−
2	8. 8. 61	0.28	+
3	1. 6. 62	0.15	+
4	26. 2. 63	0.37	+
5	23. 11. 63	0.48	+
6	20. 12. 63	0.12	+
7	10. 3. 64	0.27	−
8	6. 4. 64	0.05	−
9	3. 5. 64	0.48	−
10	12. 10. 64	0.14	−
11	8. 11. 64	0.27	+
12	5. 12. 64	0.10	−
13	5. 8. 65	0.37	+
14	18. 12. 65	0.23	+
15	5. 4. 66	0.24	+
16	29. 5. 66	0.17	+
17	24. 2. 65	0.58	−

Beside the regression coefficients ϱ_{FS}, ϱ_{SF} and mean values of $F_{10.7}$ and S_d have been computed for each 27-day interval. The dependence of the regression coefficients ϱ_{FS} on the solar activity level has been investigated. The only case of the strong correlation ($r \geqslant 0.84$) has been considered. In all 84 such cases have been found. The results do not show the evident dependence of ϱ_{FS} on S_d. The values of ϱ_{FS} vary simultaneously within wide limits (from 0.0075 to 0.0330) especially for the solar activity minimum when four values of ϱ_{FS} exceed the mentioned upper limit. The mean value of regression coefficient ϱ_{FS} is 0.019 for 27-day intervals.

The dependence of $\overline{F}_{10\,7}$ on $\overline{S}_d$ has been investigated. It has been found by least squares method that

$$\overline{F}_{10.7} = 67 + 0{,}046\,\overline{S}_d - 0{,}0000021\,\overline{S}_d^2$$

in the case of parabolic dependence.

The use of the linear dependence gives

$$\overline{F}_{10.7} = 73 + 0{,}035\,\overline{S}_d\,.$$

The mean values of the ratio $F_{10.7}/S_d$ for 27-day intervals (0.019) are different from that for 11-year cycle (0.035). This phenomenon is quite similar to that showed by satellite motions. The density variations with the solar activity are to be given by the different formulae for 27-day intervals and 11-year cycle when using $F_{10.7}$-index.

The results show the strong correlation between the two indices of the solar activity does not yet mean that these indices are in functional dependence. The same can be likely said about the connection of the solar activity indices with the UV and corpuscular radiation. Therefore, the conclusion can be drawn that it is necessary to have the data on the UV and corpuscular radiation directly when investigating the dependence of air density distribution on the solar activity because these two fluxes are the main ones to be responsible for the atmospheric density changes.

References

1. ANDERSON, A. D.: On the inexactness of the 10,7-cm flux from the sun as an index of the total extreme ultraviolet radiation. J. Atmosph. Sci. 21 (1964) 1–14.
2. Astron. Mitt. Eidgen. Sternw. Zürich, 1958–1967: NN 215, 221, 229, 237, 244, 251, 261, 267, 273, 279.
3. J. Geoph. Res., 1958–1967: 63–72.
4. SHCHIGOLEV, B. M.: Observation mathematical treatment, 1962 (in russian).
5. Solar data, 1957–1966: NN 1–12 (in russian).
6. WARD, F., SHAPIRO, R.: Decomposition and comparison of time series of indices of solar activity. J. Geophys. Res. 67 (1962) 541–554.

New Investigations of Motion of an Artificial Satellite about its Center of Mass

By

V. V. Beletskii

Institute of Applied Mathematics, Academy of Sciences, Moscow, USSR

Abstract. The last papers dedicated to the analysis of dynamics of the satellite rotational motion are reviewed. The review concerns some old fundamental investigations on which are based some new ones; the main attention should be paid to the new fundamental investigations. The paper deals with the following questions:

the analysis of the passive stabilized effects (gravitational, magnetic effects);

the role of nonlinear resonance effects in the motion of artificial and natural celestial bodies;

some new results on the motion stability theory and their role in the problems of dynamics of the satellite rotational motion;

new aerodynamic effects in the rotation of a satellite (effect of the aerodynamic "propellation");

satellite orientation determination with respect to the measurements.

Special attention has been paid to those dynamical effects of rotational motion, the analysis of which promotes the solution of problems of the upper atmosphere physics (its density, its character of module interaction with satellite surface).

The report represents the review of some papers (1964–1968) on motion of the artificial satellites about the center of mass.

I wish to simulate an interest in such papers which have character of principle as it seems to me. Analysing motion of a satellite about its center of mass one can see two trends: the first one is the analysis of the stabilizing effect of the external torques and libratory motion of a satellite in the neighbourhood of the equilibrium position in the orbital coordinate system; the second trend is the analysis of the perturbed rotational motion of the quickly rotating satellites.

1. Stabilization and Libratory Motion of a Satellite in the Gravitational Force Field

The equations of motion of a satellite in the gravitational field in the circular orbit allow the particular solution, i.e. relative equilibrium in the orbital coordinate system. In this motion conditions the main central axes of inertia of a satellite coincide correspondently with the radius-vector of the orbit, with the tangent to the orbit, and with the normal to the orbital plane. In view of the fact that this solution is very important to create the gravitational stabilization system, many papers deal with the analysis of motion in the neighbourhood of relative equilibrium.

In 1959 the author of this report published [1] the strict proof (in the Lyapunov's sense), which was received in 1956–1957, that a sufficient condition for the stability of relative equilibrium of a satellite in a circular orbit is that the major axis of the ellipsoid of inertia of the satellite should point along the radius-vector, the minor axis along the normal to the orbital plane, and the medium axis along the tangent to the orbit (Fig. 1). The analysis of small space oscillations of the satellite in the circular orbit shows that with exception of the mentioned sufficient stable condition there is a region of the inertia moment values in which the necessary stable conditions are accomplished.

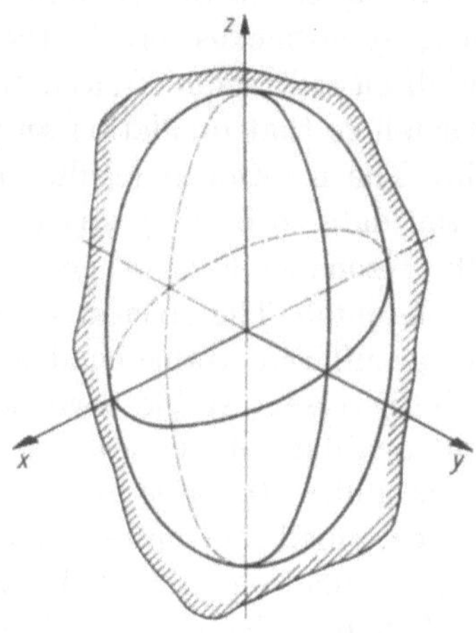

Fig. 1.

Breakwell and Pringl [2] had examined nonlinear effects in oscillations of a satellite about the position of equilibrium. Non-linearity is essential in the neighbourhood of resonance between the frequencies ω_1, ω_3 of the lateral and longitudinal oscillations of the satellite. A certain phase angular Φ the values α_3, α_1 which are proportional to the square of amplitudes of the lateral and longitudinal oscillations are introduced. Let ε be different from resonance correlation between the frequencies ω_1, ω_2. Fig. 2 illustrates the integral curves $\alpha_1(\Phi)$ describing satellite oscillations, when sufficient stable conditions are accomplished. Two cases are given: the case of the large values ε and one of the small values ε.

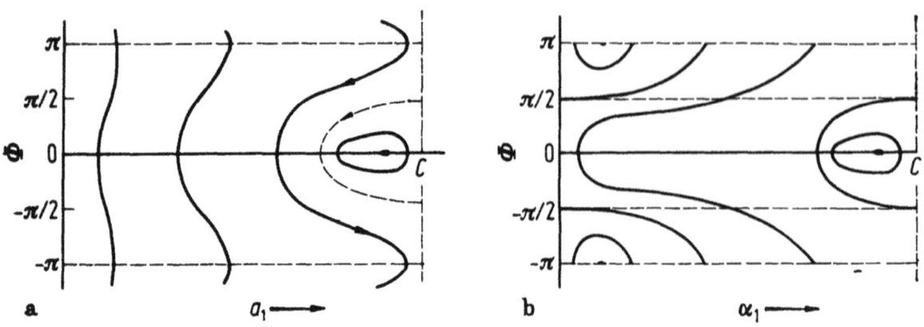

Fig. 2.

In the first case the initial small oscillations of amplitude α_1^* (and α_3) remain small. In the second case, when disorder of oscillation frequencies with resonance one is sufficiently small the initial small oscillations α_1^* become in time essentially if only $c \approx \alpha_3^0$ is not very small.

This theoretical result agrees with the numerical calculations by KANE [3] and, certainly, by no means, breaks the above proved stabilizing of motion in the discussed region. Indeed the initial data should be chosen sufficiently small so that oscillation amplitudes had not surpassed the given limits. The mentioned results show however the requirements for the initial data to be very strict. To provide moderate lateral oscillation in the resonance case we need very small initial deviation with respect to the pitch. The given case corresponds to motion in the region in which the sufficient stable conditions are accomplished. In the region, where the necessary stable conditions are accomplished the situation is different qualitatively. It is found when discrepancy $|\varepsilon|$ by resonance frequency is sufficiently large then motion can be remained bounded but when $\tilde{\varepsilon}$ is very small (but finite) there are no any bounded motions in the given approximation (Fig. 3). In addition, the angle of bank of yaw and the angle of pitch increase simultaneously. These figures illustrate the following thing. How small the value ε were it might be always found such region of the initial data where motion was stable.

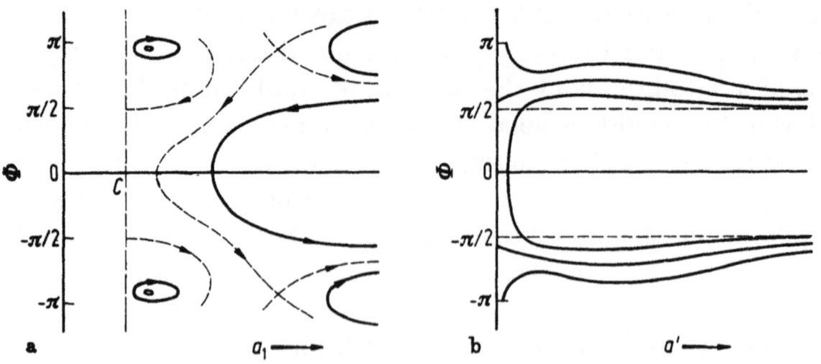

Fig. 3.

But this region decreases with $\tilde{\varepsilon}$ and when $\varepsilon = 0$ is disappeared completely. When the frequency has a resonance value – and only in this case – the motion is unstable. This fact is very interesting from the point of view of the stable theory, but it may not be considered as the strictly proved. Indeed analysing the given motion with three degrees of freedom two essential simplification were made:

1. in Hamiltonian the terms only up to the third order of smallnes (inclusively) were remained;

2. motion with such truncated Hamiltonian was considered only for the first approximation (in sense of asymptotical methods). That is why it is very important to stress that the analogous results have already proved strictly [4] for motion with two degrees of freedom. These strict results may be used in the stable analysis of stationar motion of the dynamically symmetric satellite in the gravitational field. The stationar conditions exist in the circular orbit; in addition, the axis of symmetry of the satellite remaines unstable in rotational coordinate system being normal to the radius-vector of the orbit or to the velocity vector and composing a constant (in particular, zero) angle with the normal to the orbit plane.

These stationar motions were discovered and investigated for stability [6–9]. The regions of the necessary stable conditions and of the sufficient stable conditions were found.

Recently A. P. Markeev had finished the analysis of stability of this motions [5]. Using Arnold and Moser's results he had proved that the motion was stable everywhere in the region of the necessary stable conditions, with exception of set values having zero measure. It is also interesting such fact that the existence of such parameters among this set of zero measure, which make motion be unstable, was proved.

These parameters correspond to the resonance between the typical frequencies of motion.

2. Rotational Motion in the Gravitational Field. Planet Resonances

The investigation of rotational motion of a satellite is connected with difficulties of finding of perturbations slowly accumulated in motion. It was found a fruitful idea [10] to use the following parameters as the variables in the equations of motion: modulus of the angular momentum vector, modulus of two angles describing the spatial position of this vector and modules of three Euler's angles describing the position of a satellite about the angular momentum vector. If there are no perturbations then non-perturbed motion by Euler-Puanso is described by this variables. If there are small disturbance, then these variables allow to do the effective analysis of motion by the asymptotical methods (in the first approximation – the method of averaging). During the last decade these variables in any modification are widely used for the analysis of satellite motion about its center of mass at the different disturbing factors. As a rule nonresonance case was researched. For example, invastigating motion in the gravitational field it was found that Euler-Puansò is remained in the first approximation with respect to the angular momentum vector and also the value of the angular momentum vector is remained; all the perturbations are found to be put only in slow precession of the angular momentum vector about the normal to the orbit plane at the constant (in the first approximation) angular distance from it. New results obtained recently (A. P. TORJEVSKY [11]) deal with motion in the case of resonances between Euler's frequencies and between them and the orbital revolution frequencies of a satellite. It was found that in the neighbourhood of resonance motion "swings" so that in the first approximation not only precession motion becomes essential but also its nutational motion and modulation of the angular momentum vector. Fig. 4 illustrates non resonance and resonance motions of the angular momentum vector.

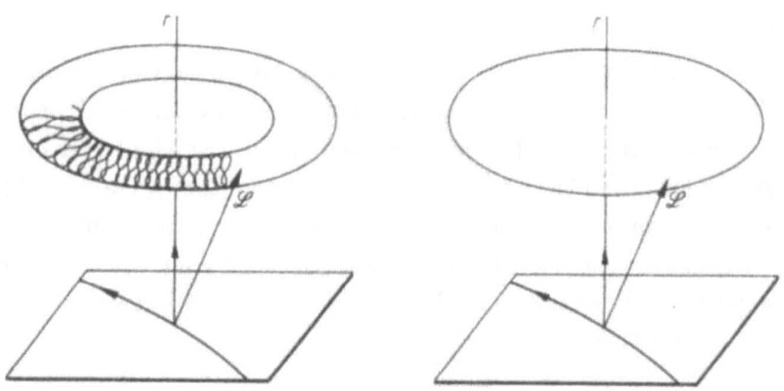

Fig. 4.

Resonances between the revolution frequency and the rotational frequency of the body in the gravitational field are of great importance. Such motion is not a rare case for the natural planets and satellites of the solar system. The Moon is a classical example of the resonance 1:1. Evidently motion of the Moon about the center of mass may be described successfully as in the terms of libratory motion and also in the terms of rotational motion. In the first case small oscillation relative to the position stable-oriented on the Earth is considered. In the second case motion about the center of mass is considered as the resonance rotation with a frequency which is equal to the orbital revolution frequency.

Radar-location of the planets Mercury and Venus gives new information of their rotation. Period of the Mercury rotation as it was found, is 2/3 from the period of its orbital circulation. Colombo [12] was the first who shows the possibility of such stable rotation of the Mercury. In addition, eccentricity ($e = 0.2$) of the Mercury's orbit plays the leading role. Colombo investigated the plane case. Fig. 5 gives the possible spa-

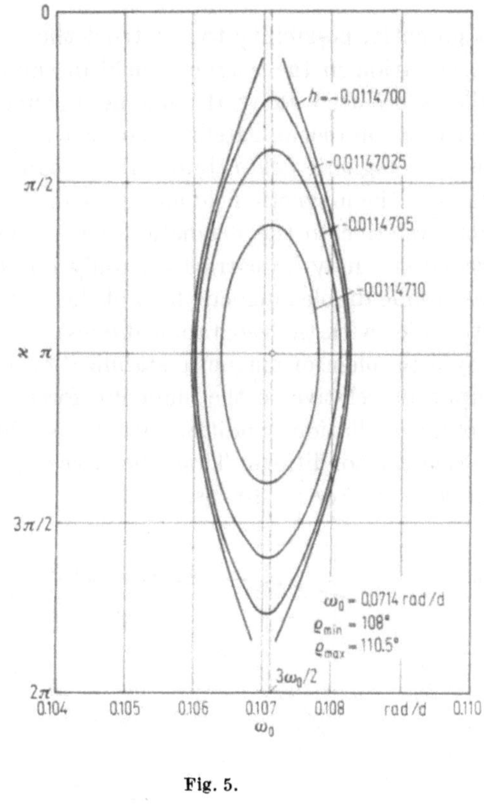

Fig. 5.

tial motion of the spin axis of the planet in the neighbourhood of resonance 2/3, where the angular velocity of rotation of the planet ω relatived to the orbital angular velocity ω_0 depending on the phase angle $\varkappa$; There is also the angle ϱ between the vector $\boldsymbol{\omega}$ and the normal to the orbital plane depending on ω. It is regarded that in the planet $\boldsymbol{\omega}$ is situated along the axis of maximum moment of inertia. Then the stationar point corresponding to $\omega = {}^3/_2\, \omega_0$ (and to the value ϱ) is really stable. When the dissipative forces exist (tidal phenomenons) motion in the process of evolution, may be reduced to such resonance motion. Resonance of the Venus rotation is quite phenomenal. The Venus was found to be turned to the Earth by the same side in every nearest positions; during the period between such two rapprochements the Venus rotating about its own axis makes four complete turns with respect to the Earth's observer. The authors of [13] try to explain this phenomenon.

3. Interaction of the Magnetized Satellite with the Earth's Magnetic Field

There is a potential possibility to construct the system of the satellite passive orientation in the magnetic field in similar manner as the compass needle is oriented along the magnetic force line. This possibility was used in a number of satellites launched. To receive accurate stabilization in the magnetic field is impossible because of the essentially irregular changes of the magnetic field along the orbit. In addition satellite's oscillations relative to the magnetic force line are essentially spatial, plane oscillations may be carried out only on the equatorial and polar orbits (assuming dipole magnetic field of the Earth and coincidence of the magnetic poles with the geographical ones).

That is why a problem of magnetic stabilization is to provide small oscillation amplitude relative to the magnetic force line; achieve it by providing satellite oscillation conditions in the neighbourhood of suitable stable periodical conditions. Thus the investigations of periodical oscillations are very interesting for us.

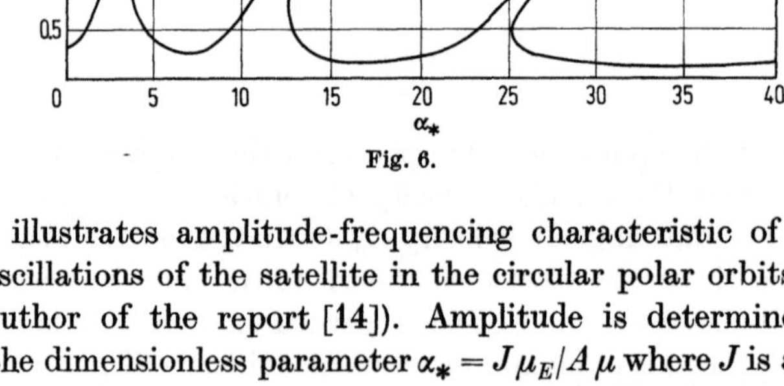

Fig. 6.

Fig. 6 illustrates amplitude-frequencing characteristic of plane periodical oscillations of the satellite in the circular polar orbits (received by the author of the report [14]). Amplitude is determined by the value of the dimensionless parameter $\alpha_* = J\mu_E/A\mu$ where J is a magnetic torque of a satellite, μ_E is a magnetic torque of the Earth, A is a moment of inertia of a satellite, $\mu = fM$ is gravitational constant (so that f is universal gravitational constant, M is the Earth's mass). Amplitude of periodical oscillations does not depend on the orbit parameters. As it is seen from the figure the more α_* the better stabilization can be attained (on the lower branches of the curves), in addition the "dangerous" resonance values α_* should be avoided. The number of periodical solutions increases with α_*; it creates new difficulties to provide the desired periodical condition of a satellite. Motion of stabilized satellite will be essentially spatial in the arbitrary orbit, and in this case the scheme of stabilization may be understood from the following arguments. In satellite orbital motion the vector of the magnetic strength $\boldsymbol{H}$ describes a certain cone closed to circular, but quite irregularly circumscribed in the translational moving system of coordinates, Fig. 7. The cone is closed in one-half of satellite orbital period. Let the regular precession be conducted to the non-magnetic dynamically-symmetric satellite with the same period of precession and with the angle of the nutation which is equal to one-half of the smallest opening angle of the cone $\boldsymbol{H}$.

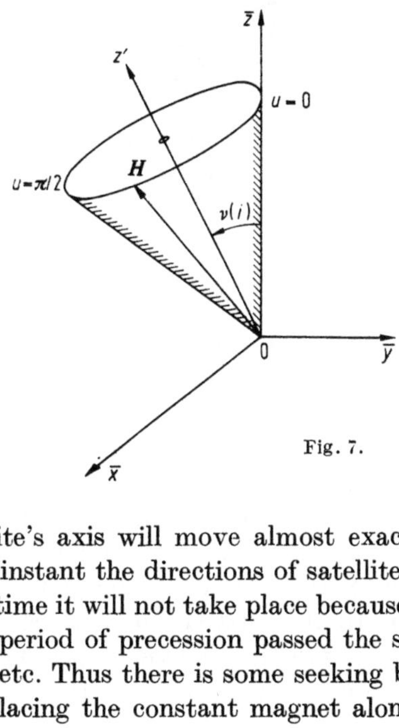

Fig. 7.

Then the satellite's axis will move almost exactly in this cone but regular. If in some instant the directions of satellite's axis and vector $\boldsymbol{H}$ coincide then next time it will not take place because of irregular motion of $\boldsymbol{H}$, but one-half period of precession passed the satellite's axis and $\boldsymbol{H}$ will coincide again etc. Thus there is some seeking by the satellite's axis of the vector $\boldsymbol{H}$. Placing the constant magnet along the satellite's axis we can make this seeking stable and decrease the amplitude of oscillation between the satellite's axis and vector $\boldsymbol{H}$.

4. New Aerodynamic Effects

The effects of "the conservative type" had studied well thanks to the experimental data of the flights of such satellites as Sputnik III, Explorer and Electron and also from the theoretical investigations. These effects retain the value of the angular momentum vector and Euler's motion near it, but they conduct precessional nutation motion to the angular momentum vector. These effects are superposed by the effects of the dissipative type changing monotoneously the value of the angular momentum vector (and the kinetic energy) and changing Euler's motion relative to the angular momentum vector. New effects were found by analysing of the motion of a satellite (Proton type [15]) about its center of mass. These effects are caused by propellating action of the highly essential serodynamic torques from the solar battery planes which are placed skew-symmetrically relative to the satellite Fig. 8. In the incident

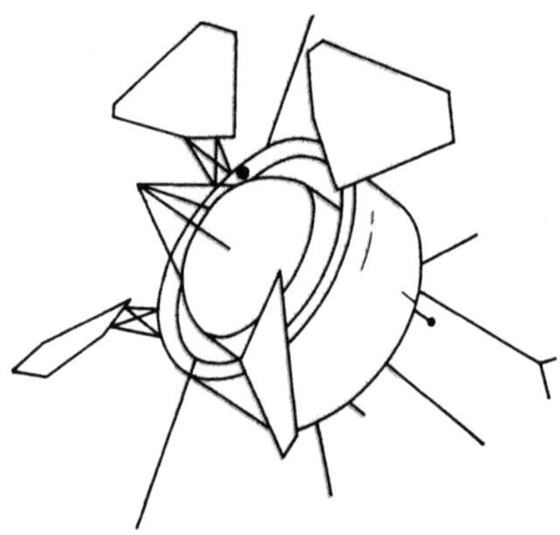

Fig. 8.

flow these planes operate as the mill wings or paddlewheel untwisting or twisting the satellite. In the elliptic orbit the majority of changes takes place in the region of the orbit perigee. We may assume in a certain premise that the propelling momentum vector lies in the plane of the velocity vector and longitudinal axis of a satellite. It causes (because of the known gyroscopic effects) motion of the angular momentum vector in the plane passing through it and the velocity vector. The change of orientation of the angular momentum vector relative to the velocity vector leads to the alternation of a sign of the propellating effect from twisting to untwisting and vice versa. As a result – the parameters of Euler's motion are subjected to the deep modulations of a long period. The angular velocity of a satellite may change several times by longperiodical way; by this the satellite to the condition of almost pure axial twisting or to the condition of tumbles. The integral curves of such motion are plotted in Fig. 9. Here L is a modulus of the angular momentum vector, ν is an angle between the longitudinal axis of the dynamically symmetric satellite and the angular momentum; Θ is an angle between the angular momentum vector and the velocity vector in the orbit perigee. The possibility of deep modulations L, ν is shown. It

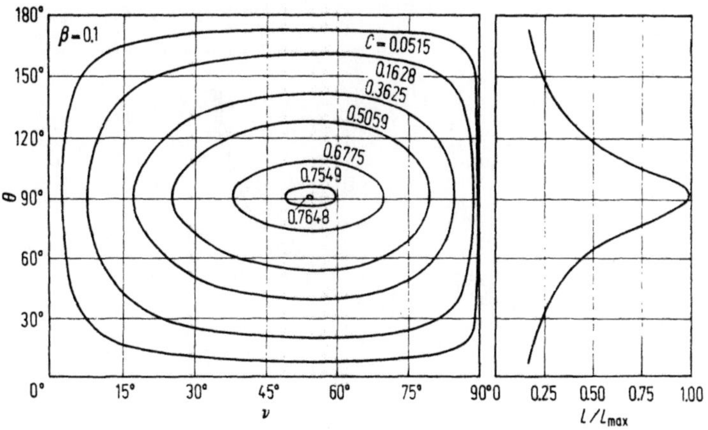

Fig. 9.

is very important that the period of these modulations does not almost depend on the initial data and depend essentially on the character of molecula reflection from the satellite surface. It gives us possibility to estimate the coefficient of accomodation according to the experimental data obtained from the rotational motion of the satellite "Proton" [16]. If the satellite provided by skew-symmetric blades twisted rapidly, then the torques of aerodynamic friction become essential. Then the damping of angular velocity of a satellite is noticeable. The value of this damping becomes also sensitive to the value of the accomodation coefficient, which may be estimated by this way [17].

5. Determination of Satellite Actual Orientation from the Measurements

Thanks to the readings of the different sensors placed on a satellite we may receive information about satellite actual orientation and actual torque of forces influenced on it. For this aim data of the following devices are used: magnetometers, sun sensors, the sensor of angle velocity, ionic traps, data of radiosignal modulation etc. Because of presence of measurement errors statistical data processing is natural method for determination of orientation.

If at the fixed moment of time a sufficient number of the different measurements is given then it makes possible to determine orientation by the local method knowing nothing concerning motion of a satellite about the centre of mass beforehand; but as a rule a sufficient number of the measurements are distributed on the considerable interval of time. In this case orientation should be determined by the integral method, using the total sum of information to construct some model of motion. In this connection the model of motion of a satellite about the center of

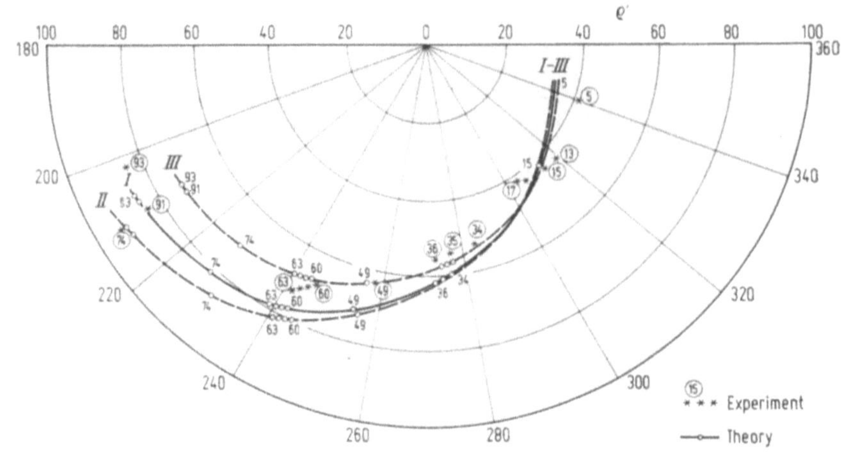

Fig. 10.

mass is of great importance. As such model we can take the non-pertubed motion, the differential equations of motion etc. The algorithms of the statistical data processing are iterative as usual. That is why the methods of receiving of zero approximation to motion of a satellite play the very important role. As a rule this zero approximation is taken from the same information which lately are used in the statistical data processing. This very difficult and complicated data processing allows, in the end, not

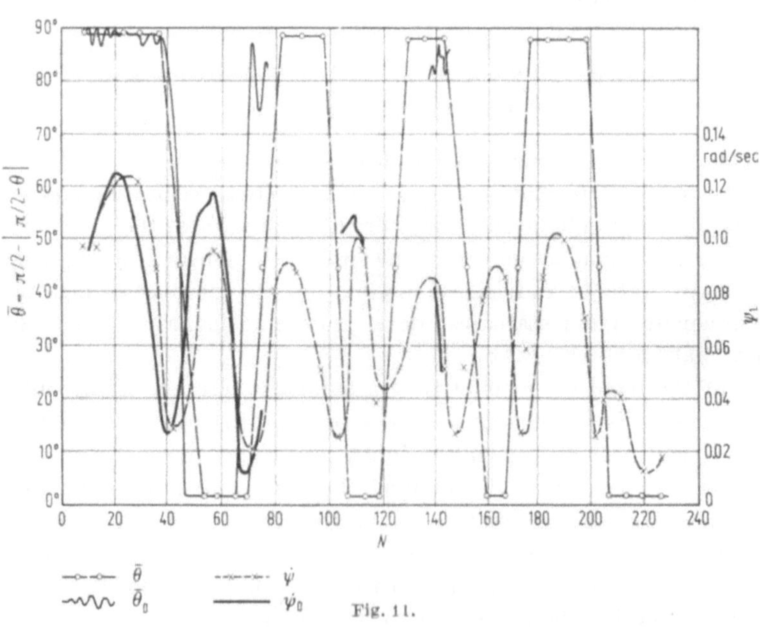

Fig. 11.

to give orientation "for exterior use" (using for interpretation of the scientific measurements) but to receive the very important data, from the point of view of a mechanician, about actual evolution of a satellite motion and the perturbing moments effected on it (see for example [18]). For instance let us point out the experimental and theoretical picture Fig. 10. of evolution of the angular momentum vector of the satellite "Electron" [19]. The known gravitational torque and the selected magnetic one explane satisfactoryly the experimental picture. Figs. 11 and 12 illustrate the experimental data concerning motion of the satellite Proton-2 about the center of mass. The effects caused by the influence of the aerodynamic propellating torque mentioned above are essentially marked. Fig. 11 illustrates the trajectory of the angular momentum vector in space. In Fig. 12 the longperiodical modulations of the value of the angular momentum vector (the angular velocity) and the character of satellite rotation (the angle θ) are plotted. It is clear that the satellite goes out from the twisting condition ($\theta = 0$) to condition of tumbles ($\theta = \pi/2$) and vice versa.

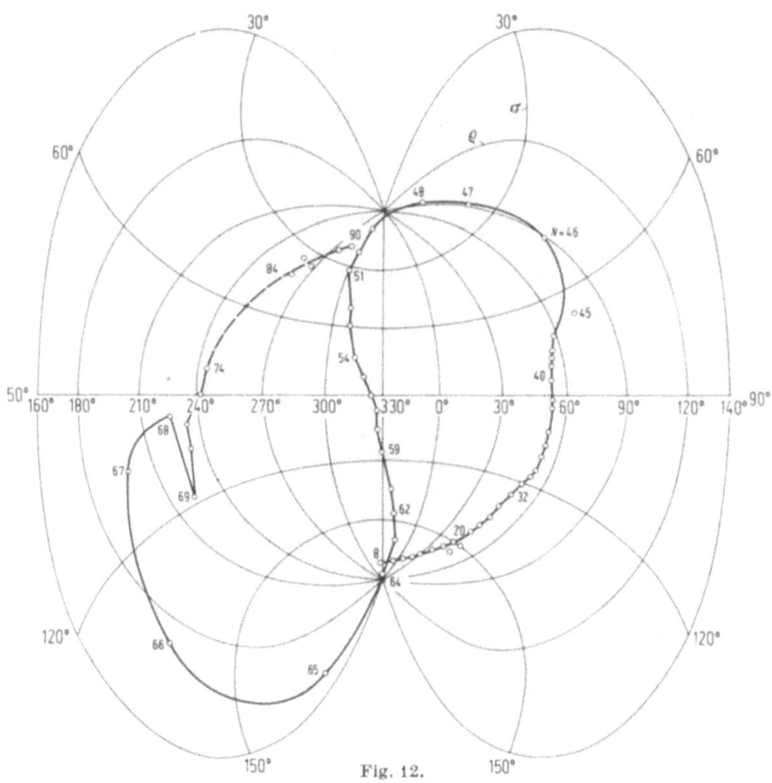

Fig. 12.

References

1. Beletskii, V. V.: On Satellite's In: The Artificial Earth Satellites, No. 3, USSR Academy of Sciences, 1959, 13–31. In English: NASA-Transl., TTF-10 (1960).
2. Breakwell, T. V., Pringl, R. Jr.: Nonlinear Resonances Affecting Gravity-Gradient Stability. Rept. XV Intern. Astronaut. Congr. Athens, 1965.
3. Kane, T. R.: AIAA J. 3, No. 4, 1965.
4. Markeev, A. P.: On Canonical System Stability with two Degrees of Freedom having the Resonance. Journal of Applied Mat. and Mech. 32, No. 4, 1968.
5. Markeev, A. P.: The Resonance Effects and Stability of the Stationar Satellite Rotation. Space Research, 5, No. 3, 1967.
6. Duboshin, G. N.: On Rotational Motion of the Artificial Celestial Body. Bul. JTA USSR Academy of Sciences, 7, ed. 7 (1960) 90.
7. Condurar, V. T.: The Particular Solutions of General Problem on the Translational Spin Spheroid Movement Spheroid under the Influence of the Balloon Attraction. Astron. J. 36, No. 5, 1959.
8. Kane, T. R., Marsh, E. L., Wilson, W. G.: Letter to the Editor. J. Astronaut. Sci. 9, No. 4 (1962) 108.
9. Chernousko, F. L.: On Stability of the Regular Satellite Precession. Institute of Applied Mathematics 28, No. 1 (1964).
10. Beletskii, V. V.: Motion of an Actificial Earth Satellite about Its Center of Mass. In: Artificial Earth Satellites, No. 1, p. 25–43. Academy of Sciences, USSR, 1958. In English: Artificial Earth Satellites, vol. I, pp. 30–54, New York: Plenum Press.
11. Torjevsky, A. P.: Motion of an Artificial Satellite about Its Center of Mass and Resonances. Astron. Acta 13 (1968).
12. Colombo, G.: Nature 208 (1965) 575.
13. Goldrech, P., Peale, S. K.: Nature 209 (1966) 1117.
14. Hentov, A. A.: Influence of the Magnetic and Gravitational Earth Fields on the Satellite Ocsillations about Its Center of Mass. Space Research 5, No. 4 (1967) 554–572.
15. Beletskii, V. V.: Evolution of Rotation and Orientation of the satellite of the type "Proton". In: Proc. of the Meeting on the General Questions of the Celestial Mechanics and Astrodynamics, Moscow, 1967, March 23–29. Beletskii, V. V., Golubkov, V. V., Hatshevitch, I. G.: Analysis of the Satellite "Proton-2" Rotation and Orientation. Proc. of the XVIII. Meeting of IAF, Belgrad, 1967; vol. I, 151–161.
16. Beletskii, V. V.: The Character Estimation of the Satellite-Aerodynamic Flow Interaction with respect to the Analysis of the Satellite Proton-2 Motion about Its Center of Mass. In: "Space Res.", 1970 No. 2.
17. Moe, O. K.: Atmospheric Densities Determined from the Spin of Explorer VI. University of California 1966 (Thesis).
18. Beletskii, V. V., Golubkov, V. V., Hatshevitch, J. G., Stepanova, E. A.: Determination of the Artificial Satellite Orientation with respect to the Measurement Data. Preprint. Part I: Method, Moscow 1967 (p. 195). Part II: Results. The Analysis of Motion, Moscow 1968 (p. 134).
19. Lavrovskii, A. K., Trushin, S. I.: Determination of the Artificial Earth Satellite. Proc. of the Meeting on the General Questions of the Celestial Mechanics and Astrodynamics. Moscow, 1967, March 23–29.

Dynamics of Spin-Stabilized Flexible Satellites of Crossed-Dipole Configuration

By

F. R. Vigneron

Communications Research Centre, Ottawa, Canada

Abstract. Recent analytical studies of the dynamics of a spin-stabilized satellite with long crossed dipoles or slender beams, such as the Canadian Alouette I and II and ISIS-I spacecraft, are described. Linearized equations of motion for the flexible satellite rotating about its mass centre and with its spin axis approximately normal to the orbit plane are found to uncouple into two sets; a first set which describes the spin rate history and deforming motions in the plane of the configuration, and a second set which describes nutational motions of the body and out-of-plane flexing of the beams. Analytical solutions obtained for the first set demonstrate the action of the gravity field in causing dynamic instability of the beams and satellite despin due to structural damping. Inspection of the second set of equations indicates that the force of gravity may also parametrically excite dynamic instability of the beams in the out-of-plane direction. Necessary ans sufficient conditions for stability of spin axis orientation are determined for the freely-spinning satellite i.e., for cases where torques and deforming forces due to gravity may be neglected. These conditions are described by extending existing criteria of rigid body stability to account for the influence of elasticity and structural damping of the system.

Introduction

The Canadian Alouette and ISIS satellites for ionospheric sounding utilize the spin-stabilized crossed-dipole structural configuration, composed of a central body and four long beams deployed in a planar array (shown schematically in Fig. 1). The satellite dimensions, characteristically 75 by 240 ft. tip to tip, are constructed from 1/2 in. diameter tubular STEM members, and are thus large and flexible.

Much of the dynamics analysis relating to design and flight performance of the Alouette satellites has been based on established theory developed for rigid satellites [1, 2]. Design information has been obtained from digital computer solutions of motion equations of a finite element model of the configuration, but these studies have yielded limited information of a general nature [2].

More recent investigations have been directed at exploring the basic characteristic features of the dynamics of flexible crossed-dipole configurations, from an analytical viewpoint. In this paper, recent studies of a configuration with four equal length beams, spin stabilized with spin axis normal to the plane of a circular orbit, are described.

Motion Equations

The spin-stabilized satellite, composed of a central body and four equal length beams of identical material and tubular cross section, is shown in Fig. 1. The reference axes $(Oxyz)$ are assigned to the satellite

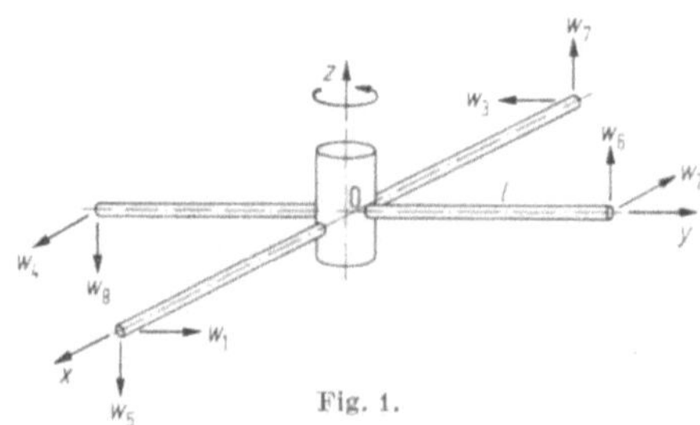

Fig. 1.

in such a way that the origin O always coincides with the mass center of the satellite, and the axes Ox and Oy are always parallel to a line tangential to the root of the beams. The absolute angular velocity is designated by the three components $(\omega_x, \omega_y, \omega_z)$, in the $(Oxyz)$ reference frame. Eight deformation functions $w_i(x, t)$, $(i = 1, 2, \ldots 8)$, where x is the length coordinate and t is the time, are referenced to the $(Oxyz)$ axes, as shown in Fig. 1.

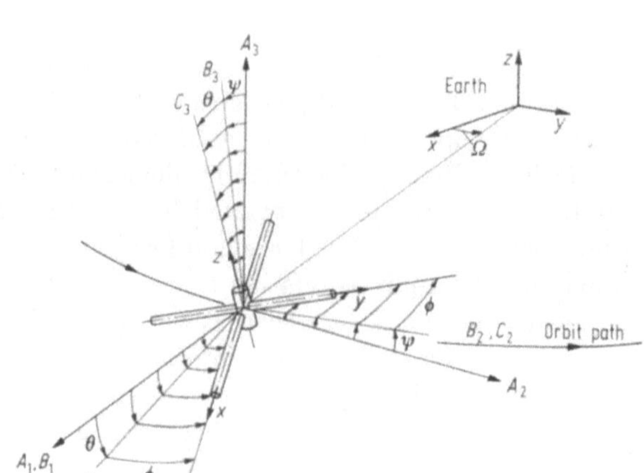

Fig. 2.

The orientation of the satellite axes ($Oxyz$) are referenced from the orbit-fixed axes (OA_1, OA_2, OA_3) by three Euler angles (ψ, θ, Φ), generated by the right-hand rotation scheme, $(\alpha)\psi$ about A_1, $(\beta)\theta$ about B_2, $(\gamma)\Phi$ about C_3, where the A_1, B_2, and C_3 axes are shown in Fig. 2.

It is convenient to represent the deflections in the form,

$$w^i(x,t) = \sum_{n=1}^{\infty} l\, q_{in}(t)\, \Phi_n(\xi),$$

where l is the length of one beam, $\xi = x/l$, $\Phi_n(\xi)$ are the set of cantilever vibrational modes of a beam of length l and resonant frequencies p_{in}, and $q_{in}(t)$ are a set of coordinate functions.

The variables θ, ψ, and $q_{in}(t)$ are "small" when the spin vector and orbit normal are approximately parallel, and beam deflections are small. Accordingly the equations of motion may be linearized in these variables. The following linearized equations of motion, derived by variational principles, are obtained:

$$\dot{\Phi} = \omega_z - \Omega \tag{1a}$$

$$C_0\dot{\omega}_z + \varrho l^3 \sum_{i=1}^{4}\sum_{n=1}^{\infty} \mathfrak{P}_n \ddot{q}_{in} = 3\Omega^2 \varrho l^3 \sum_{i=1}^{4}\sum_{n=1}^{\infty} (-1)^{i+1} q_{in} \tag{1b}$$

$$\begin{aligned} &\ddot{q}_{in} + c_{in}\dot{q}_{in} + [p_{im}^2\,\delta_{nm} + \omega_z^2(\varepsilon_{nm} - \delta_{nm}) + \frac{\Omega^2}{2}\{(\varepsilon_{nm} - \delta_{nm}) \\ &\quad + 3(\varepsilon_{nm} + \delta_{nm})\cos 2(\Phi + \Phi_0^i)\}]\, q_{in} \\ &\quad + \frac{3}{2}\mathfrak{P}_n[\dot{\omega}_z + \Omega^2 \sin(\Phi + \Phi_0^i)] = 0 \quad \text{(summation on } m\text{)} \end{aligned} \tag{1c–f}$$

where

$$i = \{1, 2, 3, 4\}, \qquad \Phi_0^i = \left\{0, \frac{\pi}{2}, \pi, \frac{3\pi}{2}\right\}.$$

$$\dot{\theta} = \omega_x \sin\Phi + \omega_y \cos\Phi - \Omega\psi \tag{2a}$$

$$\dot{\psi} = \omega_x \cos\Phi - \omega_y \sin\Phi + \Omega\psi \tag{2b}$$

$$\begin{aligned} &A_0\dot{\omega}_x + (C_0 - A_0)\,\omega_z\omega_y + \varrho l^3 \sum_{n=1}^{\infty} \mathfrak{P}_n(\ddot{q}_{6n} + \ddot{q}_{8n}) + \omega_z^2(q_{6n} + q_{8n}) \\ &\quad + \dot{\omega}_z(q_{5n} + q_{7n}) = 3\Omega^2[(A_0 - C_0)\,\theta\sin\Phi - \frac{\varrho l^3}{2} \\ &\quad \sum_{n=1}^{\infty} \mathfrak{P}_n\{(1 - \cos 2\Phi)(q_{6n} + q_{8n}) + (q_{5n} + q_{7n})\sin 2\Phi\}] \end{aligned} \tag{2c}$$

$$A_0\,\dot{\omega}_y - (C_0 - A_0)\,\omega_z\omega_x + \varrho l^3 \sum_{n=1}^{\infty} \mathfrak{P}_n\{(\ddot{q}_{5n} + \ddot{q}_{7n}) + \omega_z^2(q_{5n} + q_{7n})$$

$$- \dot{\omega}_z(q_{6n} + q_{8n})\} = 3\,\Omega^2\,[(A_0 - C_0)\,\theta\cos\Phi - \frac{\varrho l^3}{2} \tag{2d}$$

$$\sum_{n=1}^{\infty} \mathfrak{P}_n\{(1 + \cos 2\,\Phi)\,(q_{5n} + q_{7n}) + (q_{6n} + q_{8n})\sin 2\,\Phi\}]$$

$$\ddot{q}_{in} + c_{in}\dot{q}_{in} + [\delta_{nm}p_{im}^2 + \varepsilon_{nm}\omega_z^2 + \Omega^2\{(\delta_{nm} + \varepsilon_{nm}/2)$$
$$+ 3/2\,\varepsilon_{nm}\cos 2\,\Phi\}]\,q_{im} + 3\,\Omega^2\,\mathfrak{P}_n\,\theta\cos\Phi \tag{2e + f}$$
$$+ \mathfrak{P}_n(\dot{\omega}_y - \omega_z\omega_x) = 0;\; i = 5, 7$$

$$\ddot{q}_{in} + c_{in}\dot{q}_{in} + [\delta_{nm}p_{im}^2 + \varepsilon_{nm}\omega_z^2 + \Omega^2\{(\delta_{nm} + \varepsilon_{nm}/2)$$
$$- 3/2\,\varepsilon_{nm}\cos 2\,\Phi\}]\,q_{im} + 3\,\Omega^2\,\mathfrak{P}_n\,\theta\sin\Phi \tag{2g + h}$$
$$+ \mathfrak{P}_n(\dot{\omega}_x + \omega_y\omega_z) = 0;\; i = 6, 8\,.$$

In the above equations, Ω is the angular velocity of the circular orbit, ϱ is the mass per unit length of the beams, c_{in} is a structural damping parameter, and A_0 and C_0 are the transverse and spin moments of inertia of the undeformed configuration, (i.e. $C_0 = C_0' + 4\,\varrho l^3/3$, $A_0 = A_0' + 2\,\varrho l^3/3$, where C_0' and A_0' are the polar and transverse moments of inertia of the central rigid body). The superscript dots denote differentiation with time, δ_{nm} is the Kronecker delta, and $\mathfrak{P}_n$ and ε_{nm} are modal coefficients,

$$\mathfrak{P}_n = \int_0^1 \xi\,\Phi_n(\xi)\,d\xi\,, \qquad \varepsilon_{nm} = \int_0^1 \left(\frac{1-\xi^2}{2}\right)\Phi_n'(\xi)\,\Phi_m'(\xi)\,d\xi\,.$$

An important observation is that the above linearized equations are uncoupled into two sets, (1) and (2) which may be solved separately. The first set, Eqs. (1) in the variables ($\Phi, \omega_z, q_{1n}, q_{2n}, q_{3n}$ and q_{4n}) describe the spin rate changes and flexing motions of the beams in the plane normal to the spin axis. The second set, Eqs. (2) in the variables ($\omega_x, \omega_y, \theta, \psi, q_{5n}, q_{6n}, q_{7n}, q_{8n}$) define the nutational and out-of-plane flexing motions of the configuration.

Related non-linear equations have been previously obtained in connection with the study of planar motions [3], and with the study of a freely-spinning satellite [4].

Analysis of Planar Motions

For spin stabilized satellite of present design, Ω is "small" with respect to other parameters in Eqs. (1). As a consequence, one may devise a technique of solution using an asymptotic method. A suitable technique employs $q_{in} = 0$ as a generating solution for solving Eqs. (1a and 1b) and (1e and 1f) by successive iterations. Such a technique has been utilized in studying planar motion described by non-linear equations [which reduce to Eqs. (1), upon linearization in w^i] in [3].

A first step is to solve Eqs. (1a and 1b) when $q_{in} = 0$; this leads to $\dot{\Phi}$ constant. This solution is equivalent to the "rigid body" approximation.

The second step is to substitute the above result into Eqs. (1c–f) and to re-solve for the $q_{in}(t)$. Eqs. (1c–f) are thus reduced to linear second order equations with coefficients periodic in twice the spin angle, and thus are of the Mathieu-Hill category. This theory of them is well established; depending on the values of the parameters, the solution may be either bounded (stable) or unbounded (unstable). The cases of unboundedness relate, in physical terms, to dynamic buckling of the beams as a result of the action of the gravity field. The character of the solutions may be categorized in the Mathieu parameter chart shown in Fig. 3,

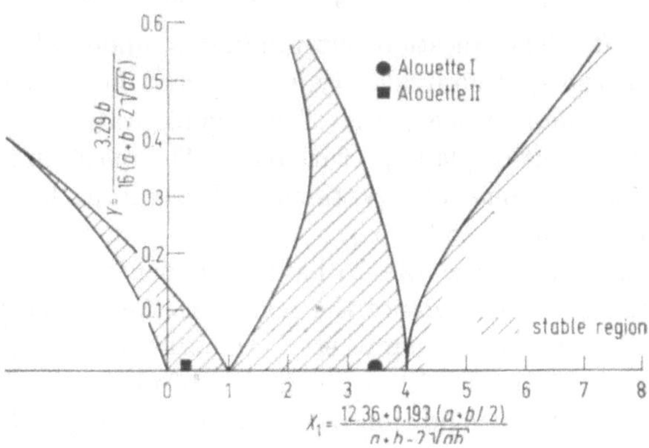

Fig. 3.

which has been deduced from Eqs. (1e–f), by truncating the system at the first mode, $n = 1$, and ignoring the structural damping. In the figure, $a = \varrho \bar{s}^2 l^4/EI$, $b = \varrho \Omega^2 l^4/EI$, where EI is the bending stiffness of the beams, and $\bar{s}$ is the spin rate of the satellite. Parameter points corresponding to satellites of the dimension of Alouette I and Alouette II are shown, and are seen to be in the stable regions of the chart. Closed-form approximate solutions of these linear equations may be obtained for cases of interest where the behaviour is known to be "stable".

A third step in the iteration procedure is performed by substituting approximate stable solutions of the equations of the preceding paragraph into Eqs. (1a and 1b) and re-solving them using the Method of Averaging [5], with Ω as a small parameter. One obtains expressions describing the long term (average) spin decay caused by structural damping combined with the gravity field. An approximate solution obtained by this technique, which is valid for sub-resonant excitation of the beams, is,

$$\bar{s} = \Omega + (\bar{s}_0 - \Omega) \exp \left[- \left\{ \frac{18 \varrho l^3 \Omega^4}{C_0} \sum_{n=1}^{\infty} \frac{\mathfrak{P}_n^2 c_n}{p_n^4} \right\} t \right]$$

where $\bar{s}_0$ is an initial average spin rate. The above expression describes an asymptotic spin decay leading to gravity capture.

Solutions obtained by the technique outlined above have been verified by comparing them with results obtained independently and directly by digital computation [3].

Stability of Nutational and Out-of-plane Bending Motions

It is evident that the solution

$$\theta = \psi = \omega_x = \omega_y = q_{5n} = q_{6n} = q_{7n} = q_{8n} = 0$$

satisfies Eqs. (2). The question outstanding is, under what conditions is this solution stable, or equivalently, under what conditions will the spin axis remain normal to the plane of the configuration after being subjected to a small out-of-plane perturbation? The question is not yet fully resolved; however some observations and advances have been made, and are reported in the following.

One observes that Eqs. (2e–h) have coefficients periodic in 2Φ; consideration of this in the light of the results of the previous section indicates that for some sets of parameters, the beams will be parametrically excited to an unstable condition in the out-of-plane direction. It is evident that this unstable condition represents dynamic buckling in the out-of-plane direction, excited by the gravity field. In addition, the well known gravity-induced features of "rigid body" instabilities of [6] and parametric resonances of [7] will be reflected in some modified form in the analyses of these nutational instabilities. Formal results of studies into this aspect of the problem are not yet available.

Results pertaining to a freely spinning satellite (i.e. one for which forces and torques external to the spinning satellite are ignored) are of great importance in satellite system design. An effectively rigid satellite which has structural damping has been studied by employing "energy sink" methods (e.g. [8]). The analogous case may be studied in the present context by exploring Eqs. (1) and (2), when $\Omega = 0$. Eqs. (1) are then solved by

$$\omega_z = S \tag{3}$$

a solution which may be shown to be infinitesimally stable (i.e. stable in the linear approximation) for all parameter values. Substitution of Eq. (3) into Eqs. (2e–h) leads to the reduced system,

$$A_0 \dot{\omega}_x + (C_0 - A_0) S \omega_y + \varrho l^3 \sum_{n=1}^{\infty} \mathfrak{P}_n \{(\ddot{q}_{6n} + \ddot{q}_{8n}) + S^2 (q_{6n} + q_{8n})\} = 0 \tag{4a}$$

$$A_0 \dot{\omega}_y - (C_0 - A_0) S \omega_x + \varrho l^3 \sum_{n=1}^{\infty} \mathfrak{P}_n \{(\ddot{q}_{5n} + \ddot{q}_{7n}) + S^2 (q_{5n} + q_{7n})\} = 0 \tag{4b}$$

$$\ddot{q}_{in} + c_{in} \dot{q}_{in} + [\delta_{nm} p_{im}^2 + \varepsilon_{nm} S^2] q_{im} + \mathfrak{P}_n (\dot{\omega}_y - S \omega_x) = 0; \quad (i = 5, 7) \tag{4c}$$

$$\ddot{q}_{in} + c_{in} \dot{q}_{in} + [\delta_{nm} p_{im}^2 + \varepsilon_{nm} S^2] q_{im} + \mathfrak{P}_n (\dot{\omega}_x + S \omega_y) = 0; \quad (i = 6, 8) \tag{4d}$$

in the variables $(\omega_x, \omega_y, q_{5n}, q_{6n}, q_{7n}, q_{8n})$.

Eq. (4) may be solved by solutions of the form, $\exp(mt)$, and infinitesimal stability may then be assessed by investigating the character of the coefficients, m. The above system has been investigated in this man-

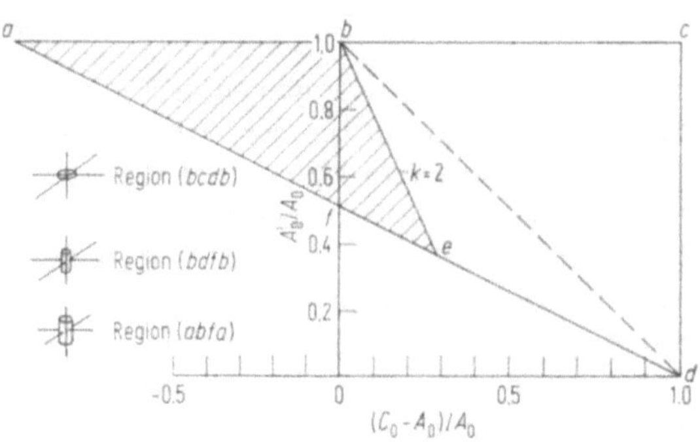

Fig. 4.

ner for the first mode approximation, $n = 1$. The character of the roots, m, was investigated by using Routh's rules [9], programmed appropriately for a CDC 3200 digital computer. The results are expressed in the form of a parameter chart shown in Fig. 4. Parameter points corresponding to physically possible configurations map onto the boundary, or into the triangular region, $(abcdefa)$. Furthermore, one finds that the points for which the central rigid body has $(C_0' - A_0')/A_0' > 0$, map into the subregion, $(bcdb)$. It is found (by analysis and computation, as discussed above) that the parameter points in the unshaded regions ($bcdeb$) denote infinitesimally stable configurations, and that points in the shaded regions, ($abefa$) denote unstable configurations. The separating line, (b–e), is given analytically by

$$A_0'/A_0 = 1 - k\{(C_0 - A_0)/A_0\} \tag{5}$$

where $k = p_{51}^2/S^2 + 1.193$, the "stiffness" of the system due to the flexural properties of the beams and the centrifugal forces due to spin. The chart is found to be valid for all values of damping greater than zero.

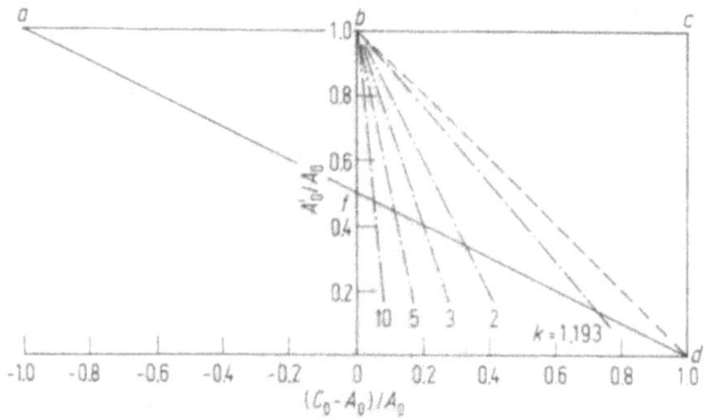

Fig. 5.

The effect of varying stiffness, k, is shown in Fig. 5. A limiting value of k equal to 1.139 corresponds to the case where the beams have no stiffness ($p_{in} = 0$) and behave like strings. Examination of the chart thus shows that a sufficient (but not necessary) condition for stability is that $(C - A)/A > 0$ for both the central rigid body and the total configuration when the damping is non-zero, since the subregion ($bcdb$) is seen to contain stable points for all values of k.

For a rigid satellite, $k \to \infty$. From Fig. 5, one observes that the necessary and sufficient condition for stability then becomes $(C_0 - A_0)/A_0 > 0$. This result agrees exactly with that obtained by the "energy sink" analysis of [8], which is based on heuristic notions concerning the role of energy dissipation in mechanical systems. By contrast, the analysis presented in this paper is based on the classical laws of dynamics, although the assumptions regarding damping are somewhat restrictive.

The above stability results for the freely-spinning satellite are based on the examination of the linearized motion equations. The analysis may be extended, by using well known theorems (due to Lyaponov) regarding the linearization procedures, to show that the stable regions are also asymptotically stable when all non-linear terms are accounted for and damping is non-zero [4].

Concluding Remarks

The problem discussed herein serves to illustrate the characteristic features of gravity-induced instabilities of spin-stabilized flexible crossed-dipole configurations of satellites. It is evident that the results present only a partial explanation of the overall dynamics of these satellites, and that a wealth of similar problem areas remain uninvestigated.

References

1. Paghis, I., Franklin, C. A., Mar, J.: Alouette I, The First Three Years in Orbit. Defence Research Board, DRTE 1159, March 1967.
2. Graham, J. D.: Dynamics of Satellites Having Long Flexible Extendible Members. International Ast. Fed., 17th International Astronautical Congress, Madrid, Spain, Oct. 9–15, 1966.
3. Vigneron, F. R.: Configuration Instability and Despin of Crossed-Dipole Satellites Due to the Earth's Gravity Field. Ph. D. Thesis, University of Illinois, Urbana, USA.
4. Vigneron, F. R.: Stability of a Freely Spinning Flexible Satellite of Crossed-Dipole Configuration. Internal Report, Defence Research Telecommunications Establishment, Ottawa, Canada.
5. Bougoliubov, N. N., Mitropolsky, Y. Q.: Asymptotic Methods in the Theory of Non-Linear Oscillations (Chap. V and VI), Delhi: Hindustan Publ. Corp. 1961.
6. Thomson, W. T.: Spin Stabilization of Attitude Against Gravity Torque. J. Ast. Sci. 9 (1962) 31–33; also Kane, T. R., et al.: Letter to the Editor 9 (1962) 108–109.
7. Markeev, A. P.: Resonance Effects and Stability of Stationary Rotation of a Satellite. Cosmic Research 5, No. 3 (1967) 318–326.
8. Thomson, W. T., Reiter, G. S.: Attitude Drift of Space Vehicles. J. Ast. Sci. VII, No. 2 (1960) 29–34.
9. Routh, E. J.: Dynamics of Rigid Bodies, Dover: 1955.

Comparison of Air Densities Derived from the Orbits of 1966-51 A, B and C

By

D. G. King-Hele and **D. W. Scott**

Royal Aircraft Establishment, Farnborough, Hants, England

Abstract. On 9 June 1966 three satellites were launched into polar orbits with perigee heights near 180 km and apogees near 3600 km. They were Secor 6 (1966-51B), a rectangular box of mass 17 kg and length 0.33 m; ORS 2 (1966-51C), a regular octahedron of side 0.23 m and mass 5 kg; and the final-stage Agena rocket (1966-51A), of length at least 6 m. Analysis of the extensive orbital data now available on these three satellites gives 260 values of air density at heights near 190 km. Comparison of these values provides a severe test of the accuracy of the orbital data and methods of analysis, and shows that the accuracy is at least as good as has previously been estimated. The comparison also reveals that the effective cross-sectional area of Secor 6 oscillated with an amplitude of about 10%, with minima when perigee was near the equator and maxima at latitudes near 70°. As a consequence, the amplitude of the semi-annual variation in air density previously obtained from Secor 6 should be reduced from 1.45 to 1.40. Other findings are that the effective cross-sectional areas of the satellites never approached their extreme possible values, and that the effective area of 1966-51A changed by less than 10% from week to week.

1. Introduction

The three satellites 1966-51A, B and C, launched on 9 June 1966, have yielded valuable information about the upper atmosphere. All three objects were in similar polar orbits, with perigees near 180 km and apogees initially near 3600 km. The decrease in inclination of all three orbits was used to determine the rotational speed of the atmosphere [1]; the orbit of 1966-51B (Secor 6) was used [2, 3] in a detailed study of atmospheric density; and the orbit of 1966-51C (ORS 2) was calculated at RAE over a short time interval, and the variations in density analysed [4].

The orbit of 1966-51A (Secor 6 rocket) has recently been determined [5] at RAE, mainly from visual observations. In the present paper, the values of orbital acceleration given in [5], together with those derived from Spacetrack bulletins for 1966-51A and C, have been used to obtain values of air density for comparison with the results of [2].

The comparison has three main purposes. First, it provides a severe test of the accuracy and reliability of determining density from Spacetrack orbital data: if the methods or the data are unreliable, the densities obtained from 1966-51B and C will differ significantly. Second, the results [2] from 1966-51B indicated an important semi-annual variation in air density, by a factor of 1.45 at 191 km: this is the only such evidence available for 1966-7 at heights below 200 km, and the value needs to be improved, if possible, because it has a crucial effect on the choice of upper-atmosphere models [6]. Third, the area/mass ratios for 1966-51B and C are known, with errors of about $\pm 10\%$, and the results can be compared with those from 1966-51A, for which the size, shape and mass are unknown, to find the area/mass ratio of 51A and its variations, if any.

2. Calculation of Atmospheric Density

2.1. Orbital Decay Rate

The method used to obtain values of density is generally the same as that described in [2]; i.e. the Spacetrack values of mean motion n were differenced to give values Δn, which were then divided by the time interval Δt to obtain values of $\dot{n} = \Delta n/\Delta t$, which were assumed to apply at the midpoint of the time interval. Figs. 1–3 show the values of $\dot{n}$ for 1966-51B, C and A respectively, from Spacetrack and other sources.

The only difference in treatment here is that up to MJD 39347.8 (1966 Aug. 10) the values of $\dot{n}$ given by the RAE orbit of 1966-51A were used instead of Spacetrack values of $\Delta n/\Delta t$. After this date the RAE values were at irregular intervals and the Spacetrack values were used. As Fig. 3 shows, there were no significant differences between the RAE and Spacetrack values when both were available, so there should be no discontinuity at the change-over point.

2.2. Perigee Height

The variation of perigee height for a polar orbit is given by

$$y_p = K - 9.2 \sin \omega + 21.4 \sin^2 \omega - \frac{H}{2} \ln \frac{e_0}{e} \text{ km}, \qquad (1)$$

where ω is the argument of perigee, e the eccentricity, and H (taken as 32 km) the density scale height at a height $3H/2$ above perigee. The constant K was chosen to give the best fit to observed values of y_p; the values used were 176.0 km, 174.4 km and 180.0 km for 1966-51A, B and C respectively.

The "observed" values of y_p were taken from RAE orbits for 51A

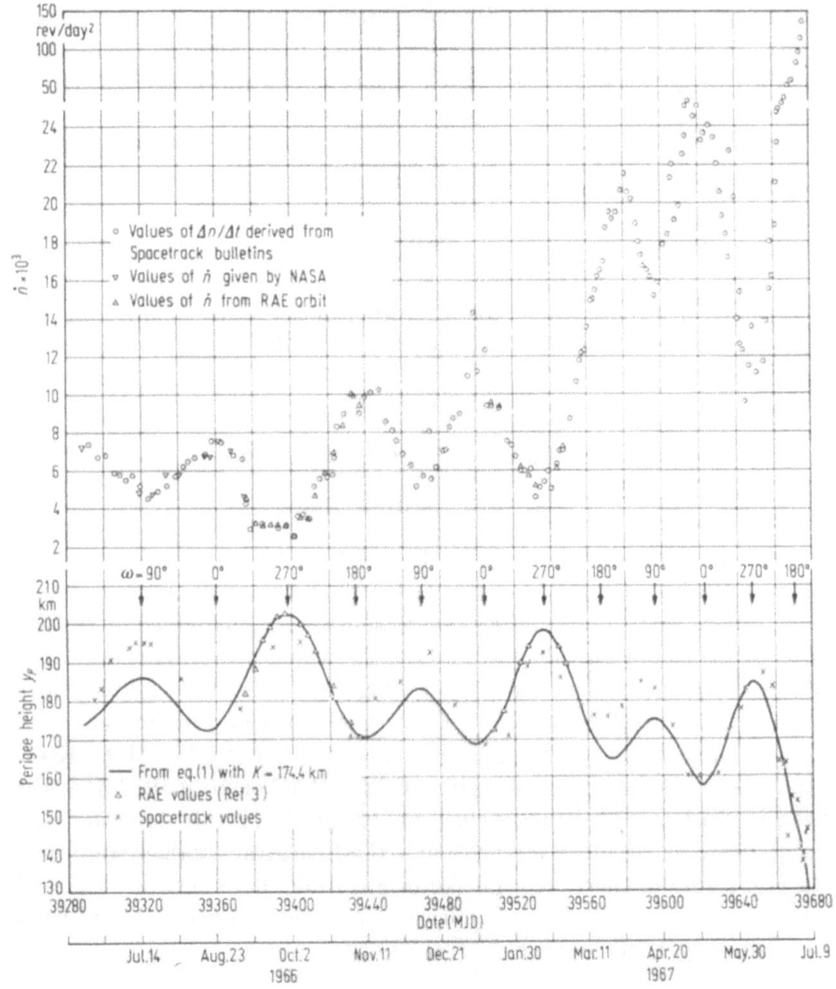

Fig. 1. Orbital decay rate, $\dot{n}$, and perigee height, y_p, for 1966-51 B (Secor 6 rocket).

and B, and from Spacetrack bulletins for 51C. The 3 RAE values [4] of y_p and the 12 NASA values shown in Fig. 2 indicate that our chosen value of K for 51C is probably not more than 2 km in error. It should be noted that the Spacetrack values of y_p in Figs. 1 and 2 require a correction of about $-7 \sin\omega$ km to bring them into agreement with the usual definition of perigee height (for further explanation see [2 and 7]). In Fig. 3 the value of K is chosen so that the curve passes through the group of values between MJD 39330 and 39350, which are much more accurate than the others [5].

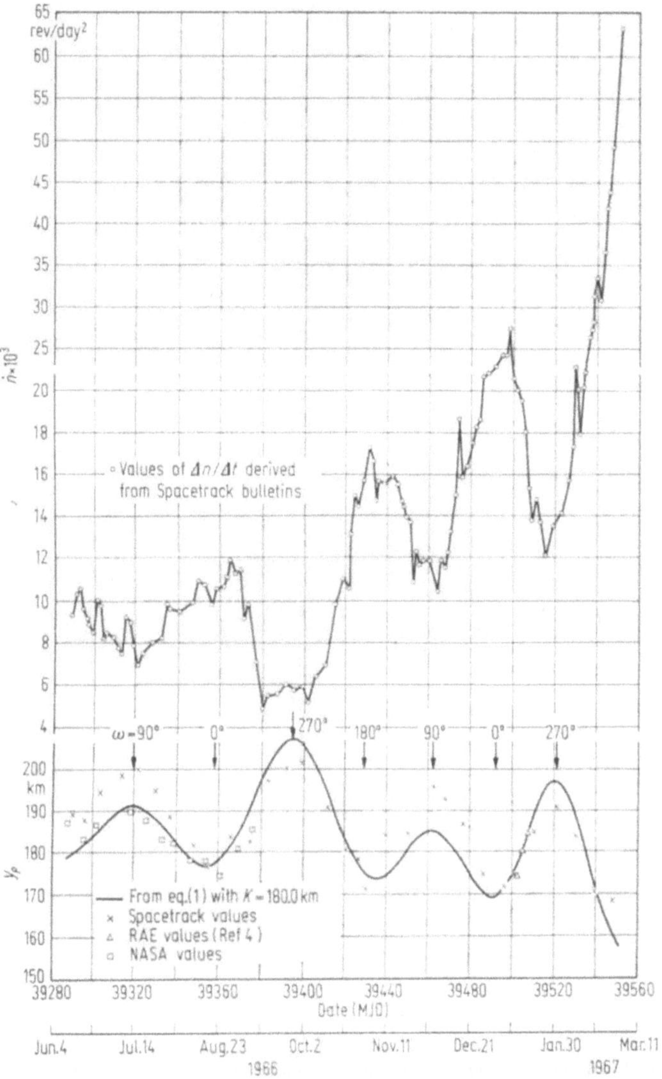

Fig. 2. Orbital decay rate, $\dot{n}$, and perigee height, y_p, for 1966-51 C (ORS 2).

2.3. Area/Mass Ratios

The main difference between the three orbits arises from the different area/mass ratio of the three satellites, giving lifetimes of 6, 13 and 9 months for 1966-51A, B and C respectively. The orbital acceleration is proportional to $\delta = FSC_D/m$, where $F = 1$ for a polar orbit, S is the effective cross-sectional area, C_D is the drag coefficient, and m is the mass of the satellite.

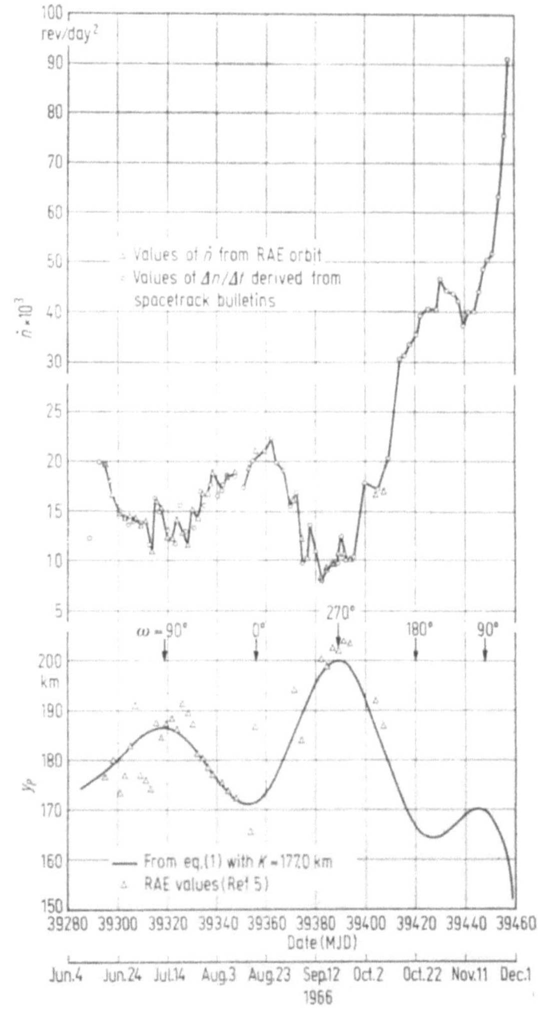

Fig. 3. Orbital decay rate, $\dot{n}$, and perigee height, y_p, for 1966-51 A (Secor 6 rocket).

For 1966-51B, a rectangular box measuring 33 × 28 × 23 cm, the effective cross-sectional area, assuming $C_D = 2.2$, was between 0.078 m² and 0.115 m², and the mean value of S was taken [2] as 0.10 m², giving $\delta = 0.0129$ m²/kg with an error of ± 10% (s.d.).

The shape of 1966-51C (erroneously described in [4] as "disc-like") is that of a regular octahedron of side 0.23 m, and choosing a value for S posed an interesting geometrical problem. The four principal projections of a regular octahedron of unit side are illustrated in Figs. 4a–d:

heavy lines represent edges which are perpendicular to the line of sight andtherefore of unit length; other visible edges are represented by continuous light lines, and invisible edges by broken lines. The maximum and minimum values of effective cross-sectional area are shown by Figs. 4a and 4b, giving possible extreme values of S for 1966-51C of 0.0529 m² and 0.0374 m² respectively, if the satellite were stabilized in either of these directions. It is more likely that the satellite was spinning, about

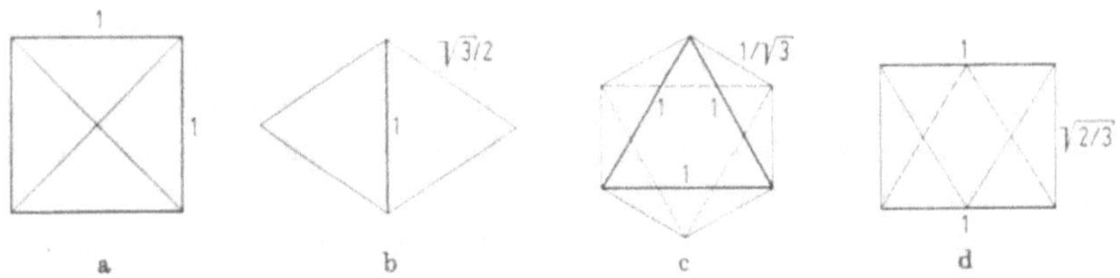

Fig. 4a–d. Visible areas A of a regular octahedron with edges of unit length, along four principal lines of sight: a) along axis of symmetry ($A = 1$); b) along plane of symmetry ($A = 1/\sqrt{2} = 0.707$); c) perpendicular to face ($A = \sqrt{3/2} = 0.866$); d) parallel to face ($A = \sqrt{2}/\sqrt{3} = 0.816$).

an unknown axis, so that the best value of S is probably that given by Fig. 4c, i.e. 0.0458 m², which is the same as for spin with random orientation, as well as being close to the mean of the two extremes quoted above. We therefore took $S = 0.0458$ m², with an error of $\pm 8\%$ (s.d.). With $C_D = 2.2$ and $m = 5.0$ kg, this gives $\delta = 0.0202$ m²/kg.

As the size and weight of the rocket, 1966-51A, were not accurately known (it exhibited much sharper variations in brightness than a normal Agena rocket), it seemed best to choose a value of δ to give values of density consistent with those from the other two satellites. The value of δ used was 0.031 m²/kg, with a standard deviation which will be assessed in Sec. 3.2.

2.4. Calculation of Density

Values of density were calculated using the equations

$$\varrho_A = \frac{0.157\,(1000\,\dot{n})}{10^6\,n^2\,\delta}\left(\frac{e}{a H^*}\right)^{1/2}\frac{(1-e)^{1/2}}{(1+e)^{3/2}}\left\{1 - \frac{H^*}{8ae}\left(1 - 8e + \frac{7H^*}{16ae}\right) + \frac{0.00335\cos 2\omega}{e}\right\} \tag{2}$$

$$\varrho_B = \varrho_A \exp\left(\frac{y_p - \bar{y}_p}{H'}\right) \tag{3}$$

where n is the mean angular motion in rev/day, a is the semi-major axis in km, and H^* is the value of H at perigee; $\bar{y}_p$ is a chosen "mean" perigee height and H' is the value of H at a height $\bar{y}_p + {}^1/_2 H^*$. Eq. (2) gives ϱ_A

with an error of 1% (s.d.) if, as here, $e > 3H^*/a$. For all three satellites $\bar{y}_p$ was chosen as 177.5 km, while H^* and H' were taken as 27 km and 30 km respectively. The resulting values of density, ϱ_B apply to a height $\bar{y}_p + {}^1/_2 H^* = 191$ km, and will be written as ϱ_{191}.

3. Results

3.1. Values of Density

The resulting values of density at a height of 191 km, ϱ_{191}, are shown in Fig. 5, a complicated diagram which calls for extensive explanation.

Each of the 260 values of density obtained is plotted either at the epoch of the orbit determination (for the RAE orbit), or at the mid-point of the time interval over which it applies (for Spacetrack values): squares, circles and triangles indicate 51A, B and C respectively. The points have been joined, by unbroken lines, in histogram form: since the Spacetrack values of $\dot{n} = \Delta n/\Delta t$ represent an average over the time interval Δt, the values of ϱ_{191} are also averages over the time interval Δt, and a histogram is a more appropriate presentation than joining the points. The values of $\dot{n}$ from the RAE orbit are nominally "instantaneous" values, at the epoch of the orbit determination; but in fact the variation of $\dot{n}$ with time was taken as linear in most of the orbit determinations of 1966-51A, so that the "instantaneous" value is in effect a mean value, over a time interval whose end points may conveniently be taken midway between the successive epochs of the orbit determinations.

Comparison between the values of density given by 1966-51A, B and C in Fig. 5 is difficult, because the times at which the values are plotted are different, and so are the time intervals over which they are averaged. In view of these differences and the "spiky" nature of the variations in density linked with geomagnetic activity – the peaks at MJD 39474 offer a clear example – comparisons are bound to be inexact. On the other hand, we are unlikely ever to have any better sets of data, unless satellites are specially launched for this purpose, so the comparisons are worth making.

3.2. Cross-sectional Area of 1966-51A

We concentrate first on comparing the mean of the values from 51B and C with the values from 51A, to expose any large variations in the cross-sectional area of 51A. Taking the mean of B and C is easier said than done: we must average the values of ϱ_{191} over time intervals that are, as nearly as possible, the same for both satellites. The values have therefore been combined into groups, as shown by the broken lines in Fig. 5. The grouping was made with the aim of ensuring that the end

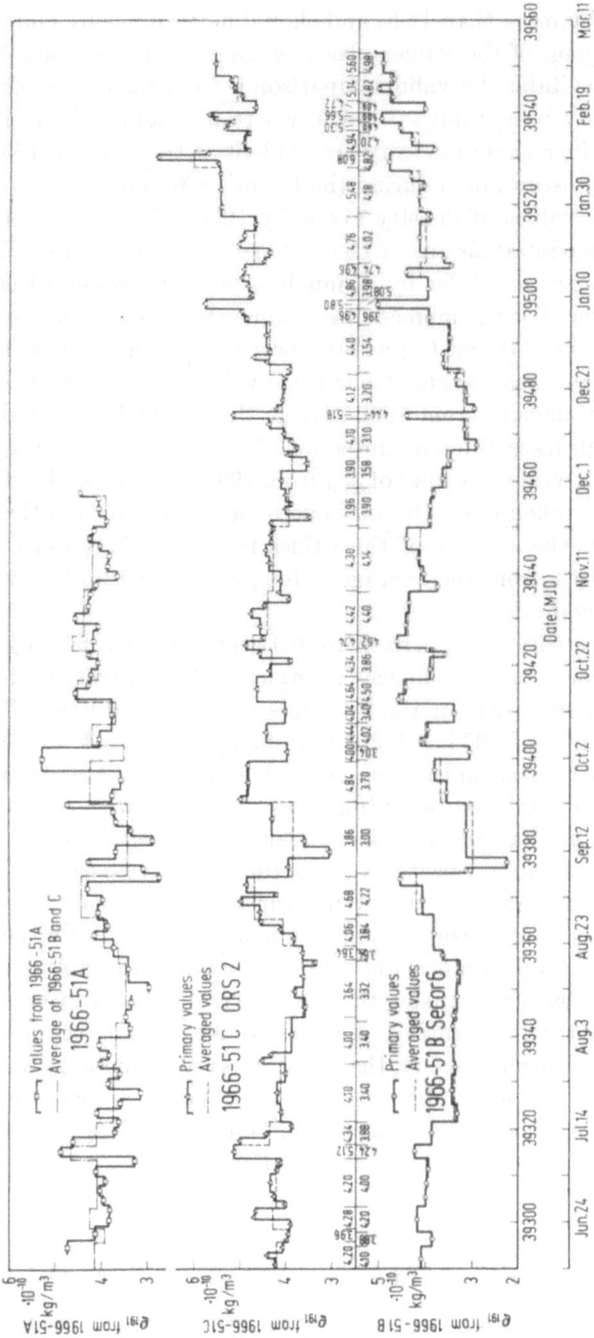

Fig. 5. Values of ϱ_{191}, air density at a height of 191 km, given by 1966-51 A, 51 C and 51 B.

Fig. 5. Values of ϱ_{191}, air density at a height of 191 km, given by 1966-51 A, 51 C and 51 B.

points of the corresponding time intervals for 51B and C should if possible not differ by more than 12 h, and should never differ by more than 24 h. This grouping of the values obscures some of the fine detail: but such a loss is inevitable if a valid comparison is to be made; and the fine detail has already been analysed in [2], where it is noted, for example, that significant increases in density are evident at the times of 17 of the 19 geomagnetic disturbances during the life of 1966-51B.

The average values of density given by 1966-51B and C over each of the chosen time intervals are written between the two graphs in Fig. 5. These values sometimes differ by as much as 30%, for reasons that will be discussed later. For the moment we assume that the arithmetic mean of the values from 51B and C gives the best indication of the values of density, and these mean values are superposed as a dot-dash histogram on the values of density from 1966-51A at the top of Fig. 5. Although the time intervals have different endpoints, the dot-dash histogram agrees remarkably well with the values of ϱ_{191} from 1966-51A, shown by the histogram with unbroken lines. On only one occasion, between MJD 39400 and 39403, does the average of the values from 1966-51B and C differ by more than 10% from the average value given by 1966-51A over the same time interval.

This unexpectedly good agreement between satellites moving quite independently provides a conclusive empirical demonstration that (a) the errors inherent in the method are less than 10%, (b) the effective cross-sectional area of 1966-51A did not change from week to week by as much as $\pm$ 10%, and (c) the assumed value of δ for 51A, 0.031 m^2/kg, corresponding to $S/m = 0.0141$ m^2/kg, was correct to within $\pm$ 10%. If 51A had been a normal Agena rocket, of length 8 m and diameter 1.5 m, the expected maximum variations from the mean cross-section would have been $\pm$ 13%, and since the actual variations are smaller, it seems very probable that any extra equipment aboard, responsible for the satellite's rather flashy appearance, did not greatly affect the cross-sectional area; i.e. the flat surfaces causing the flashes were probably integral with the body rather than on protruding vanes. If the size and shape of 51A were similar to a normal Agena, the mean cross-sectional area would be about 10 m^2 and the mass about $10/0.0141 \triangleq 700$ kg, which is close to the usual mass of an empty Agena rocket. Thus the flashing of 1966-51A was probably caused by fairly trivial alterations to the basic design of the Agena rocket.

3.3. Comparison of Densities from 1966-51B and 51C

A comparison of the pairs of values of density obtained from 1966-51B and C, written on Fig. 5, shows that the values from 51C are on average

16% greater than those from 51B. The possible sources of this bias are errors in (a) perigee height, (b) cross-sectional area and (c) mass. The perigee height of 1966-51B should be accurate to ± 1 km; the perigee height of 51C is based on rather more scattered observational values (see Fig. 2), and the standard deviation is estimated as $1^1/_2$ km, corresponding to an error (s.d.) of 5% in ϱ_{191}. The estimated errors (s.d.) in the cross-sectional areas of 1966-51B and C have already been quoted as 10% and 8% respectively. Recent data on the masses of the satellites suggest that the values used should not be in error by more than 3%. These three sources of error are enough to account for the 16% average difference between the densities found from 1966-51B and C. It is neither necessary nor possible to specify how much each of the three errors contributes to the total; but perhaps the most likely explanation is that the area taken for 1966-51B was about 10% too large, and that the remaining 6% difference comes from errors in the perigee height, cross-section and mass of 51C.

3.4. Cross-sectional Areas of 1966-51B and C

The variations in the ratio of the density given by 1966-51C, ϱ_{51C} say, to the density given by 1966-51B, ϱ_{51B} say, provide a record of the variations in the relative cross-sectional areas of the satellites, though we have to remember that the record may be slightly contaminated by error because the time intervals over which the density is evaluated are not exactly the same for the two satellites. We write

$$\mu = \frac{\varrho_{51C}}{1 \cdot 16\, \varrho_{51B}}, \tag{4}$$

where the factor 1.16 has been inserted so that μ has a mean value of 1; then, since ϱ_{51C} is inversely proportional to the assumed cross-sectional area S_{51C}, and similarly for 51B,

$$\mu = K \frac{S_{51B}}{S_{51C}} = \frac{S_{51B}/\bar{S}_{51B}}{S_{51C}/\bar{S}_{51C}} \tag{5}$$

where K is a constant which it is convenient to write in terms of the mean values of S_{51B} and S_{51C}, namely $\bar{S}_{51B}$ and $\bar{S}_{51C}$ respectively.

The 41 values of μ obtained from the values of ϱ_{51C} and ϱ_{51B} in Fig. 5 are plotted as circles in Fig. 6. The variations in μ – between 0.86 and 1.14 – are smaller than might be expected. For if $\bar{S}_{51B}$ was taken 10% too large, as we suspect, we have $\bar{S}_{51B} = 0.090\,\text{m}^2$, and from [2] the limits for S_{51B} are $0.078 < S_{51B} < 0.115\,\text{m}^2$. So, if $\bar{S}_{51C} = 0.0458\,\text{m}^2$, as assumed, and $0.0374 < S_{51C} < 0.0529\,\text{m}^2$, the extreme limits for μ are $0.75 < \mu < 1.57$. Since in fact μ remains between 0.86 and 1.14, our first conclu-

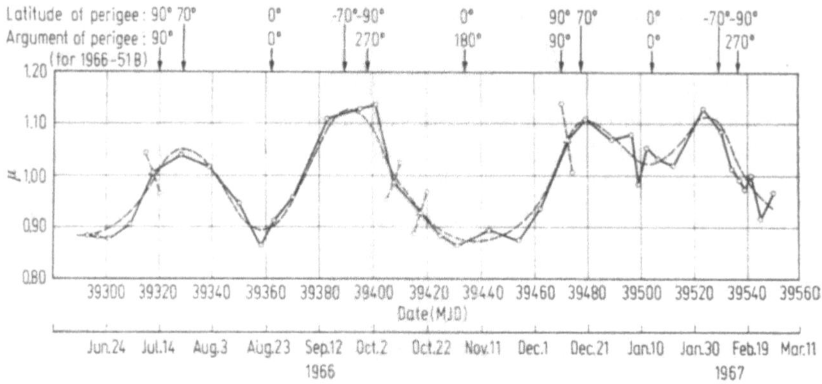

Fig. 6. Values of μ.

sion from Fig. 6 is that it provides strong observational support for the hypothesis that the cross-sectional areas of spinning satellites very rarely reach their extreme values. Indeed Fig. 6 suggests that the variations in cross-sectional area are much less than the estimated bias error in the mean area, and that the s.d. of the variations may be taken as approximately 1/3 of the difference between the mean and extreme values, instead of about 1/2, as we usually assume. Taking the factor 1/3 would give variations (s.d.) of 7% in S_{51B} and 6% in S_{51C}, i.e. a variation of 9% in μ: the actual variation (s.d.) in μ is 8%, so the factor 1/3 provides an adequate margin of error.

The variations in μ in Fig. 6 cannot be dismissed as mere scatter: they evidently have a significant structure. To remove distracting jaggedness, four pairs of points in Fig. 6 have been averaged, and these average values, indicated by triangles, have been used when joining up the points with the unbroken lines. If we accept that the variations in μ, i.e. in the ratio of the cross-sectional areas of 1966-51B and C, have a significant structure, the variation in μ will be regular and continuous if, as is likely, the satellites were spinning and the changes in cross-section are caused by regular precession and nutation. A tentative curve has been drawn in Fig. 6 to suggest a likely form for the variation. Our second conclusion from Fig. 6 is, therefore, that it reveals a regular variation in S_{51B}/S_{51C}, as indicated approximately by the broken curve.

The third conclusion to be drawn from Fig. 6 is concerned with the remarkably small scatter of the points about the broken curve, about 2% rms. It must be remembered that the values plotted in Fig. 6 represent the ratios of values of density obtained at slightly different times, averaged over different time intervals, from satellites moving in independent orbits and having variable cross-sectional areas. Since the corresponding time intervals for 51B and C often differ by as much as 20%, and the change in density between one interval and the next is also often as much as 20%, errors in μ of 4% could arise from this source alone; since the actual scatter in μ is only about 2%, errors inherent in the method of density determinations must be less than 2%.

To illustrate the contribution of differences in the time intervals to the scatter, consider the point furthest from the broken curve in Fig. 6, that at MJD 39470. As Fig. 5 shows, the value of ϱ_{51C} at this point is slightly "contaminated" by the beginning of the magnetic storm at MJD 39478, whereas the effect of this storm on ϱ_{51B} is concentrated in the *subsequent* value of density. Averaging of the two successive values of μ should noticeably reduce the scatter – and does so, as shown by the triangle in Fig. 6 at day 39472. Examination of other points not very near the curve leads to similar conclusions.

3.5. Discussion of Variations in μ

In considering the variations of μ, the latitude of perigee is relevant. The scales at the top of Fig. 6 show the values of ω, the argument of perigee for 51B, and the latitude of perigee, Φ_p. The variation of μ shows a consistent pattern: the four maxima in the broken curve occur at $\Phi_p \triangleq \pm 70°$ and the three minima occur when perigee is near the equator ($\Phi_p \triangleq 0$).

Details of the spin behaviour of the two satellites are not known, but it is believed that 1966-51C was spinning at rates close to its design value of 12 rev/min. If so, its cross-sectional area should not change by more than a few per cent, and the variations in μ would be primarily due to variations of S_{51B}.

Certainly 1966-51B, a rectangular box measuring $33 \times 28 \times 23$ cm, had more opportunities to change its cross-sectional area. The satellite was designed to be magnetically stabilized; but it entered an incorrect orbit on which the aerodynamic forces at perigee were about 10^8 times greater than on the intended orbit. On the actual orbit the magnetic stabilization is unlikely to have been fully successful; but it may been partially effective, for the limited evidence from radar, radio and optical tracking is all consistent with a slow rotation rate, which may have been an angular oscillation rather than true spin. If so, the minimum cross-sectional area might have occurred when perigee was near the equator, when the longest side (33 cm) would be horizontal if it was aligned with the magnetic field; S_{51B} could [2] then be as low as 0.078 m^2, as compared with the probable average value (see Sec. 3.4) of 0.090 m^2. This may explain the minima in μ when perigee is near the equator. The maxima in S_{51B} at $\Phi_p \triangleq \pm 70°$ cannot so easily be accounted for: the north and south magnetic poles are at latitudes of 74°N and 68°S; if they are responsible for the maxima, however, there should also be maxima at $\omega = 110°$ and $250°$, when perigee latitude is 70°. It is true that there are maxima in μ at MJD 39317 ($\omega \triangleq 100°$) and MJD 39401 ($\omega = 260°$), though they might not be recognized except by those who were looking for them. Despite its deficiencies, this "magnetic explanation" may well be correct.

If the variations in μ are caused by variations in the cross-sectional area of 1966-51B, the values of air density obtained from that satellite should be corrected to allow for the variations in S_{51B}. If this correction was made, the maxima and minima of the semi-annual variation in density (averaged over 30 days) would be as follows:

1966 July	minimum	3.7×10^{-10} kg/m³
1966 October	maximum	4.5×10^{-10} kg/m³
1967 January	minimum	3.3×10^{-10} kg/m³
1976 March	maximum	5.0×10^{-10} kg/m³ (unchanged).

In fact we cannot be sure that this full correction should be made, and the safest procedure is to take the mean of these and the original values: the amplitude of the semi-annual variation in density would then be reduced from the value of 1.45 given in [2] to 1.40.

We also suspect (see Sec. 3.4) that the mean value of S_{51B} should be rather smaller than was assumed in [2]. If so, all the values of density in [2] should be slightly increased. We cannot specify the increase exactly, but it is unlikely to exceed 10%.

4. Conclusions

1. The effective cross-sectional area of the Agena rocket 1966-51A changed by less than 10% from week to week, and the average area/mass was 0.014 m²/kg. The size and mass were probably similar to a normal Agena, and the rather flashy appearance was probably caused by fairly minor alterations to the basic shape.

2. The quantity μ plotted in Fig. 6 represents the (normalized) ratio of the effective cross-sectional area of 1966-51B and 51C, i.e. S_{51B}/S_{51C}, as deduced from the variations in orbital decay rates for the two satellites. Although theoretically capable of ranging between 0.75 and 1.57, μ remains between 0.86 and 1.14, thus confirming the usual assumption that the effective cross-sectional areas of satellites rarely approach their extreme values.

3. Fig. 6 shows that μ undergoes regular oscillations with an amplitude of about 10%. This should probably be interpreted as a variation in S_{51B}, since S_{51C} is likely to have remained fairly constant. If so, S_{51B} had minimum values when perigee was at the equator and maximum values at latitudes near 70°. The most likely explanation is that the in-

tended magnetic stabilization of 1966-51B, a rectangular box, was partially effective and that the maxima and minima in $S_{51\mathrm{B}}$ occur near the magnetic equator and poles.

4. If $S_{51\mathrm{B}}$ varied in the manner indicated by μ, some small amendments should be made to the values of air density deduced in [2] (in which $S_{51\mathrm{B}}$ was assumed constant). In particular the amplitude of the semi-annual variation in air density in 1966-7 at a height of 190 km should be changed from 1.45 (as given in [2]) to 1.40.

5. The values of μ in Fig. 6 are obtained from the ratio of orbital decay rates of 1966-51B and C, evaluated over slightly different time intervals, and errors of $\pm 3\%$ (s.d.) would be expected from this source, in addition to any errors due to inaccuracies in orbital data and the theory used. In fact the scatter of the values of μ about the curve is 2% (s.d.): this severe test provides conclusive empirical confirmation that the errors inherent in the data and method of analysis give rise to errors of less than 2%.

Acknowledgement

We thank Doreen Walker for calculating the values of $\dot{n}$ for 1966-51C.

References

1. King-Hele, D. G., Scott, D. W.: Planet. Space Sci. 15 (1967) 1913.
2. King-Hele, D. G., Hingston, J.: Planet. Space Sci. 16 (1968) 675.
3. Gooding, R. H., Milner, R.: RAE Tech. Rpt. 67283 (1967).
4. Scott, D. W., Cook, G. E.: RAE Tech. Rpt. 67273 (1967).
5. Scott, D. W.: J. Brit. Interplan. Soc. 22 (1969) 123.
6. Cook, G. E.: Annales de Géophys. 25 (1969) 451.
7. King-Hele, D. G., Walker, D. M. C.: Planet. Space Sci. 17 (1969) 985.

The Use of AES Orbits Evolution Data for Refinement of Some Geophysical Parameters

By

P. E. Elyasberg

Institute of Space Research, Academy of Sciences, Moscow, USSR

Abstract. Evolution satellites orbit data contain information about values of some geophysical factors such as coefficients of the expansion of the terrestrial gravitational field in a spherical harmonic series and parameters of the earth's atmosphere.

A good many works are devoted to the topic of the determination of these factors by using the data of satellites movements. However there are discrepancies between the results obtained by authors. For this reason the estimation of accuracy of results obtained by different methods is the problem of great importance.

The present paper deals with the estimation of the errors appearing in the determination of atmosphere density according to the approximate expression of the type:

$$\delta T = \varrho_n \sqrt{H}\, \mathfrak{F}(a, e, i, H),$$

as well as with some problems of more precise determination of the zonal harmonic coefficients derived from the long-term changes of the orbital elements.

It is shown that the formulae of the type given above may lead to considerable errors which determine the range of application of these formulae. It is also shown that the limited number of parameters which provide the agreement of evolution orbit data with the whole set of the gravitational potential expansion factors may be used for the prediction of the movement of the satellite.

This problem has been solved for the case of the long-term prediction of satellite movement along the quasi-circle orbit with the account of the zonal harmonics.

The determination of different geophysical factors using the results of AES (Artificial Earth Satellite) orbital motion observations is based on the introduction of some agreement parameters q_i defining with sufficient accuracy the numerical values of the considered geophysical factors and affecting significantly the AES orbit evolution. These parameters are obtained by means of the correlation between the theoretical calculations results and the experimental AES motion data. It should be noted that the quantity of parameters q_i increases indefinitely in case of the absolutely precise problem statement. In doing so the formulated problem becomes unsolvable as it is based on the using of finite number of measurements. In that case if the number of parameters is large enough (on the order of several scores) the solution of the given problem meets with the significant calculation difficulties and the influence of random errors of measurements noticeably increases. In this connection the reasonable selection of agreement parameters set and a technique for its determination plays essential role. The given report devotes to the investigation of this problem as applied to the problems of refinement of the upper atmosphere parameters and zonal harmonic coefficients in gravitational potential expansion. The final approximate formulas establishing the relationship between the parameters q_i and the orbital elements evolutions are derived.

In determining the upper atmosphere parameters one usually uses the formulas of the kind [1, 2]:

$$\delta P = - \varrho_n \sqrt{H_n}\, \mathfrak{F}(a, e, i, H_n)\,, \tag{1}$$

where δP is an orbital period change per one revolution obtained experimentally; ϱ_n and H_n are air density and scale height in the orbit perigee region respectively; a, e and i are major axis, eccentricity and orbital inclination respectively; $\mathfrak{F}(a, e, i, H_n)$ is some function of the forementioned parameter that is slightly dependent on the scale height H_n.

By this approach the problem of the atmosphere parameters refinement is solved in two stages. At the first stage the suitable value of the agreement parameter-product $\varrho_n \sqrt{H_n}$ (or density ϱ at the altitude that is somewhat more than the perigee altitude h_n [3]) is calculated for each of available definitions of the value δP. At the second stage all available definitions of the value $\varrho_n \sqrt{H_n}$ corresponding to different perigee conditions (different values of altitude h_n, angle ψ_n, solar activity indices, geomagnetic indices, etc.) are used. Such approach to the solution of the formulated problem is good because of its simplicity. However as it will be shown below such simplicity can be responsible in many cases for the significant distortion of the final calculation results.

It is connected with the fact that the derivation of the equations of the form (1) is essentially based on hypotheses about spherical or spheroidal symmetry of the atmosphere. and the diurnal effect influence in the atmosphere density distribution is taken into account only by the determination of the values ϱ_n and H_n as applied to the orbit perigee conditions. To estimate errors occurring in this case it is necessary to obtain approximate euqations for the calculation of the value δP with the diurnal effect accounting.

We will ignore the influence of the noncentrality of the Earth's gravitational field and the nonsphericity of its surface (the effect of the Earth's surface nonsphericity can be additionally taken into account by the known technique [2]) and also we will assume that the air density in the orbit perigee region is determined by the model analogous to that suggested by JACCHIA [4]:

$$\varrho = \varrho_n^* \exp\left(-\frac{h-h_n}{H^*}\right)\left\{1 + A_n \exp[c(h-h_n)]\cos^m\frac{\psi}{2}\right\}, \tag{2}$$

where ϱ_n^* is minimum ("night") air density at the perigee altitude h_n; H^* is corresponding scale height; h is instantaneous altitude; A_n is amplitude of the diurnal effect at the altitude h_n; c is a constant coefficient determining the change of the diurnal effect amplitude with the altitude; ψ is an angle between the directions from the Earth's center to the instantaneous point and to the point with the maximum atmosphere density; m is some constant.

Using this model we can find the expression for air density at any orbit point that is defined by the value of eccentric anomaly E. In so doing we will let $m = 6$ [4] (analogous results can be obtained for any even m).

After some elementary transformation we obtain:

$$\varrho = \varrho_0 + \varrho_1, \tag{3}$$

where:

$$\varrho_0 = \varrho_n^*\left[\exp(-\nu + \nu\cos E) + A_n \exp(-\nu_1 + \nu_1 \cos E)\cos^6\frac{\psi_n}{2}\right] \tag{4}$$

$$\begin{aligned}\varrho_1 = \varrho_n^* A_n \exp(-\nu_1 + \nu_1\cos E)\Bigg[f(E) &- \frac{3(1+e)}{2}\varphi_1\frac{1-\cos E}{1-e\cos E}\\ &+ \frac{3(1+e)^2}{4}\varphi_2\left(\frac{1-\cos E}{1-e\cos E}\right)^2 - \frac{(1+e)^3}{8}\varphi_3\left(\frac{1-\cos E}{1-e\cos E}\right)^3\\ &+ \frac{3(1-e^2)}{4}\varphi_4\left(\frac{\sin E}{1-e\cos E}\right)^2 - \frac{3(1-e^2)(1+e)}{8}\frac{\sin^2 E(1-\cos E)}{(1-e\cos E)^3}\Bigg],\end{aligned} \tag{5}$$

$$\nu = \frac{ae}{H^*}, \qquad \nu_1 = \frac{ae}{H_1}, \qquad H_1 = \frac{H^*}{1-cH^*}, \tag{6}$$

$$\begin{aligned}&\varphi_1 = \cos^4\frac{\psi_n}{2}\cos\psi_n, \quad \varphi_2 = \cos^2\frac{\psi_n}{2}\cos^2\psi_n, \quad \varphi_3 = \cos^3\psi_n,\\ &\varphi_3 = \cos^2\frac{\psi_n}{2}\sin^2\psi_n\cos^2\alpha, \quad \varphi_5 = \sin^2\psi_n\cos\psi_n\cos^2\alpha.\end{aligned} \tag{7}$$

Here $f(E)$ is an odd trigonometrical function E; ψ_n is the value of angle ψ at the orbit perigee; α is the angle between the orbital plane and the plane passing through the Earth's center, perigee and the point of maximum air density.

Notice that the first term of the right-hand side of the Eq. (3) can be rewritten as follows:

$$\varrho_0 = \varrho_n^* \left[\exp\left(-\frac{h-h_n}{H^*}\right) + A_n \exp\left(-\frac{h-h_n}{H_1}\right)\cos^6\frac{\psi_n}{2}\right].$$

This expression corresponds to the spherical atmosphere with the such altitude density distribution as for the assumed model (2) in the perigee region. As far as the second term ϱ_1 is concerned it is determined by the atmosphere nonsphericity.

In the all further analysis it will be assumed that

$$\bar{\bar{c}} H^* < 1\,, \qquad H_1 > 0\,. \tag{8}$$

As the calculations based on CIRA-65 atmosphere data show these conditions are fulfilled almost everywhere except for the altitudes close to the altitude at which the diurnal effect changes its sign.

Now make use of the known expression for the orbital period change per one revolution [1]:

$$\delta P = -\frac{6\pi}{\sqrt{\mu}}\bar{\bar{s}}\, a^{5/2}\int_0^{2\pi}\varrho\,\frac{(1+e\cos E)^{3/2}}{\sqrt{1-e\cos E}}\,dE\,, \tag{9}$$

where μ is gravitational parameter of the Earth; $\bar{\bar{s}} = \frac{c_x F_m}{2\mathfrak{M}}$; c_x is air drag coefficient; F_m is master cross section area of the satellite and $\mathfrak{M}$ is its mass.

Substitute the relationships (3), (4) and (5) into the right-hand side of this expression. Now it is necessary to note that the integral of the expression containing $f(E)$ is equal to zero (as the integral of odd periodic function over its period). Expanding the rest integrand into a power series in e we obtain finally the expression δP in terms of Bessel functions. In the following considerations we restrict ourselves to the orbits with average eccentricity i.e. we will assume that:

$$\nu, \nu_1 \gg 1\,, \qquad e \ll 1\,. \tag{10}$$

The using the known asymptotic expansions of the Bessel function into series in the negative powers of argument [1] and neglecting terms of order of ν^{-4}, ν_1^{-4}, $\left(\frac{H^*}{a}\right)^2$, $\left(\frac{H_1}{a}\right)^2$, $e\frac{H^*}{a}$, $e\frac{H_1}{a}$ as well as some terms of order of $\frac{H^*}{a}\nu^{-1}$ and $\frac{H_1}{a}\nu_1^{-1}$ (with the small coefficients at them) after some transformation we obtain that:

$$\delta P = \delta_0 P + \delta_1 P, \tag{11}$$

where:

$$\delta_0 P = -\frac{6\pi}{\sqrt{\mu}} s \varrho_n^* a^{5/2} \left\{ \sqrt{\frac{2\pi}{\nu}} \left[f_0(\nu) \frac{(1+e)^{3/2}}{\sqrt{1-e}} - \frac{H^*}{a} \right] + A_n \cos^6 \frac{\psi_n}{2} \sqrt{\frac{2\pi}{\nu_1}} \left[f_0(\nu_1) \frac{(1+e)^{3/2}}{\sqrt{1-e}} - \frac{H_1}{a} \right] \right\}, \tag{12}$$

$$\delta_1 P = -\frac{6\pi}{\sqrt{\mu}} s \varrho_n^* A_n \sqrt{\frac{2\pi}{\nu_1}} \left[\frac{3(1+e)^{5/2}}{4(1-e)^{3/2}} \Theta_1 - \frac{9(1+e)}{2} \frac{H_1}{a} \Theta_2 \right], \tag{13}$$

$$\Theta_1 = (-\varphi_1 + \varphi_4)\nu_1^{-1} + \frac{3}{4}\left(-\frac{1}{2}\varphi_1 + \frac{1+e}{1-e}\varphi_2 - \frac{1}{2}\varphi_4 - \frac{1+e}{1-e}\varphi_5 \right)\nu_1^{-2} + \frac{5}{16}\left[-\frac{9}{8}\varphi_1 + \frac{3}{2}\frac{1+e}{1-e}\varphi_2 - \left(\frac{1+e}{1-e}\right)^2 \varphi_3 - \frac{3}{8}\varphi_4 - \frac{3}{2}\frac{1+e}{1-e}\varphi_5 \right]\nu_1^{-3} \tag{14}$$

$$\Theta_2 = \left[-\frac{3}{4}\varphi_1 + (1-e)\varphi_4 \right]\nu_1^{-1} + \frac{5}{4}\left[-\frac{3}{8}\varphi_1 + (1+e)\varphi_2 - \frac{1}{2}(1-e)\varphi_4 - \frac{5}{4}(1-e^2)\varphi_5 \right]\nu_1^{-2},$$

$$f_0(\nu) = 1 + \frac{1}{8}\nu_1^{-1} + \frac{9}{128}\nu_1^{-2} + \frac{75}{1024}\nu_1^{-3} + \ldots$$

In the approximate calculations and at sufficiently large values of ν and ν_1 the terms proportional to $\frac{H^*}{a}$, $\frac{H_1}{a}$, ν_1^{-2}, ν_1^{-3} can be neglected in the given equations and also one can assume that $f_0(\nu) \cong f_0(\nu_1) \cong 1$. As a result we obtain the simpler relationship

$$\delta P = \delta_0' P + \delta_1' P, \tag{15}$$

where:

$$\left.\begin{aligned} \delta_0' P &= -\frac{6\pi}{\sqrt{\mu}} s \varrho_n^* a^{5/2} \frac{(1+e)^{3/2}}{\sqrt{1-e}} \left(\sqrt{\frac{2\pi}{\nu}} + A_n \cos^6 \frac{\psi_n}{2} \sqrt{\frac{2\pi}{\nu_1}} \right), \\ \delta_1' P &= -\frac{6\pi}{\sqrt{\mu}} s \varrho_n^* a^{5/2} \frac{(1+e)^{5/2}}{(1-e)^{3/2}} A_n \sqrt{\frac{2\pi}{\nu_1^3}} \\ &\times \frac{3}{4} \cos^2 \frac{\psi_n}{2} \left(\sin^2 \psi_n \cos^2 \alpha - \cos^2 \frac{\psi_n}{2} \cos \psi_n \right). \end{aligned}\right\} \tag{16}$$

It should be noted that the first terms in the right-hand sides of the Eqs. (11), (15) are obtained by the substitution the value ϱ_0 determined by the Eq. (4) for the density ϱ in the right-hand side of the expression (9). In other words, these terms correspond to the "perigee" spherical model of the atmosphere. As far as the second terms in the right-hand sides of the expressions (11) and (15) are concerned they characterize the influence of the unsphericity of the accepted atmosphere model.

The analysis shows that the Eq. (12) for $\delta_0 P$ can be in some cases represented with the sufficient accuracy as follows (1):

$$\delta_0 P \cong -6\pi \sqrt{\frac{2\pi}{\mu}} \frac{a^2}{\sqrt{e}} s \varrho_n \sqrt{H_n} \left[f_0(\nu) \frac{(1+e)^{3/2}}{\sqrt{1-e}} - \frac{H_n}{a} \right], \tag{17}$$

where according to the accepted atmosphere model (2)

$$\varrho_n = \varrho_n^* \left(1 + A_n \cos^6 \frac{\psi_n}{2} \right),$$

$$H_n = -\frac{\varrho_n}{\frac{d\varrho}{dh}(h_n, \psi_n)} = \frac{1 + A_n \cos^6 \frac{\psi_n}{2}}{H_1 + H^* A_n \cos^6 \frac{\psi_n}{2}} H_1 H^*,$$

$$\nu_n = \frac{H_n}{ae}.$$

The results of the estimation of the maximum relative error ξ that occurs in passing from the Eq. (12) to the expression (17) are presented as a function of the value cH^* in the Table 1.

Table 1.

$\bar{c}H^*$	−1	−0.5	0	0.1	0.2	0.3	0.4	0.5	0.6	0.7	0.8
max $\lvert\xi\rvert$ [%]	4.7	1.6	0	0.1	0.4	1.2	2.6	4.7	8	11	30

It can be seen from Table 1 that when $-1 \leq cH^* \leq 0{,}5\ \lvert\xi\rvert < 5\%$. As the calculations based on the CIRA-65 atmosphere show the presented inequalities are satisfied at the solar intensity at the wave 10.7 cm when $F = 65 \times 10^{-22}$ W/m²Hz for the altitudes 300 km $\leq h_n \leq$ 800 km and when $F = 275 \times 10^{-22}$ W/m²Hz – for the altitudes 400 km $\leq h_n \leq$ 800 km. In this case max$\lvert\xi\rvert$ doesn't exceed $1 \div 2\%$ for the greatest part of the mentioned altitude intervals. For the values h_n less than the lower limits of these intervals max$\lvert\xi\rvert$ increases; it is equal to 12.5% when $F = 65 \times 10^{-22}$ W/m²Hz, $h_n = 200$ km and to 6.5% when $F = 275 \times 10^{-22}$ W/m²Hz, $h_n = 300$ km. For the case $F = 275 \times 10^{-22}$ W/m²Hz, $h_n = 200$ km the value of max$\lvert\xi\rvert$ was not determined as under these conditions the requirements (8) are not satisfied.

Turn now to the estimation of the errors which are caused by the neglecting of the influence of the nonsphericity of the accepted atmosphere model that is equivalent to droping of the second terms in the right-hand sides of the expressions (11) and (15). Using the relation (6),

(7) and (16) one can show that the corresponding relative error η is determined by the equation:

$$\eta = \frac{\delta_1' P}{\delta_0' P} = \frac{3(1+e)}{4(1-e)} \nu_1^{-1} \mathfrak{F}(K, \psi_n, \alpha), \tag{18}$$

where:

$$K = \frac{A_n}{\sqrt{1 - cH^*}},$$

$$\mathfrak{F}(K, \psi_n, \alpha) = K \frac{\cos^2 \frac{\psi_n}{2} \left(\sin^2 \psi_n \cos^2 \alpha - \cos^2 \frac{\psi_n}{2} \cos \psi_n \right)}{1 + K \cos^6 \frac{\psi_n}{2}}. \tag{19}$$

Denote maximum $|F(K, \psi_n, \alpha)|$ with fixed K and variable ψ_n, α as $\Phi(K)$.

Then it follows from the relationship (18) that

$$|\eta| \leq \frac{3(1+e)}{4(1-e)} \Phi(K) \nu_1^{-1}. \tag{20}$$

The values of $\Phi(K)$ are listed in the Table 2.

Table 2.

K	0	0.5	1	2	3	4	5	6	7	8	9	10	11	12
$\Phi(K)$	0	0.32	0.49	0.80	1.10	1.37	1.60	1.82	2.02	2.20	2.36	2.53	2.70	2.86

The values max $|\eta|$ calculated by Eq. (20) as applied to the CIRA-65 atmosphere are given in the Table 3 for different values F, e and h_n.

It is obvious from the relations (6) and (20) as well as from the Tables 2 and 3 that max $|\eta|$ increases with decreasing of the orbit eccentricity e and with increasing of the parameters K and H_1.

Table 3.

$F\left[\frac{\mathrm{W}}{\mathrm{m^2Hz}}\right]$	65×10^{-22}			275×10^{-22}		
e	0.05	0.07	0.10	0.05	0.07	0.10
h_n[km]						
200	9%	6%	4.5%	—	—	—
300	9%	7%	5%	14%	10%	7.5%
500	34%	25%	18%	26%	19%	13%
800	22%	16%	11%	63%	46%	33%

In this case the satisfactory calculation accuracy ($|\eta| < 10\%$) is achieved for the CIRA-65 atmosphere model during the time of minimum solar activity for $200\text{ km} \leq h_n \leq 300\text{ km}$ for orbits with the eccentricity $e > 0.05$. At the higher solar intensity and the higher perigee altitudes the same calculation accuracy can be assured only for orbits with larger eccentricities (this is connected with increasing of the values K and H_1).

At the determination of the atmosphere parameters the relative calculation error in the value δP causes the same relative error in the value of $\varrho_n \sqrt{H_n}$ product. Further it will result in errors in the final values of the atmosphere density ϱ during the atmosphere characteristic definition on the base of all available measured values $\varrho_n \sqrt{H_n}$. It follows from the relations (13) and (19) that these errors essentially depend on an illumination of the perigee region. Maximum of relative errors in the density prevails on the day side of the Earth and achieves the values given in the Table 3. Such an irregularity of the error distribution leads to the appropriate errors in the value of diurnal effect amplitude A. If the latter is determined by the equation

$$A = \frac{\varrho(\psi = 0) - \varrho^*}{\varrho^*},$$

then using the relation (11) one can obtain the following approximate expression for the estimation of the relative error arising during the amplitude A calculation:

$$\left.\begin{aligned} \zeta = \frac{A - \tilde{A}}{A} = \frac{3(1+e)}{4(1-e) f_0(\nu_1)} \Big\{ \nu_1^{-1} - \frac{3}{4}\left(\frac{1+e}{1-e} - \frac{1}{2}\right)\nu_1^{-2} \\ + \frac{5}{16}\left[\frac{9}{8} - \frac{3}{2}\frac{1+e}{1-e} + 2\left(\frac{1+e}{1-e}\right)^2\right]\nu_1^{-3}\Big\}, \end{aligned}\right\} \tag{21}$$

where A is an "accurate" value of the amplitude found by the Eq. (11) and $\tilde{A}$ is an approximate value calculated with the use from relationships of type (1).

For the approximate estimation of the value ζ the simplified equation can be used:

$$\zeta = \frac{3(1+e)}{4(1-e)} \nu_1^{-1}. \tag{22}$$

The values ζ calculated by the Eq. (22) with the initial data used for the calculation of Table 3 are listed in the Table 4.

It follows from the Eq. (21) and Table 4 that the calculation by the equations of type (1) results finally in the significant underestimation of the diurnal effect amplitude value. Particularly large errors will occur for small eccentricity orbits and for high solar activity times.

Table 4.

$F\left[\frac{\mathrm{W}}{\mathrm{m^2Hz}}\right]$	65×10^{-22}			275×10^{-22}		
e	0.05	0.07	0.10	0.05	0.07	0.10
h_n[km]						
200	23%	17%	12%	—	—	—
300	12%	8%	6%	32%	24%	17%
500	15%	11%	8%	28%	21%	15%
800	22%	16%	11%	32%	23%	17%

To eliminate these errors it is desirable to use more accurate relationships such as (11) or to apply another calculation technique (for example, the numerical integration of the satellite motion equations with accounting the diurnal effect).

It can be seen from the expressions (11) that for the equations considering the diurnal effect the proportionality between δP and $\varrho_n \sqrt{H_n}$ that is specific to the equations of type (1) doesn't exist. For the fixed (or slightly varying) perigee altitude h as well as for the constant values of both the solar activity and the geomagnetic disturbance indices the following non-linear dependence between orbit elements and the atmosphere parameters exists:

$$\delta P = \delta P(a, e, i, \psi_n, \alpha, \varrho_n^*, H^*, A_n, c, m) .$$

The values δP, a, e, i, ψ_n and α appearing in this dependence are defined by measurement data but the values ϱ_n^*, H^*, A_n, c, m can be employed as the agreement parameters which must be determined simultaneously by the results of several measurements of δP value.

Turn to the question of the agreement parameters choice permitting to consider the influence of zonal harmonics in the gravitational potential expansion on the orbit evolution. The principle difficulty arising from investigation of the gravitational field influences on AES motion is concerned with the fact that the Earth's potential expansion in a series in spherical functions has a poor convergence and the coefficients by the high order harmonics are not well known enough. That's why it is reasonable to make an attempt of replacing of larger number of harmonic coefficients by the limited amount of the equivalent agreement parameters during the AES motion computation.

In this report the similar substitution is produced for the case of the computation of nearly circular orbits evolution in long time intervals. In this case secular and long-periodic variations of the elements at the ascending nodes of the orbit that are determined generally by the zonal harmonics are considered [5].

As it was shown before [6] these variations (subject to the elimination of the orbits with critical inclinations) can be described by the following equations:

$$\left.\begin{aligned}
l &= \frac{(B+C)l_0+A}{B+C}\cos\left[\sqrt{B^2-C^2}\,(N-N_0)\right] - \\
&\quad - \frac{(B-C)h_0-D}{\sqrt{B^2-C^2}}\sin\left[\sqrt{B^2-C^2}\,(N-N_0)\right] - \frac{A}{B+C}, \\
h &= \frac{(B-C)h_0-D}{B-C}\cos\left[\sqrt{B^2-C^2}\,(N-N_0)\right] + \\
&\quad + \frac{(B+C)l_0+A}{\sqrt{B^2-C^2}}\sin\left[\sqrt{B^2-C^2}\,(N-N_0)\right] + \frac{D}{B-C}, \\
T_\Omega &= T_{\Omega_0} + \frac{3C_{20}T(5\mu-4)}{2\bar{a}^2}\left[(l-l_0) - \frac{7}{4}(l^2-l_0^2) - \frac{1}{4}(h^2-h_0^2)\right], \\
t_\Omega &= t_{\Omega_0} + T_{\Omega_0}(N-N_0) - \frac{3C_{20}T(5\mu-4)}{2\bar{a}^2}\left\{\frac{A}{B+C} + \frac{7}{8}\left[\frac{(B+C)l_0+A}{B+C}\right]^2 + \right. \\
&\quad + \frac{7}{8}\left[\frac{(B-C)h_0-D}{\sqrt{B^2-C^2}}\right]^2 + \frac{7}{4}\left(\frac{A}{B+C}\right)^2 + \frac{1}{8}\left[\frac{(B-C)h_0-D}{B-C}\right]^2 + \\
&\quad \left. + \frac{1}{8}\left[\frac{(B+C)l_0+A}{\sqrt{B^2-C^2}}\right]^2 + \frac{1}{4}\left(\frac{D}{B-C}\right)^2 + l_0 - \frac{7}{4}l_0^2 - \frac{1}{4}h_0^2\right\}(N-N_0) + \\
&\quad + \frac{3C_{20}T(5\mu-4)}{2\bar{a}^2}\left\{(h-h_0)\left[\frac{1}{B+C} + \frac{21A}{8(B+C)^2} + \frac{A}{8(B^2-C^2)}\right] + \right. \\
&\quad \left. + (l-l_0)\left[\frac{7D}{8(B^2-C^2)} + \frac{3D}{8(B-C)^2}\right] + (hl-h_0l_0)\left[-\frac{7}{8(B+C)} + \frac{1}{8(B-C)}\right]\right\} \\
\Omega &= \Omega_0 + (N-N_0)\left[K_\Omega + \frac{1}{2}H_{\Omega_2}\left(h_0^2 + \frac{B+C}{B-C}l_0^2\right) + \frac{1}{2}L_{\Omega_2}\left(l_0^2 + \right.\right. \\
&\quad \left.\left. + \frac{B-C}{B+C}h_0^2\right)\right] + (l-l_0)\left[-\frac{H_\Omega}{B-C} - \frac{3H_{\Omega_2}D}{2(B-C)^2} - \frac{L_{\Omega_2}D}{2(B^2-C^2)}\right] + \\
&\quad + (h-h_0)\left[\frac{L_\Omega}{B+C} + \frac{H_{\Omega_2}A}{2(B^2-C^2)} - \frac{3L_{\Omega_2}A}{2(B+C)^2}\right] + \\
&\quad + (hl-h_0l_0)\left[-\frac{H_{\Omega_2}}{2(B-C)} + \frac{L_{\Omega_2}}{2(B+C)},\right. \\
i &= i_0 - (l-l_0)\left[\frac{H_i}{B-C} - \frac{AH_{hl}}{B^2-C^2}\right] + (h-h_0)\frac{L_i}{B+C} + (h^2-h_0^2)\frac{H_{hl}}{2(B+C)}, \\
\bar{a} &= \bar{a}_0 - \frac{C_{20}}{\bar{a}_0}\left[3(l-l_0) + 6(l^2-l_0^2) + 3(h^2-h_0^2)\right],
\end{aligned}\right\} \quad (23)$$

where

$$l = e\cos\omega; \qquad h = e\sin\omega,$$

$$\mu = \sin^2 i, \qquad \bar{a} = \frac{a}{a_e}. \qquad \bar{p} = \bar{a}(1-e^2),$$

a_e is an average equational radius of the Earth; N is the number of the revolution; index "0" denotes the initial values of the appropriate parameters; $A, B, C, D, K_\Omega, H_\Omega, L_\Omega, H_{\Omega_2}, L_{\Omega_2}, H_i, L_i, H_{hl}$ are constants that are functions of the average values of the orbit elements and of the coefficients C_{n0} by zonal harmonics.

Using the expression of the Earth's potential in terms of the orbit elements function [5] it can be obtained the expressions for these constants involving arbitrary number of zonal harmonic coefficients:

$$\left.\begin{aligned}
A &= \frac{3\pi}{4p^4}(5\mu - 4)(3 - 2\mu) C_{20}^2,\\
B &= \frac{3\pi}{2p^2}(5\mu - 4) C_{20} + \frac{3\pi}{32p^4}(380\mu - 445\mu^2) C_{20}^2 +\\
&\quad + \frac{15\pi}{16p^4}(16 - 62\mu + 49\mu^2) C_{40}(1 + \beta_B),\\
C &= \frac{3\pi}{16p^5}(-48 + 46\mu + 15\mu^2) C_{20}^2 + \frac{15\pi}{16p^4}(6\mu - 7\mu^2) C_{40}(1 + \beta_c),\\
D &= \frac{3\pi\sin i}{4p^3}(4 - 5\mu) C_{30}(1 + \alpha_D),\\
K_\Omega &= \frac{3\pi\cos i\cdot C_{20}}{p^2} + \frac{3\pi\cos i(3 - 20\mu)}{4p^4} C_{20}^2 +\\
&\quad + \frac{15\pi\cos i(7\mu - 4)}{8p^4} C_{40}(1 + \beta_{K\Omega}),\\
H_\Omega &= \frac{3\pi\,\mathrm{ctg}\, i(15\mu - 4)}{4p^3} C_{30}(1 + \alpha_{H\Omega}),\\
L_\Omega &= \frac{6\pi(2 - 5\mu)\cos i}{p^4} C_{20}^2,\\
H_i &= \frac{3\pi\sin 2i(5\mu - 4)}{4p^4} C_{20}^2,\\
L_i &= \frac{3\pi\cos i(5\mu - 4)}{4p^3} C_{30}(1 + \alpha_{Li}),\\
H_\alpha &= \frac{9\pi(5\mu - 4)}{2p^3} C_{20}^2,\\
H_{hl} &= \frac{3\pi\sin 2i(14 - 15\mu) C_{20}^2}{16p^4} - \frac{3\pi\sin 2i(30 - 35\mu)}{16p^4} C_{40}(1 + \beta_{Hhl}),\\
H_{\Omega_2} &= \frac{15\pi\cos i(2 - 7\mu) C_{20}^2}{16p^4} - \frac{15\pi\cos i(18 - 35\mu)}{16p^4} C_{40}(1 + \beta_{H\Omega_2}),\\
L_{\Omega_2} &= \frac{3\pi\cos i(25\mu - 18) C_{20}^2}{16p^4} - \frac{15\pi\cos i(6 - 7\mu) C_{40}(1 + \beta_{L\Omega_2})}{16p^4}.
\end{aligned}\right\}\quad (24)$$

The coefficients α and β appearing in the expressions (24) are defined by the equations:

$$
\left.
\begin{aligned}
&\beta_B = \frac{16}{15(16 - 62\mu + 49\mu^2)} \sum_{\substack{n=6\\(n=2K)}}^{n\max} \left[\frac{C_{n0}}{\bar{p}^{n-4} C_{40}} \right.\\
&\qquad \left. \times \sum_{t=0}^{\frac{n}{2}} \frac{(-1)^t (2n - 2t)!\, \mu^{\frac{n}{2} - t - 1} [\mu(n^2 + 3n - 4t) - 2n + 4t]}{t!\,(n-t)!\, 4^{n-t} \left[\left(\frac{n}{2} - t\right)!\right]^2} \right],\\
&\beta_C = -\frac{16}{15(6 - 7\mu)} \sum_{\substack{n=6\\(n=2K)}}^{n\max} \left[\frac{(n-1)(n-2) C_{n0}}{\bar{p}^{n-4} C_{40}} \right.\\
&\qquad \left. \times \sum_{t=0}^{\frac{n}{2} - 1} \frac{(-1)^t (2n - 2t)!\, \mu^{\frac{n}{2} - t - 1} (n - 2t)}{t!\,(n-t)!\, 4^{n-t} (n + 2 - 2t) \left[\left(\frac{n}{2} - t\right)!\right]^2} \right],\\
&\beta_{K\Omega} = \frac{16}{15(7\mu - 4)} \sum_{\substack{n=6\\(n=2K)}}^{n\max} \left[\frac{C_{n0}}{\bar{p}^{n-4} C_{40}} \right.\\
&\qquad \left. \times \sum_{t=0}^{\frac{n}{2} - 1} \frac{(-1)^t (2n - 2t)!\, (n - 2t)\, \mu^{\frac{n}{2} - t - 1}}{t!\,(n-t)!\, 4^{n-t} \left[\left(\frac{n}{2} - t\right)!\right]^2} \right],\\
&\qquad \beta_c = \beta_{H_{hl}} = \beta_{L_{\Omega_2}},\\
&\beta_{H_{\Omega_2}} = -\frac{8}{15(18 - 35\mu)} \sum_{\substack{n=6\\(n=2K)}}^{n\max} \left[\frac{(n-1)(n-2) C_{n0}}{\bar{p}^{n-4} C_{40}} \right.\\
&\qquad \left. \times \sum_{t=0}^{\frac{n}{2} - 1} \frac{(-1)^t (2n - 2t)!\, (n - 2t)(2n - 4t + 2)\, \mu^{\frac{n}{2} - t - 1}}{t!\,(n-t)!\, 4^{n-t} (n - 2t + 2) \left[\left(\frac{n}{2} - t\right)!\right]^2} \right],\\
&\alpha_D = -\frac{8}{3(4 - 5\mu)} \sum_{\substack{n=5\\(n=2K-1)}}^{n\max} \left[\frac{(n-1) C_{n0}}{\bar{p}^{n-3} C_{30}} \right.\\
&\qquad \left. \times \sum_{t=0}^{\frac{n-1}{2}} \frac{(-1)^t (2n - 2t)!\, \mu^{\frac{n-1}{2} - t}}{t!\,(n-t)!\, 4^{n-t} \left(\frac{n+1}{2} - t\right) \left[\left(\frac{n-1}{2} - t\right)!\right]^2} \right],\\
&\alpha_{H\Omega} = -\frac{8}{3(4 - 15\mu)} \sum_{\substack{n=5\\(n=2K-1)}}^{n\max} \left[\frac{(n-1) C_{n0}}{\bar{p}^{n-3} C_{30}} \right.\\
&\qquad \left. \times \sum_{t=0}^{\frac{n-1}{2}} \frac{(-1)^t (2n - 2t)!\, (n - 2t)\, \mu^{\frac{n-1}{2} - t}}{t!\,(n-t)!\, 4^{n-t} \left(\frac{n+1}{2} - t\right) \left[\left(\frac{n-1}{2} - t\right)!\right]^2} \right],\\
&\qquad \alpha_{Li} = \alpha_D.
\end{aligned}
\right\} \quad (25)
$$

These coefficients considered as "corrections" to both the third and the fourth zonal harmonics can be calculated before and set during prediction together with the initial conditions of the motion or they can be included in the number of unknowns which are refined according to the data of orbit evolution in long time intervals. For these coefficients charts can be also plotted having two inputs – the inclination and the orbit parameter. These charts can be computed using known and more reliable groups of the zonal harmonics coefficients and they can be refined as far as the appropriate processing and the analysis of the available orbits are carried out.

It follows from the expressions (25) that for the prediction of the nearly circular orbits it is well enough to use 6 correction coefficients. The set of the agreement parameters can involve only 4 of them: β_B, β_C, $\beta_{K\Omega}$ and α_D.

The uncertainties in $\alpha_{H\Omega}$ and $\beta_{H\Omega 2}$ values have not a noticeable influence on the prediction accuracy so far as the long-periodic disturbances at the ascending node position and the secular drift in Ω of the order e^2 for nearly circular orbits are sufficiently small. It also follows from the relationships given above that during the refinement of the zonal harmonics coefficients according to the variations of the elements of several orbits the obtained values C_{n0} being the agreement coefficients for the accepted prediction technique can take different values for various orbit elements (the dropped harmonics have an unequal effect upon the various orbit elements). That is, may be, one of the reasons of the difference between the existing coefficients C_{n0} resulting from using of various orbit elements.

In conclusion it should be noted that for some "critical" inclinations the third and the fourth zonal harmonics have not an influence on the orbits evolution.

For such inclinations the coefficients α and β derived above cannot serve any longer as the agreement parameters. In these cases corrections to the values of zonal harmonic coefficients of higher order are to be taken as the agreement parameters.

References

1. Elyasberg, P. E.: Introduction to the Theory of the Flight of Artificial Satellites of the Earth. Moscow 1965 (in russian).
2. King-Hele, D.: Theory of the Orbits of Artificial Satellites in the Atmosphere. Moscow 1966.
3. Elyasberg, P. E., Yastrebov, V. D.: Determination of the Density of the Upper Atmosphere using Observations of the Motion of the Third Artificial Satellite of the Earth. In: The artificial satellites of the Earth, No. 4, Moscow 1960 (in russian).

4. JACCHIA: Model of an Atmosphere with a Variable Density obtained from Air-drag on Satellites. In: Investigations of the Upper Atmosphere with the Aid of Rockets and Satellites. Moscow 1961.
5. KAULA, V.: Cosmic geodesy. Moscow 1966.
6. KOUGAENKO, B. V., ELYASBERG, P. E.: Evolution of Nearly Circular Orbits of Artificial Satellites of the Earth under the Influence of Zonal Harmonics. Cosmic Research 6, No. 2, 1968, Moscow.

Analysis of the Atmospheric Drag of the Echo 1 Satellite Using the PERLO Orbital Period Determination Program

By

I. Almár, A. Horváth and **E. Illés**

Konkoly Observatory of the Hungarian Academy of Sciences, Budapest, Hungary

Abstract. The PERLO program determines the quasi-nodal period of a satellite. Using visual observations of the decaying Echo 1 the density at 800 km was deduced. During April 1968 the solar flux $S_{10.7}$ did not follow the 27-day cycle but the density ϱ_0 continued to show this variation.

1. Introduction

A new computer program PERLO has been completed at the Konkoly Observatory, Budapest, and the Satellite Tracking Station, Miskolc, in order to deduce the orbital period of satellites with a high time resolution. The procedure is based on A. M. Lozinsky's suggestion [1] to determine the quasi-nodal period of a satellite from its consecutive transits through the topocentric celestial equator of a tracking station. Using the PERLO program, the orbital acceleration of Echo 1 has been studied and correlated with changes in the solar activity in 1968. The results are presented in the following sections.

2. Method of Analysis

The PERLO program permits the determination of the crossing time of the topocentric celestial equator with considerable accuracy, if observations with topocentric declinations $|\delta| < 10°$ are used. Frequent positions on both sides of the equator are preferred, but not indispensable. Every observed position (in horizontal or in equatorial system) is the starting point of an arithmetic calculation proposed by M. Ill [2] and of another by I. Almár [3] in order to derive approximate values of the $t(\delta)$ function at $\delta = 0$. The arithmetic mean and the standard deviation are given in both cases. As these methods are based on different assumptions, they supplement and check each other for the case of incomplete series of observations. A further check of the mean values is provided by a least squares solution. The adopted values of all topocentric crossing times are transformed to a common reference latitude using a simple formula requiring only approximate values of the orbital elements [4]. The program is written in ALGOL 60 language and contains 1120 statements.

Changes in the quasi-nodal period have been derived graphically from the $T(t)$ curves, and also numerically using the O–C method [5]. We could not find in the present material any sudden change in $\dot{T}$, therefore we confined ourselves to a time resolution of 1 day. (The PERLO program allows a better time resolution up to one orbital period, if necessary.) The acceleration caused by solar radiation pressure was removed by the usual formulae [6]. Atmospheric densities were computed by means of the formula [7]

$$\varrho_\lambda = -\left(\frac{\Phi \dot{T}}{3\pi a \delta}\right) \frac{\exp(c \cos 2\omega)}{1 + \frac{2 e I_1(z^*)}{I_0(z^*)} + \frac{c I_2(z^*)}{I_0(z^*)} \cos 2\omega}$$

converted to values at a common height of 800 km

$$\varrho_{800} = \varrho_\lambda \exp\left(\frac{y - 800}{H}\right)$$

where H is the density scale height obtained from the CIRA 1965 model atmosphere [8] and $\delta = 288.2\ \text{cm}^2/\text{g}$ based on the results of earlier investigations [9].

Fig. 1. Distribution of observing stations.

3. Observational Material

About 636 visual observations of 104 transits of Echo 1 were reduced by the PERLO program using an ICT-1905 computer. The observations were made at 19 Hungarian, Soviet, French and English tracking stations (see Fig. 1). The estimated average angular accuracy was 0.1°. The mean timing error deduced from independent observations of the same crossing over the same reference latitude but made at different stations proved to be about 1–2 sec. Echo 1 was an unfavourable satellite from this point of view, crossing the topocentric equator of the European stations at a small angle.

4. Results and Conclusions

The deduced values of air density at a height of 800 km are shown as circles in Fig. 2. They are joined by continuous lines where the accuracy of the determination permitted continuous monitoring of the variations. The local time of the sub-perigeepoint changed but little, 12^h being a good average value over the whole time interval.

In order to display the variation connected with solar radiation more clearly, we need to correct ϱ_{800} for the geomagnetic effect. At the very bottom of Fig. 2 the variation of the planetary geomagnetic index a_p is

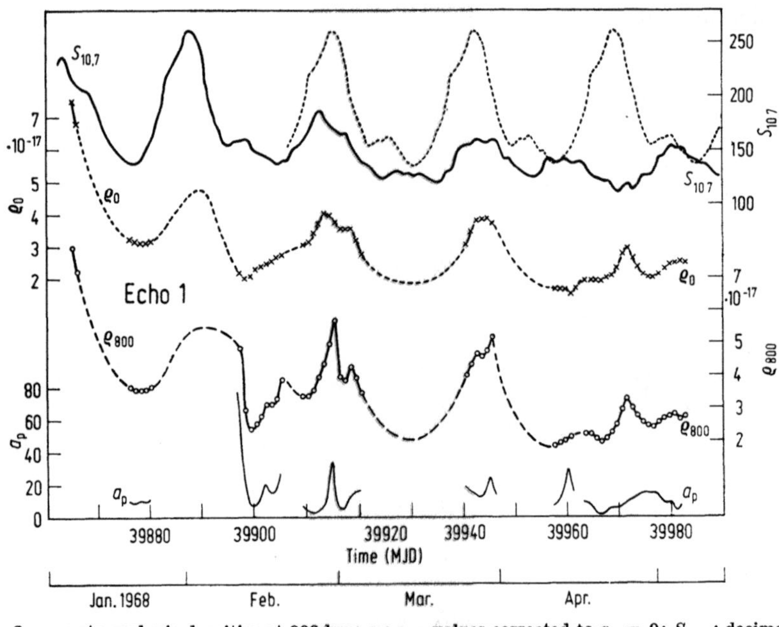

Fig. 2. ϱ_{800}: atmospheric densities at 800 km; ϱ_0: ϱ_{800} values corrected to $a_p = 0$; $S_{10,7}$: decimetric flux of solar radiation in 10^{-22} Wm^{-2} Hz^{-1} (continuous line), and repetition of the January–February cycle (dashed line).

shown during the time covered by the observations. Density values were reduced to $a_p = 0$ by [10]

$$\varrho_0 = \frac{\varrho_{800}}{1 + 0.015\, a_p}$$

making the assumption of a time delay of $0^{\mathrm{d}}.5$, and are represented by crosses in Fig. 2. The transformation yields a curve revealing an obvious correlation from January till March with the variation of the measured $S_{10.7}$ intensity – represented by the continuous line at the top of Fig. 2 – with a time lag of about two days. The dates of occurrence of the extreme values of ϱ_0, however, continue to repeat the 27 day cycle in April as well, whilst the actual $S_{10.7}$ variation was significantly disturbed. This fact is demonstrated by projecting the form of the undisturbed January–February cycle of solar activity repeatedly as a dashed line on later cycles. The correspondence in time is extremely good.

This interesting result is consistent with previous conclusions by King-Hele and Walker [11]: "During June 1967 $S_{10.7}$ failed to show the strong 27-day variation apparent in May and July, but the density *did* show this 27-day variation. We must conclude that the 10.7 cm radiation is an imperfect index of the extreme ultraviolet radiation from the Sun ... values of density are in this instance a better index of solar activity than $S_{10.7}$."

The discrepancy between the variation of $S_{10.7}$ and ϱ_0 in April 1968 probably occurs because the phase of the 27-day cycle of the 10.7 cm radiation suddenly changed in April by about 180°, but the density variation continued to follow the same run.

It is to be mentioned that the ϱ_0 curve – not reduced to a common $S_{10.7}$ value – shows a subsidiary minimum between 10 and 15 April corresponding to that revealed earlier by M. Ill and F. Barlier by means of different methods. Their paper was presented at this symposium.

References

1. Lozinsky, A. M.: In: Trajectories of Artificial Celestial Bodies, Proc. of COSPAR-Symp. Paris 1965, Berlin/Heidelberg/New York: Springer 1966.
2. Ill, M., Kassimenko, T. V.: In: Observations of Artificial Satellites of the Earth No. 7 (Sofia 1968) p. 87.
3. Almár, I.: In: Observations of Artificial Satellites of the Earth No. 7 (Sofia 1968) p. 111.
4. Horváth, A., Szabó, G.: In: Observations of Artificial Satellites of the Earth No. 5 (Bucuresti 1967) p. 176.
5. Illés-Almár, E., Almár, I.: Mitt. Budapest-Svabhegy No. 59 (1965).
6. Horváth, A.: In: Observations of Artificial Satellites of the Earth No. 8 (Budapest 1969) p. 187.

7. King-Hele, D. G.: Theory of Satellite Orbits in an Atmosphere, London: Butterworth 1964.
8. CIRA 1965 COSPAR International Reference Atmosphere, Amsterdam: North-Holland 1965.
9. Zadunaisky, P., Shapiro, I., Jones, H.: SAO Special Report No. 61, March 1961.
10. Stergin, C. G.: In: Handbook of Geophysics and Space Environments, p. 3–40, New York: McGraw-Hill 1965.
11. King-Hele, D. G., Walker, D.: Planet. Sci. 17 (1968) 197.

Correspondence to: Dr. I. Almár, Konkoly Observatory Budapest 114 Pf. 67.

The Upper Atmosphere and its Influence on Satellite Orbits

By

D. G. King-Hele

Royal Aircraft Establishment, Farnborough, Hants, England

Abstract. The chief variations in upper-atmosphere density are described, and it is shown how each of them affects the orbits of close Earth satellites. The main features discussed are:

1. the irregular variations in density dependent on solar activity, which may cause great changes in a few days, as well as a 27-day recurrence tendency;
2. the day-to-night and semi-annual variations in density, which cause nearly-periodic changes in orbits: for the day-to-night variation the period is usually between 2 months and 2 years, while the semi-annual effect leads to maxima of density in April and October, and minima in January and July; and
3. the large changes in density during a sunspot cycle, which give rise to a 10- or 11-year recurrence tendency.

The changes in orbital inclination caused by the rotation of the atmosphere are also briefly described.

1. Introduction

The analysis of changes in satellite orbits has enabled us to obtain a fairly complete picture of upper-atmosphere density and temperature at heights between 150 and 1200 km. The main results are embodied [1] in the COSPAR International Reference Atmosphere 1965, and in a revised version of CIRA soon to be prepared. In this paper, I shall describe the main features of the upper atmosphere and their effect on satellite orbits, as measured by the rate of change of orbital period, $\dot{T}$, which is also often called the decay rate. Air drag does, of course, affect elements other than the period (or semi-major axis), and particularly the eccentricity. But the changes in these other orbital elements can be related to the decay rate by means of orbital theory [2, 3] and will not be discussed here.

After a brief description of the upper atmosphere in Sec. 2, the changes in decay rate caused by variations in the atmosphere are discussed in Sec. 3, 4 and 5. The effects are distinguished by means of their time-scales – days, months or years. In Sec. 6 the effect of atmospheric rotation is described.

2. The Upper Atmosphere

In the upper atmosphere at heights between 150 and 1000 km, the maximum density and temperature are attained in the daytime during years when the Sun is active, and the minima occur at night in years when the Sun is inactive. Figs. 1 and 2 show likely maximum and minimum values of temperature and density, as given in Jacchia's 1969 model atmosphere [4]. The curves shown are for exospheric temperatures of 1600 °K, 1000 °K and 700 °K. The exospheric temperature of 1600 °K corresponds to maximum daytime values when the solar 10.7-cm radiation energy, $S_{10.7}$, is $220\times10^{-22}\,\mathrm{W\,m^{-2}Hz^{-1}}$, which was the average solar activity during the high solar maximum in 1958. The exospheric temperature of 700 °K corresponds to low solar activity, $S_{10.7} = 70\times10^{-22}\,\mathrm{W\,m^{-2}Hz^{-1}}$, and minimum night-time conditions. Also shown is an "average atmosphere", with an exospheric temperature of 1000 °K.

Fig. 1 shows that the temperature increases rapidly between 100 and 200 km height, and then levels off. The definition of temperature becomes dubious at heights above about 500 km, but in practice this causes little difficulty because the "temperature" is interpreted as the quantity that would represent temperature if the ordinary gas laws

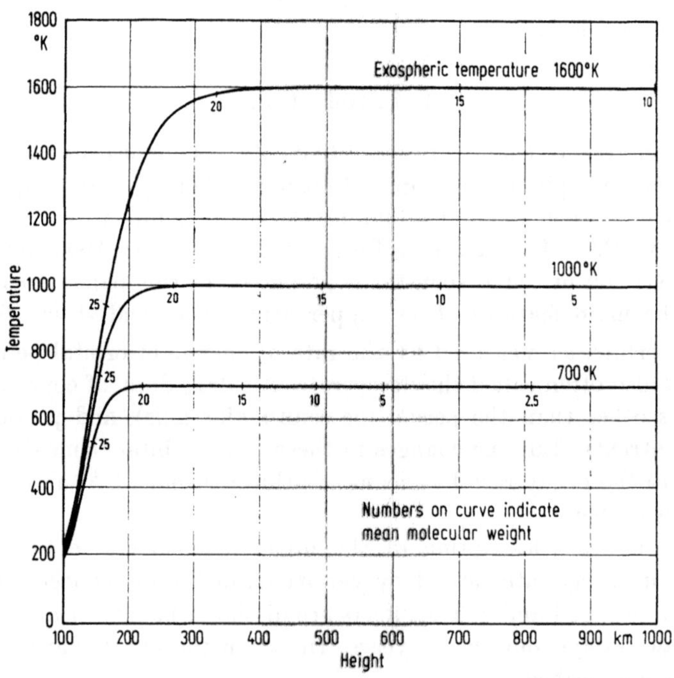

Fig. 1. Variation of temperature with height as given by Jacchia's model atmosphere[4].

applied. Also marked on the curves in Fig. 1 are the values of the mean molecular weight: at high exospheric temperatures atomic oxygen is dominant from 300 km up to nearly 1000 km; but at low exospheric temperatures helium becomes important at heights above 500 km and hydrogen is a major constituent of the atmosphere at heights above 800 km.

Fig. 2 shows the corresponding values of density given by the model and also, for comparison, some observational values of density obtained from analysis of satellite orbits [5, 6]. Fig. 2 is useful in showing the immense range of densities that may be encountered. The density decreases from approximately 10^{-6} kg/m³ at 100 km height to less than

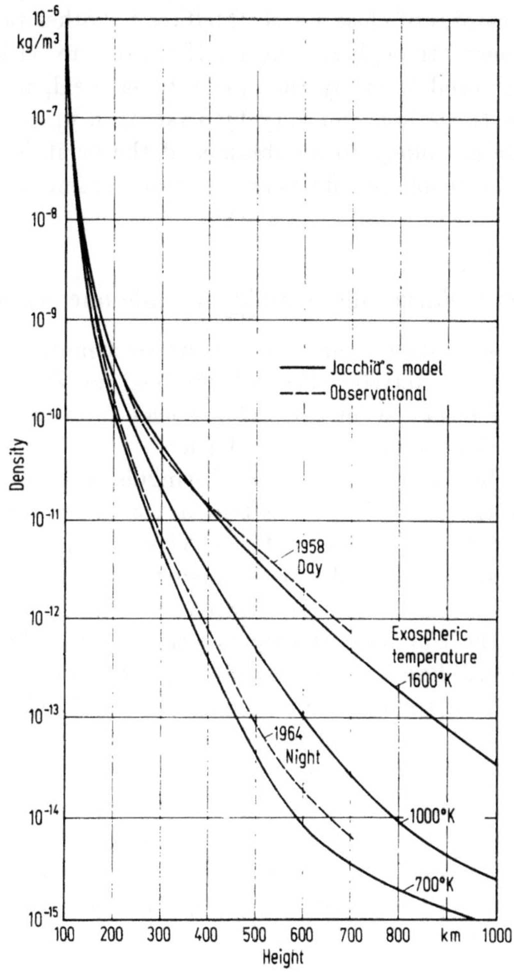

Fig. 2. Variation of air density with height given by Jacchia's model atmosphere[4], with observational values from analysis of satellite orbits.

10^{-14} kg/m^3 at 1000 km. At any particular height there are also very wide variations between the minimum night-time conditions at very low solar activity and the maximum daytime values at a strong solar maximum: at 500 km height there is a difference by a factor of 200. The slope of the lowest curve in Fig. 2 is less steep at heights above 500 km because of the greatly decreased molecular weight in these conditions, and the consequence is that the maximum changes in the atmosphere tend to occur at heights between 500 and 700 km. (The differences between the model and the observational values in Fig. 2 are probably not significant, because the observational values may not correspond to strictly constant exospheric temperatures.)

Because of the great variations in atmospheric density, everyone who studies the dynamics of close Earth satellites for other purposes should be aware of these atmospheric effects. If a satellite with a perigee of 500 km is being used to study the gravitational field, for example, the atmospheric perturbations can vary by a factor of up to 200, and more accurate results are likely to be obtained if the orbit is analysed at a time when the atmospheric effects are at their minimum.

3. Short-term Variations, with Time-scale one Month or Less

The major short-term variations in upper-atmosphere density are attributable to solar activity, through the heating of the upper atmosphere caused by either (a) streams of particles from the Sun, of (b) increases in the solar far ultra-violet radiation.

These two effects, (a) and (b), are distinguishable. The effect of the particles from the Sun shows itself in a correlation between upper-atmosphere density and geomagnetic activity. The peaks in density are closely paralleled by the peaks in the planetary indices of geomagnetic disturbances, K_p or A_p, which are controlled primarily by the influx of particles from the Sun into the upper atmosphere. The link between density and A_p is shown in Fig. 3, for Explorer 9 (1961 δ1) at 650 km [7], and for Secor 6 (1966-51B) at 190 km [8]. At a height of 650 km the effect is remarkably clear and large; the decay rate of Explorer 9 increased by a factor of 3 at the times of geomagnetic storms. (The three-hour geomagnetic index a_p has been used for Explorer 9 because of the excellent time resolution of the data.) At a height of 190 km the effect is smaller, with increases of density by up to about 40%, but the correlation is still quite obvious. The changes in density lag behind the changes in the geomagnetic index by about 6 h [9], and it is noticeable that after a strong magnetic disturbance the density often falls to a lower value than that prevailing before the disturbance.

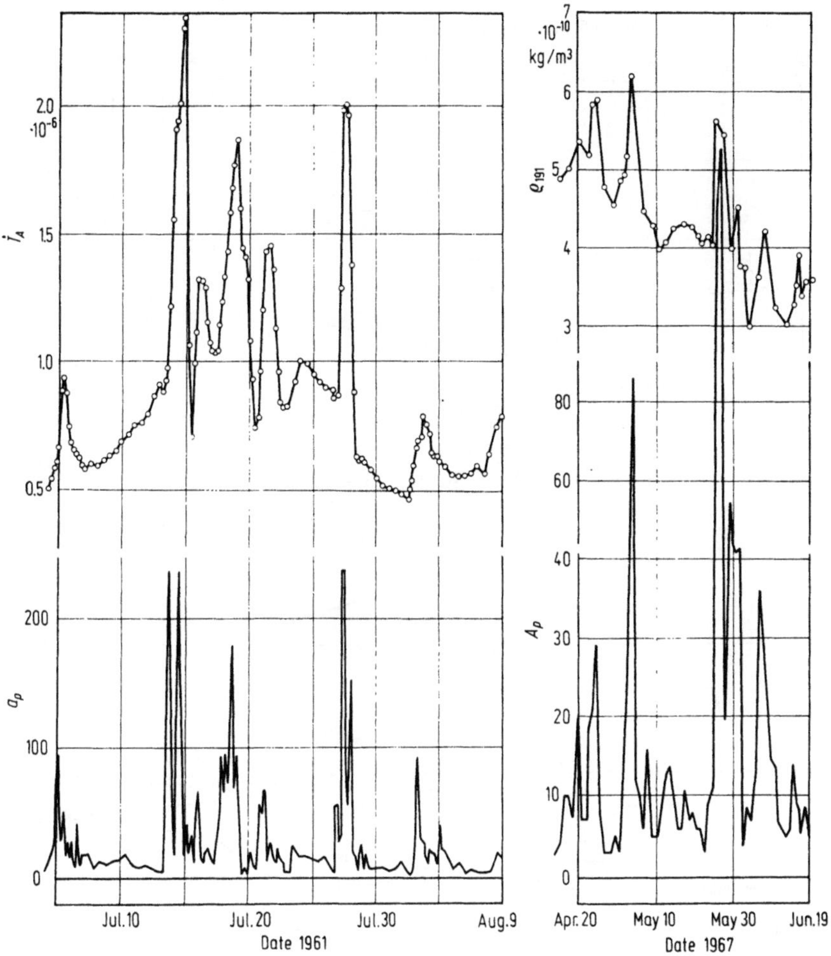

Fig. 3. Comparison of orbital decay rate or air density with geomagnetic indices a_p (3 hourly) or A_p (daily). (a) Orbital decay rate due to air drag, $\dot{T}_A$, for Explorer 9 (Ref 7); (b) Air density at 191 km height, ϱ_{191}, from Secor 6 (Ref 8).

The second main effect of the Sun on the upper atmosphere is the influence of solar radiation, chiefly at wavelengths between 100 and 1000 Å. Until this radiation can be accurately and regularly monitored by instruments aboard spacecraft, we have to rely on ground-level measurements of quantities that are believed to be correlated with extreme ultraviolet radiation. The best index has proved to be the solar 10.7-cm radiation, as measured by the N.R.C., Ottawa. The variations in air density and $S_{10.7}$ are frequently similar. In particular the 27-day recurrence tendency of the 10.7-cm radiation is well paralleled by the density. The response to variations with a time-scale shorter than 27 days is neither

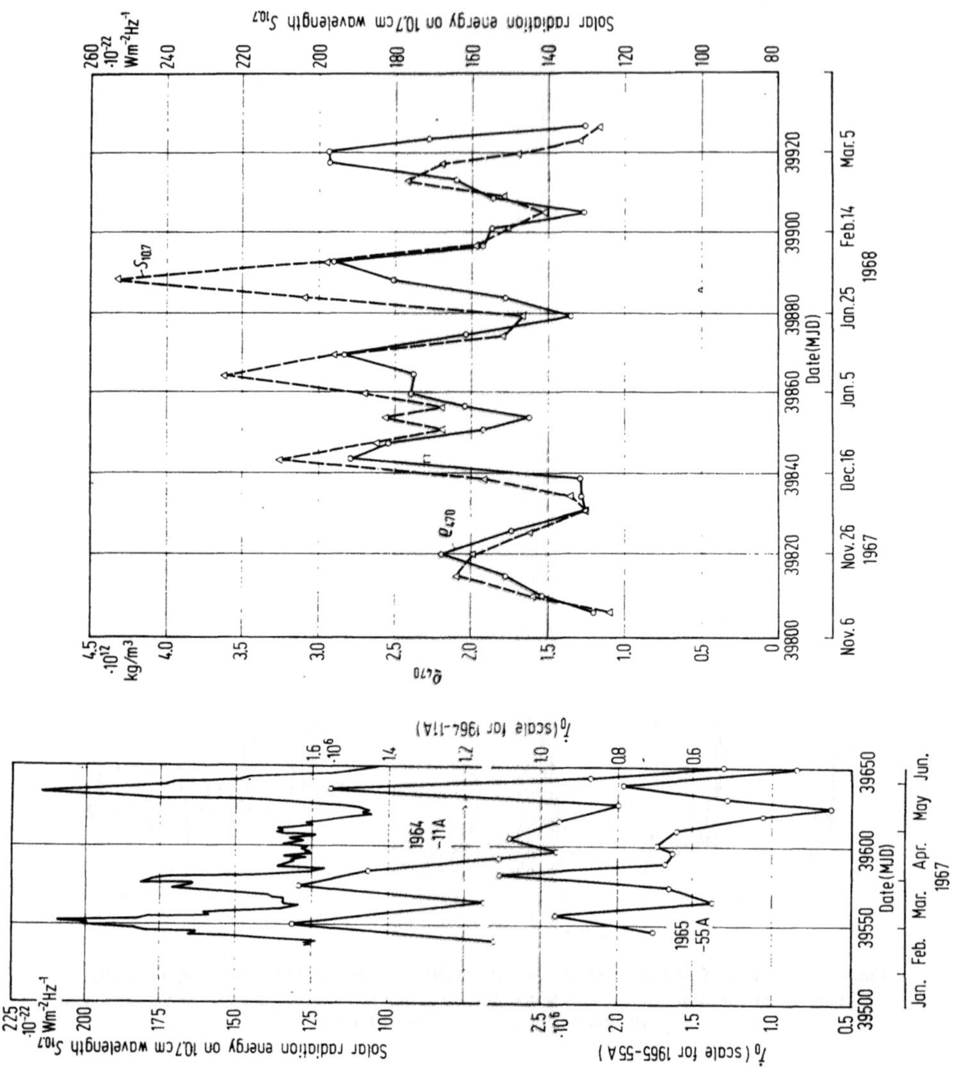

Fig. 4. Resemblance between solar activity ($S_{10.7}$) and orbital decay rate ($\dot{T}_0$) or air density (ϱ_{470}).
(a) $\dot{T}_0$ from satellites 1963-27 A and 1965-55 A (Ref 10); (b) Air density at 470 km, ϱ_{470}, from 1966-118 A (Ref 11).

so strong nor so consistent, but is still usually recognisable. Examples of variations correlated with solar activity, for satellites in near-circular orbits [10, 11] at heights near 500 km, are shown in Fig. 4. The increases in solar radiation quite often cause increases in density by a factor of 2, and the response to the solar radiation is somewhat variable, not only in magnitude but also in time. As Fig. 4b shows, the density follows $S_{10.7}$ with scarcely any time lag on some occasions, e.g. between MJD 39838 and 39860, but sometimes there is a time lag of up to 3 or 4 days. The average time lag is between 1 and 2 days [12]. The 27-day recurrence is particularly clear in Fig. 4b.

For the analysis of satellite orbits the important features of both Figs. 3 and 4 are that large changes in the decay rate, by a factor of 2 or more, can often occur within a few days. In the largest disturbance recorded (during November 1960), variations by a factor of 8 within one day were measured [14] at heights near 650 km. Outbursts such as this are rare and irregular, but the 27-day recurrence tendency, caused by rotation of the Sun, is present more often than not. The air density responds to solar activity at all heights [7–17] between 150 and 1200 km, but, as with the long-term influence, the maximum effect occurs at about 600 km height.

4. Variations with Time-scale between 2 Months and 5 Years

4.1. Semi-annual Variation

Two quite different features of the upper atmosphere give rise to "medium-term" variations with a period between 2 months and 5 years. The first is the semi-annual variation in density [17–19]: it is found that the density, after correction for the effects of other sources of variation, exhibits a semi-annual oscillation with maxima of density in early April and late October, and minima in mid-January and late July. The October maximum is usually higher than that in April; and the July miuimum is usually lower than that in January.

A good example of the semi-annual variation is shown in Fig. 5, obtained from analysis of Echo 2 (1964-O4A) at a height of 1130 km in 1964-5, when the effect was large. The values have been corrected to a fixed level of solar activity ($S_{10.7} = 100 \times 10^{-22}\,\mathrm{W\,m^{-2}\,Hz^{-1}}$), and have been smoothed by taking the running means of 3 successive values. Fig. 5 has one unusual feature: the April maximum is slightly higher than that in October; but otherwise it is a particularly clear example of the semi-annual variation, with the density changing by a factor of about 2 between the maximum and minimum. [19] provides a thorough recent review of the semi-annual variation, and further examples may be found in [8, 10, 11, 13, 25].

The cause of the semi-annual variation is still uncertain, but it extends down to heights as low as 90 km and seems likely to originate from variations in the lower atmosphere. The ratio R of maximum-to-minimum density in the semi-annual oscillation varies with height and from year to year: in a "strong" year, like 1967, R may increase from about 1.5 at 200 km to 3 at 500 km, and then decrease slowly to about 2 at 1000

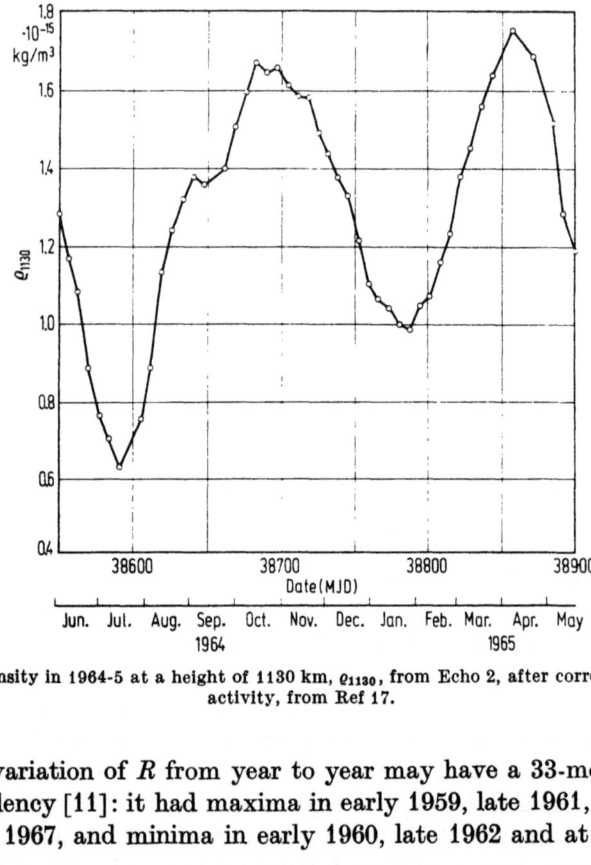

Fig. 5. Air density in 1964-5 at a height of 1130 km, ϱ_{1130}, from Echo 2, after correction for solar activity, from Ref 17.

km. The variation of R from year to year may have a 33-month recurrence tendency [11]: it had maxima in early 1959, late 1961, early 1964 and early 1967, and minima in early 1960, late 1962 and at the end of 1965.

Because of the presence of the semi-annual variation, it may sometimes be unwise to choose for analysis a satellite for which the perigee has a rotation period of 6 months, since the semi-annual variation may partly mask the effects of the motion of perigee.

4.2. Day-to-night Variation

The second feature of the upper atmosphere that gives rise to medium-term variations is the change in density between day and night. At heights between 200 and 1000 km the density increases during the morning as a result of solar heating, reaches a maximum between 14 and 16 ^{h} local time, then declines fairly steadily until midnight and reaches a flat minimum between midnight and dawn, before beginning its morning increase again [21, 22, 25]. The effect is small at a height of 200 km, increases to a high maximum at about 600 km, and then decreases again. Fig. 6 shows the day-to-night variation given by CIRA 1965 for medium

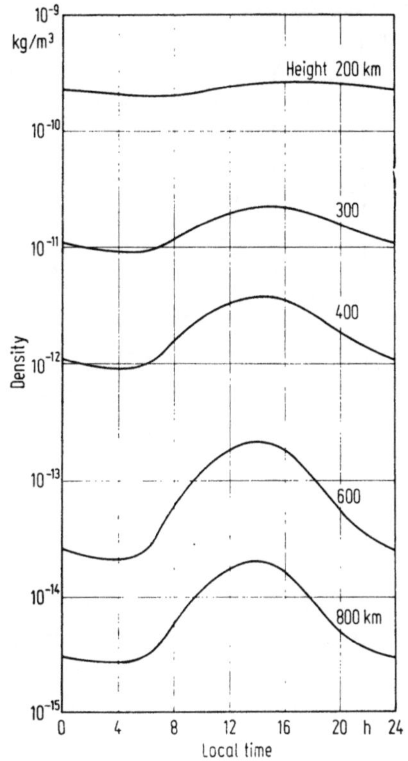

Fig. 6. Variation of upper-atmosphere density during the course of the day, given by CIRA 1965, model 3.

solar activity. CIRA 1965 indicates that the maximum daytime density exceeds the minimum night-time by a factor of 8 at a height of 600 km, though the actual day-to-night is now known to be somewhat smaller: Jacchia's recent model [4] gives a factor of 5 at 600 km for medium solar activity.

The perigee point of a satellite orbit usually takes between 2 months and 2 years to move through a complete day-to-night cycle; so the day-to-night variation in density shows itself as an oscillation having a period which is usually between 2 months and 2 years, but can be as long as 5 years if the satellite is close to the inclination (near 41°) for which perigee remains stationary relative to the Sun [23]. Fig. 7 shows the variation of orbital period for Vanguard 1 (1958 β 2, perigee 650 km) between 1958 and 1960. The effect of the day-to-night variation is very striking, although in fact it is somewhat intensified by the decline in solar activity between 1958 and 1960.

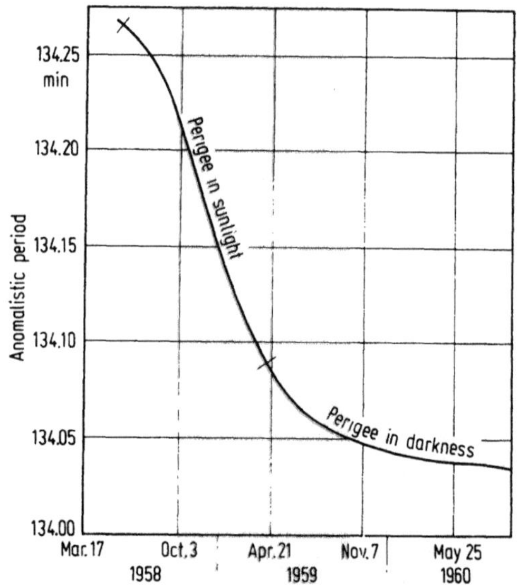

Fig. 7. Orbital period of Vanguard 1 between 1958 and 1960.

The day-to-night variation is a major influence on orbits with perigee heights between 300 and 800 km, and in analysing the orbit of Vanguard 1, for example, it would clearly be easier to recognise other perturbations if the analysis was made during 1959-60 rather than in 1958.

5. Long-term Variations, with Time-scale 10 Years

Since the atmosphere responds to solar activity, the 10- or 11-year sunspot cycle exercises a clear influence over the changes in satellite orbits. This source of variation, which for the last two solar cycles has had a period of almost exactly 10 years, is both the longest and the largest of the atmospheric variations. As Fig. 2 shows, the density in 1958 by day (exospheric temperature about 1600 °K) exceeded the density in 1964 by day (exospheric temperature about 900 °K) by a factor of about 20 at 500 km height.

The combined influence of the sunspot cycle and the day-to-night variation is nicely illustrated by Explorer 1 (1958α), which was launched

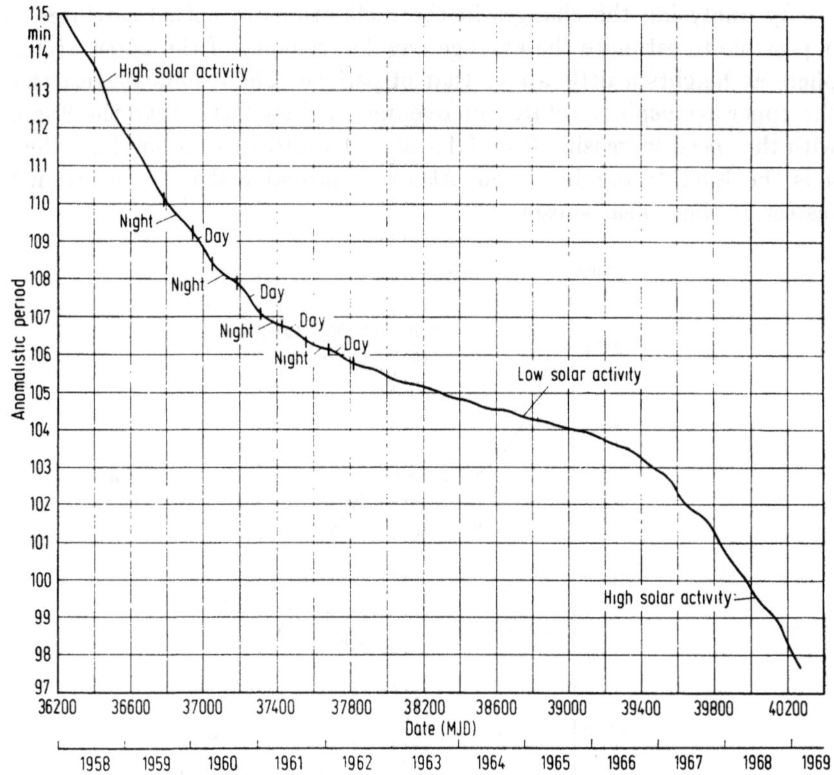

Fig. 8. Orbital period of Explorer 1 (1958 α) between 1958 and 1969.

in February 1958: Fig. 8 shows the variation of period for this satellite between 1958 and 1969. The period decreased rapidly during the years of high solar activity, 1958-60, then much more slowly during the low solar activity in 1962-65, and finally faster again in the recent solar maximum of 1968-9. Superposed on this general trend is a smaller oscillation due to the day-to-night variation in density, which at the perigee height of Explorer 1 (350 km) is considerably smaller than with Vanguard 1 (650 km) in Fig. 7.

6. The Effect of Atmospheric Rotation

If the Earth and atmosphere did not rotate, the aerodynamic force acting on a satellite would generally be tangential to the orbit and would not alter the orientation of the orbital plane. In fact the atmosphere rotates and exerts a small force normal to the orbital plane, which has the effect of slightly decreasing the inclination of the orbit to the equator (for an eastward-moving satellite).

By analysing the changes in the inclinations of actual satellites, it is possible to estimate the average angular velocity of the upper atmosphere at heights a little above that of perigee. The results suggest that the upper atmosphere rotates, on overage, slightly faster than the Earth, with the speed increasing from $1.1w$ at 210 km to $1.5w$ at 350 km, where w is the Earth's angular speed. Also it is probable that the rotation is fastest at times near sunset.

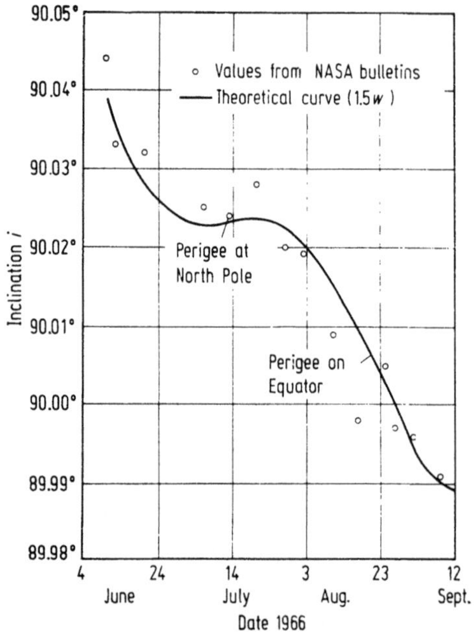

Fig. 9. Orbital inclination of 1966-51 C.

The changes in inclination i, caused by atmospheric rotation are proportional to the change in orbital period and also to $\sin i$, and are therefore greatest for polar orbits that suffer a large change in orbital period. An example of the magnitude of the effect is shown in Fig. 9, which gives values of inclination for ORS 2 (1966-51C) between June and September 1966, with the theoretical curve for an atmospheric rotational speed of $1.5w$. This is the highest rotational speed that has been obtained and occurs at a time when perigee was close to the sunset line. Fig. 9 shows how the effect of atmospheric rotation on i is much stronger when perigee is near the equator. Fig. 9 also gives an impression of the order of magnitude of the changes in i: for 1966-51C the change was only 0.05°, during a time when the orbital period decreased from 124 min to 116 min. Very few satellites exhibit a change of more than 0.1° in inclination as a result of atmospheric rotation.

References

1. CIRA 1965: COSPAR International Reference Atmosphere 1965, Amsterdam: North-Holland Publ. Co. 1965.
2. KING-HELE, D. G.: Theory of satellite orbits in an atmosphere, London: Butterworths 1964.
3. COOK, G. E., KING-HELE, D. G.: Phil. Trans. Roy. Soc. 259 (1965) 33–67.
4. JACCHIA, L. G.: Smithsonian Astrophys. Obs. Spec. Rpt. (1969).
5. KING-HELE, D. G., QUINN, E.: Journ. Atmos. Terr. Phys. 27 (1965) 197–209.
6. KING-HELE, D. G., QUINN, E.: Planet Space Sci. 14 (1966) 1023–1033.
7. JACCHIA, L. G., SLOWEY, J.: Smithsonian Astrophys. Obs. Spec. Rpt. No. 125 (1963).
8. KING-HELE, D. G., HINGSTON, J.: Planet Space Sci. 16 (1968) 675–691.
9. ROEMER, M.: Space Research VII, pp. 1091–1099; Amsterdam: North-Holland Publ. Co. 1967.
10. COOK, G. E.: Planet Space Sci. 16 (1968) 1161–1176.
11. KING-HELE, D. G., WALKER, D. M. C.: Planet Space Sci. 17 (1969) 197–215.
12. ROEMER, M.: Forschungsber. d. Astronom. Inst. Bonn No. 68-08 (1968).
13. COOK, G. E.: The semi-annual variation in the upper atmosphere during 1967 and 1968. R.A.E. Technical Report 69254, 1969.
14. JACCHIA, L. G.: Space Research II, pp. 747–750, Amsterdam: North-Holland Publ. Co. 1961.
15. KING-HELE, D. G., WALKER, D. M. C.: Planet. Space Sci. 17 (1969) 1539–1566.
16. PAETZOLD, H. K., ZSCHORNER, H.: Space Research II, pp. 958–973. Amsterdam: North-Holland Publ. Co. 1961.
17. COOK, G. E., SCOTT, D. W.: Planet-Space Sci. 15 (1967) 1933–1956.
18. JACCHIA, L. G., SLOWEY, J. W., CAMPBELL, I. G.: Planet. Space Sci. 17 (1969) 49–60.
19. COOK, G. E.: The Semi-annual Variation in the Upper-Atmosphere: A Review. Annales de Géophys., 25 (1969) 451–469.
20. HARRIS, I., PRIESTER, W.: Space Research III, pp. 53–75, Amsterdam: North-Holland 1963.
21. MAROV, M. YA.: Kosmicheskie Issledovaniya 6 (1968) 110–118.
22. JACCHIA, L. G., SLOWEY, J.: Planet Space Sci. 16 (1968) 509–524.
23. KING-HELE, D. G., WALKER, D. M. C.: Space Research II, pp. 918–957, Amsterdam: North-Holland 1961.
24. KING-HELE, D. G., SCOTT, D. W.: Planet Space Sci. 15 (1967) 1913–1931.
25. MAROV, M. YA.: Space Research X (to be published).

Radiation Pressure Effects in the Motion of Artificial Satellites

By

L. Sehnal

Astronomical Institute, Czechoslovak Academy of Sciences, Prague, ČSSR

Abstract. The problem of the perturbation analysis of the radiation pressure effects has its difficulties of both physical and mathematical nature. The process of transfer of the radiation momentum on a satellite, the changing position of a satellite body, the properties of its surface and other circumstances must be taken into account in detailed analysis. Knowing the physical conditions, the computing of the changes of satellite motion is made difficult by several discontinuities, the effect of the Earth's shadow being the most important. This can be described by a special shadow function, giving satisfactory results at least by computing the short-periodic perturbations.

The effect of the re-radiation of the Earth is still described best by numerical integration regarding the uneasy description of the radiation sources. The dynamical effect of the direct radiation (Poynting-Robertson effect) can be computed with sufficient accuracy by approximative analytical methods, since its magnitude does not reach significant values.

The interest in the study of the effects of the radiation pressure was evoked mainly after the launching of baloon satellites and now one can say that it is possible to describe with relatively good precision the changes of elements caused by this effect, provided that we know the physical constants, which determine the magnitude and direction of the disturbing force.

According to the origin of the radiation incident on a satellite, we can more or less separately study the action of the direct solar radiation, the effects of the radiation scattered or emmitted from the Earth and the so called Poynting-Robertson effect.

The Direct Solar Radiation

The mathematical description of the pressure of the direct solar radiation is not too complicated, if we restrict ourselves to the cases in which the satellite does not penetrate the Earth's shadow. The method of deriving the perturbational equations can be different: The derivation of the perturbation of different vectorial elements [MUSEN, 1960], the modification of the Hamiltonian function [BROUWER, 1962], [HORI, 1966], the usage of the Kryloff-Bogolyuboff method [BRYANT, 1961], using the von Zeipel's method [VINTI, 1964] or a simple establishing of the magnitudes of the disturbing acceleration [KOZAI, 1961]. Let us make a use of the last method to describe shortly the simple mathematical description of the whole phenomenon.

The Lagrangian equation of the changes of the orbital Keplerian elements have generally this form:

$$\frac{d\sigma}{dt} = K_\sigma \cdot f(S, T, W).$$

The Greek letter σ stands for an arbitrary element, K_σ is a constant depending on the physical conditions and S, T, W are the rectangular components of the disturbing acceleration (S – the radial component, T – the component perpendicular to the radius-vector, W – the binormal component, perpendicular to the orbital plane). Those components can be expressed as functions of the orbital elements and a time, or an anomaly. E.g. the radial component S can be written as

$$S = S_0 \cos v + T_0 \sin v \qquad (v\text{—true anomaly})$$

where S_0 depends on the mutual position of the satellite's orbit and the Sun, and can be expressed by formula:

$$S_0 \sim \frac{\sin}{\cos} i \frac{\sin}{\cos} \varepsilon \frac{\sin}{\cos} (\omega \pm \Omega \pm \lambda) \tag{1}$$

where we have i for the orbital inclination, ε for the obliquity of the ecliptic, ω for the argument of perigee, Ω for the longitude of the ascending node and λ for the mean longitude of the Sun.

The Resonance Effects

Provided the three quantities in the bracket (ω, Ω, λ) are constant, the equation can be easily solved, but owing to the Earth's asphericity, ω and Ω change relatively fast and even the apparent motion of the Sun described by λ cannot be neglected. However, we can with good precision take the speed of the changes ($= \dot{\omega}, \dot{\Omega}, \dot{\lambda}$) as a constant, and thus we have here the resonance cases, which arise if some of the combinations $\dot{\omega} \pm \dot{\Omega} \pm \dot{\lambda}$ approaches zero.

Since the speed of the changes of the argument of perigee and the ascending node is caused mainly by the second harmonics in the development of the Earth's potential, we can find such a combination of elements, in which a resonance occurs, putting the expression in the bracket in Eq. (1) equal zero. Expressing the changes of the elements by the combination of the initial elements itself, we obtain a relation, which Hori [1966] writes as

$$\left(\frac{a}{R}\right)^{7/2}(1-e^2)^2 = j\,\frac{3}{4}\,\frac{J_2}{\nu_0}(1 + k\cdot 2\cos i - 5\cos^2 i)\,,$$

J_2 being the second harmonics of the Earth's potential, a, e, i are the semi-major axis, the eccentricity and the inclination of the satellite orbit respectively, R is the mean equatorial radius of the Earth, ν_0 depends on the mean longitude of the Sun and k and j are the parameters which equal either 1 or zero. If we take $\cos i$ as the ordinate, we can plot a graph Fig. 1 consisting of six parabols, which correspond to different

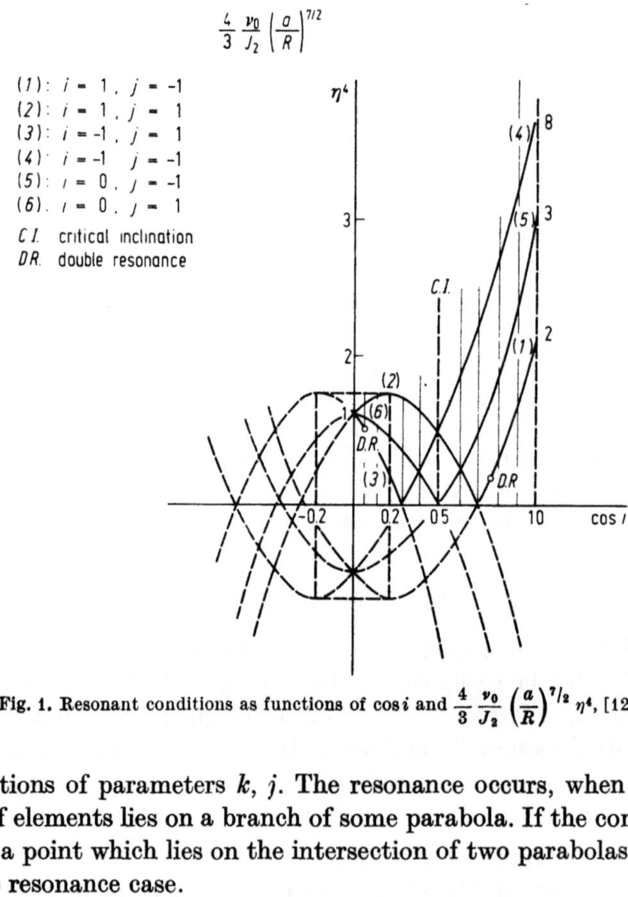

Fig. 1. Resonant conditions as functions of $\cos i$ and $\frac{4}{3}\frac{\nu_0}{J_2}\left(\frac{a}{R}\right)^{7/2}\eta^4$, [12].

combinations of parameters k, j. The resonance occurs, when a combination of elements lies on a branch of some parabola. If the combination gives us a point which lies on the intersection of two parabolas, we have a double resonance case.

Each parabola can be again divided into several graphs, in which we can immediately find out the resonance case directly from the values of the elements, Such graphs were given by Polyachova [1963], and

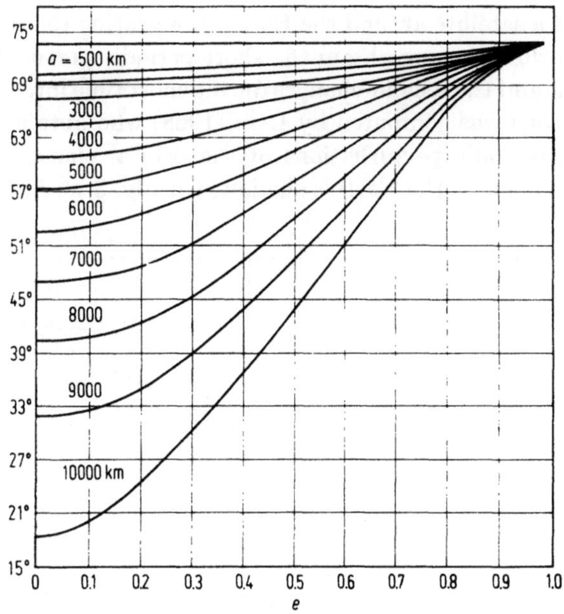

Fig. 2. Critical inclinations for resonant condition $\dot{\omega} + \dot{\Omega} = \dot{\lambda}$, [23].

from them, on Fig. 2, the probably most important case of resonance, given by $\dot{\omega} + \dot{\Omega} = \dot{\lambda}$, is shown. In this case the perigee of the orbit follows the Sun. Its influence can change over the lifetime of the satellite, sometimes by a factor 10, as was shown by PARKINSON [1960] et al.

The analytical solutions of the resonance effects were given in several papers, e.g. by MUSEN [1960], who proceeded a similar method as BROWN in his planetary theory or by COOK [1963], who solved the problem using his own simplyfied theory for the low-eccentricity orbits. All results were confirmed in praxis, probably most conspicuously in the case of the orbits of West-Ford dipoles, as was shown by SHAPIRO [1964, 1966] and his collaborators.

The satellites of a Moon will be also affected by the radiation pressure effects; the only resonance for their orbits is possible right only in case of the equality of the change of the longitude of the perigee with Sun's motion. But, as was shown by POLYACHOVA [1964] this occurs only if the satellite is farther than 10 Moon's radii from its surface, and this comes in the region of instable orbits.

The Short-periodic Perturbations

The values of the orbital elements and the position of the Sun can be, however, considered as constant during a short time interval of one revolution of a satellite around the Earth. Integrating the perturbational equations in this case, we obtain the short-periodic perturbations of the satellite orbit and the secular changes during short time interval. This kind of perturbations was computed by LÁLA [1968] who obtained the curves of the short-periodic perturbations of the orbital elements, neglecting again the influence of the Earth's shadow. On Fig. 3 we have the change

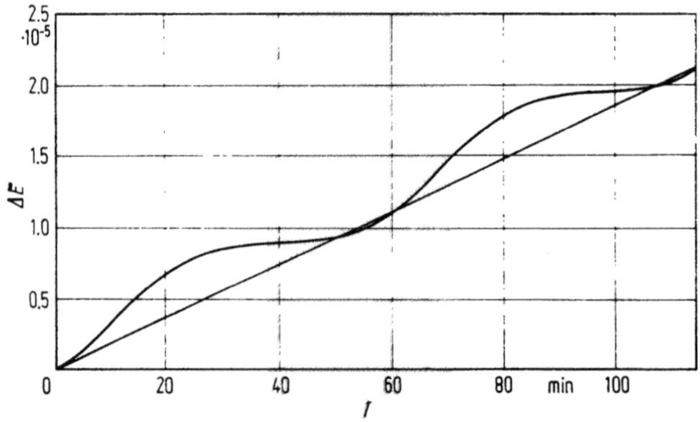

Fig. 3. Short-periodic perturbations of eccentricity during one orbital period for Echo 1 (epoch – 38445 MJD). Shadowing effect is not included [15].

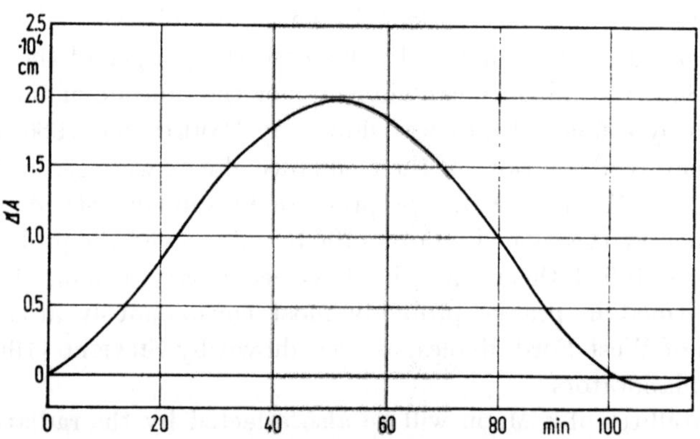

Fig. 4. Short-periodic perturbations of semi-major axis during one orbital period for ECHO 1 (epoch – 38445 MJD). Shadowing effect is not included [15].

of the eccentricity of the orbit of satellite Echo 1 (epoch 38445 MJD) during one revolution of its orbit. Fig. 4 shows the change of semi-major axis of the same orbit. This results were confirmed by comparing with the numerical integration of the equation of motions, computed e.g. by BARLIER [1964], and by comparison the secular changes for those cases with observed values.

The Effect of the Earth's Shadow

Investigating analytically the radiation pressure effects, we meet on great obstacle, caused by the Earth's shadow. Actually, we can determine the shadow boundary solving a quartic, and then we can obtain the changes of the elements by numerical integration, ignoring the part of the radiation pressure on the orbit which lies in the Earth's shadow. But as was shown by WYATT [1961] or KOZAI [1961] the equation for the shadow boundary cannot be simply expressed in terms of an orbital anomaly, even by development in series, so that an analytical solution of the problem is not possible in that way.

However, the influence of the Earth's shadow can be described by another method, introducing the so called shadow function. This function, which equals one if the satellite is illuminated by the Sun, and is zero if the satellite is in the shadow, was introduced by FERRAZ-MELLO [1963, 1964], If we multiply by such a function the right sides of the disturbing equations, we obtain a good analytical description of the whole effect. This function can be considered as being periodic over one revolution of the satellite, thus being developable in Fourier series. FERRAZ-MELLO gives the theoretical derivation of the equations, but if we want to use his development in praxis, we find out that the series converges very slowly and the computation is very labourious. Another method of forming such a shadow function was used by LÁLA and SEHNAL [1969], which gives practically applicable results. Thus, we can obtain the courses of the short-periodic changes of the orbital elements, taking into account the Earth's shadow. As an example, we have on Fig. 5 the comparison of the short-periodic changes of the eccentricity of the satellite Echo 1 (epoch 38445 MJD), without the shadow (dashed line), and taking the shadowing effect into account (full line). Moreover, one obtains also the secular changes of the semi-major axis and thus the changes of period, which vanish of we ignore the shadow. Investigating analytically the changes of period during one revolution of the satellite, we can see that the change depends also on the mutual position of the satellite's orbit and the Sun. WYATT [1961] e.g. distinguishes several typical cases and shows, that in case that the apsidal line goes through the centre of the part of the orbit, which lies in the shadow, this being the symmetrical case, the change of period vanishes. The same holds, of course, for a circular orbit.

The Terrestrial Radiation

The disturbing acceleration in this case is, of course, substantially smaller than that, caused by the direct solar radiation. The highest esti-

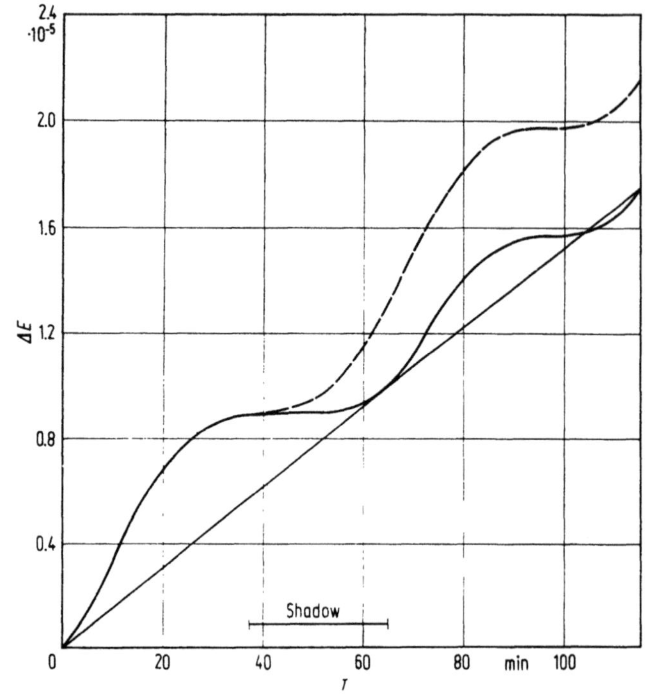

Fig. 5. A comparison of short-periodic perturbations of eccentricity with (full line) and without (dashed line) shadowing effect. Position of the shadow is for orientation only [16].

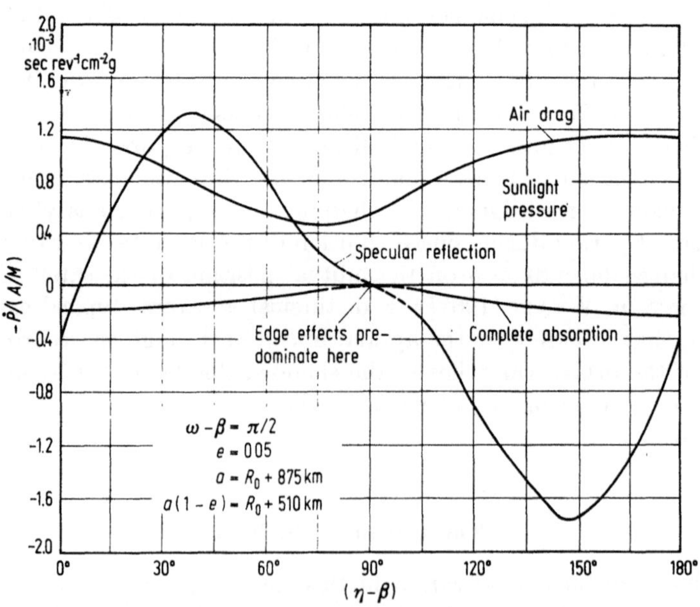

Fig. 6. Changes in the orbital period of a spinning disk as a function of its orientation [34].

mates give its value as to be 25% of the direct radiation acceleration. Generally, we can the origin of the radiation divide into three parts: 1. the thermal radiation, 2. the specularly reflected component of the solar radiation, 3. the diffusly reflected component.

The thermal radiation of the Earth is believed to be of lower importance mainly because of the symmetry of the radiation. The resulting force acts like diminishing the Earth's gravitational attraction by a negligible amount.

The specular component of the reflected radiation is obviously also small; the usual estimates give its value as 10% of the total reflection. Besides, it is very difficult tractable mathematically, since its value depends to a great extent on the changing reflectivity properties of the Earth's surface.

Thus, the diffusely reflected radiation remains as the main component of the disturbing forces of the terrestrial radiation. The problem was studied mainly by numerical integration of disturbing equations [Baker, 1965, Shapiro, 1962] and several authors [Wyatt, 1962, Sehnal, 1962, 1964, 1965] tried to describe it analytically, establishing some more or less realistic models of the albedo radiation. The radiation is usually supposed to behave according to the Lambert's law, even if this is not exactly correct. Another problem arises here with the description of the magnitude and direction of the disturbing acceleration, when the satellite moves partly above the illuminated and partly above the dark hemisphere of the Earth. The quantitative results show us, that the change of period, caused by that effect, does not go even by balloon satellites higher the 10% value of the effect, caused by the pressure of direct solar radiation. However, by precise orbital analysis the effect of terrestrial radiation pressure must be taken into account. E. G. Newton [1968] in his detailed analysis of satellite motion found the changes of the longitude of ascending node and of the inclination, which are, of course, caused by the binormal component of the disturbing acceleration. He found the perturbations with a half-year period and an amplitude smaller than 1″. In those cases, the satellites were those with relatively small area-to-mass ratio.

Generally one can say, that up till now, there was found no separate component of observed changes of orbital elements, which could be explained by the action of the terrestrial albedo radiation.

The Poynting-Robertson Effect

There is one more effect, which must be ascribed to the action of radiation, it is the so called Poynting-Robertson effect, which is conditioned on the nonvanishing velocity of the satellite in a given reference frame and has a relativistic background [ROBERTSON, 1937]. The disturbing acceleration can be described mathematically relatively easily.

Generally, the effect caused by this disturbing acceleration is very small, even in case of the Echo 1 satellite the change of the semi-major axis during one revolution of the satellite does not surpass 1 cm and the changes of the eccentricity are of the order 10^{-8}. In a longer period of time this effect could cause the displacement of the orbit in the direction from the Sun, thus forming the perigee on the sunlit side of the Earth.

The Physical Conditions of the Radiation Effects

Now, from what has been said here, we can see, that it is possible to describe relatively precisely the main effects caused by the radiation pressure, if we know, however, the initial physical conditions. Some of the presumptions are only approximations, and this refers mainly to the time invariance of some physical quantities, such as:

1. solar flux;
2. the distance Sun – satellite;
3. the direction Sun – satellite;
4. Earth's albedo;
5. satellite reflection properties;
6. the A/M ratio (area-to-mass ratio of the satellite).

But especially the phenomena, connected with the transfer of the momentum of the incident photons on the satellite's body, are not too simple. Usually, it is supposed that the radiation pressure acts in the direction from its origin. Actually, we have more components of the total pressure acceleration:

a) The absorption component – in the direction from the source;

b) the emmission component – in the direction of the normal to the satellite's surface;

c) the reflection component – depends on the type of the reflection and on the reflectivity properties of the surface.

The detailed explanation of those effects can be seen e.g. in BAKER, Astrodynamics [1967].

Thus, the precise determination of the magnitude and direction of the disturbing acceleration depends on the precise knowledge of the properties of the satellite's surface, its shape and orientation in space. All this is, however, simplified in case of spherical satellites, where the disturbing acceleration acts always in the direction from the Sun. The influence of the orientation of non-spherical satellite was studied by SHAPIRO [1963] by numerical integration, on the examples of a desk and a cylinder. E.g., from one of his graphs (Fig. 6) we can see, that the orientation, given by angle $(\eta - \beta)$, can be neglected in case of absorption, but when specular reflection arises, the influence of the orientation becomes substantial.

The effects of radiation pressure were up till now studied mainly from a somewhat negative point of view, e.i. as an effect, which must be subtracted from the observed orbital changes, to obtain most precisely the values of physical constants, which determine the other disturbing phenomena – the shape of the Earth, the air density. Let us express a hope, that may be now it will be able to determine from the observed orbital perturbation also the values of those physical quantities, which determine the radiation pressure effects themselves, such as the solar flux, Earth's albedo, etc.

References

BAKER, R. L. M.: Astrodynamics, p. 186, New York/London: Academic Press 1967.

BAKER, R. L. M.: Radiation on a Satellite in the Presence of Partly Diffuse and Partly Specular Reflecting Body. In: Trajectories of Artif. Celestial Bodies. Symp. Paris 1965; Berlin/Heidelberg/New York: Springer 1966.

BARLIER, F.: Recherches sur la détermination et l'utilisation des éléments osculateurs dans la mécanique des satellites proches. Bull. Astr. Tome XXV, Fasc. 4 (1965) 326.

BROUWER, D.: Analytical Study of Resonance Caused by Solar Radiation Pressure. Dynamics of Satellites, IUTAM Symp., Paris 1962, p. 34; Berlin/Göttingen/Heidelberg: Springer 1963.

BRYANT, R., BAILIE, A., MUSEN, P.: Perturbations in Perigee Height of Vanguard 1. Astron. J. 65 (1969) 342.

BRYANT, R. W.: The Effect of Solar Radiation Pressure on the Motion of an Artificial Satellite. Astron. J. 66 (1961) 430.

BRYANT, R. W.: A Comparison of Theory and Observations of the Echo 1 Satellite. J. Geoph. Res. 66 (1961) 3066.

COOK, G. E., HUGHES, JANICE M.: The Orbits of Needle Satellites. Planetary and Space Sci. 9 (1962) 153.

COOK, G. E.: The Use of Simplified Orbital Theory for Satellites of Large Area-to-Mass Ratio. Planetary and Space Sci. 11 (1963) 1289.

FERRAZ-MELLO, S.: Action de la Pression de Radiation sur le Mouvement d'un Satellite Artificial de la Terre, p. 41, XIVe Congress IAF, Paris 1963.

FERRAZ-MELLO, S.: Sur le problème de la pression de radiations dans la théorie des satellites artificiels. C. R. Acad. Sc. Paris, t. 258, 463–466 (1964).

HORI, G. I.: The Effect of Radiation Pressure on the Motion of an Artificial Satellite. Space Math. Part. 3. Amer. Math. Sec. 1966, 167–168.

JASTROW, R., BRYANT, R.: Variations in the Orbit of the Echo Satellite. J. Geoph. Res. 65 (1960) 3512.

KOZAI, Y.: Effect of Solar Radiation Pressure on the Motion of an Artificial Satellite. Smithsonian Astroph. Observ. Spec. Rep. 56 (1961).

LÁLA, P.: Short-periodic Perturbations of the Satellite Orbits Caused by Solar Radiation Pressure. Bull. Astr. Czech. 19 (1968) 233.

LÁLA, P., SEHNAL, L.: The Earth's Shadowing Effects in the Short-Periodic Perturbations of Satellite Orbits. Bull. Astr. Czech. 20 (1969) 327.

LEVIN, E.: Solar Radiation Pressure Perturbations of Earth Satellite Orbits. AIAA J. 6 (1967) 120.

MUHLEMAN, D. O., HUDSON, R. H., HOLDRIDGE, D. B., CARPENTER, R. L., OSLUND, K. C.: Observed Solar Pressure Perturbations of Echo 1. Science 132 (1960) 1487.

MUSEN, P.: The Influence of the Solar Radiation Pressure on the Motion of an Artificial Satellite. J. Geoph. Res. 65 (1960) 1391.

NEWTON, R. R.: A Satellite Determination of Tidal Parameters and Earth Deceleration. Geophys. J. 14 (1968) 505.

PARKINSON, R. W., JONES, H. M., SHAPIRO, I. I.: Effects of Solar Radiation Pressure on Earth Satellite Orbits. Science 131 (1960) 920.

PARKYN, D. G.: Analysis of Satellite Data. Nature 191 (1961) 896.

POLYACHOVA, E. N.: Radiation Pressure and the Motion of Earth's Satellites. Bulletin of Institute of Theoretical Astronomy, Leningrad (in Russian) 9 (1963) 15.

POLYACHOVA, E. N.: Radiation Pressure and the Motion of Lunar Satellites (in Russian). Bull. Inst. Theoret. Astron. Leningrad 9 (1964) 440.

ROBERTSON, H. P.: Dynamical Effects of Radiation in the Solar System. M. N. 97 (1937) 423.

SEHNAL, L.: The Dynamical Effects of the Solar Radiation Pressure in the Motion of Artificial Satellites. Bull. Astr. Czech. 14 (1963) 133.

SEHNAL, L.: The Influence of the Re-radiation of the Earth on the Motion of the Artificial Satellites. In: The Theory of Orbits in the Solar System and in Stellar Systems. Symposium IAU No. 25, Thessaloniki, 1964 (Academic Press).

SEHNAL, L.: The Perturbations of the Orbital Elements Caused by the Pressure of the Radiation Reflected from the Earth. Trajectories of Artificial Celestial Bodies. COSPAR-Symp. Paris 1965; Berlin/Heidelberg/New York: Springer 1966.

SEHNAL, L.: The Use of Artificial Satellites for Geodesy. Symposium, Washington, 1962. Amsterdam: North-Holland Publish. Comp.

SHAPIRO, I. I., JONES, H. M.: Perturbations of the Orbit of the Echo Ballon. Science 132 (1960) 1484.

SHAPIRO, I. I., JONES, H. M.: Loss of Mass in Echo Satellite. Science 133 (1961) 579.

SHAPIRO, I. I., JONES, H. M.: Science 134 (1961) 973.

SHAPIRO, I. I.: The Prediction of Satellite Orbits. In: Dynamics of Sattelites, IUTAM-Symp. Paris 1962; Berlin/Göttingen/Heidelberg: Springer 1963.

SHAPIRO, I. I.: Effects of Sunlight Pressure on Air Density Determinations Involving Cylindrical Satellites. J. Geophys. Res. 68 (1963) 5349.

SHAPIRO, I. I., JONES, H. M., PERKINS, C. W.: Proceedings IEEE 52 (1964) 496.

SHAPIRO, I. I.: Last of the West Ford Dipoles. Science 154 (1966) 1445.

SHAPIRO, I. I.: Effects of a Constant Force on a Keplerian Orbit. In: The Theory of Orbits in the Solar System and in Stellar Systems; Sympos. IUA No. 25, Thessaloniki (1964) 355, New York: Academic Press.

WYATT, S. P.: The Effect of Radiation Pressure on the Secular Acceleration of Satellites. Smithsonian Astroph. Observatory Special Report 60 (1961).

WYATT, S. P.: The Effect of Terrestrial Radiation Pressure on Satellite Orbits. In: Dynamics of Satellites, IUTAM-Sympos. Paris 1962, p. 180; Berlin/Göttingen/Heidelberg: Springer 1963.

Air Drag and Solar Radiation Pressure Effects on Close Earth Satellites

By

L. Sehnal

Astronomical Institute, Ondřejov, Czechoslovakia

and

E. M. Gaposhkin

Smithsonian Institution, Astrophysical Observatory, Cambridge, MA, USA

Abstract. Air drag and radiation pressure cause perturbations on satellites which must be evaluated before the most accurate ephemeris can be calculated. The results of using an analytical development for these perturbations will be given. The development uses Keplerian elements and a seven parameter Jacchia atmosphere density model.

This paper will be published in full in one of SAO Special Reports.

Effets de L'Atmosphère et de la Pression Réémise a Haute Altitude

By

M. Rapaport

Observatoire de Bordeaux, Floirac, France

and

M. Rousseau

Observatoire de Bordeaux, Floirac, France

and

Mme Lauret

Observatoire de Meudon, Meudon, France

Abstract. We study the effects of the atmosphere and of the radiation pressure reflected by the Earth, on very high satellites (altitude 2500 to 3000 km). PAGEOS A and 63.30D are the satellites that have been used. We have chosen times when the direct solar radiation pressure has no effect and we try to estimate respectively the influence of the atmosphere and of the reflected radiation pressure, noticing that the second of these factors may have a total effect which vanishes on part of the interval of time which is considered.

I.

Les effets gravitationnels ne produisent pas d'effets séculaires ou à longue période sur la période d'un satellite artificiel.

L'atmosphère et la pression de radiation solaire produisent des effets séculaires ou à longue période sur la période, ce qui facilite leur mise en évidence.

Tant que l'altitude du périgée reste inférieure à 800–1000 km, c'est l'effet de l'atmosphère qui est le plus important. En plus des deux phénomènes cités, il y a la pression de radiation réémise par la terre, qui peut atteindre 15% de la pression directe. Jusqu'à 800 km, on a donc par effet décroissant sur la période, l'atmosphère, la pression directe, la pression réémise.

Quand l'altitude du périgée augmente, l'ordre se modifie, et aux altitudes que nous allons considérer, la pression de radiation directe est le phénomène prédominant, les deux autres sont plus faibles, et ont des ordres de grandeur équivalentes.

Il y a une circonstance favorable à l'étude des deux derniers effets, à savoir que, lorsque le satellite reste constamment éclairé, la pression de radiation directe n'agit pas sur la période orbitale. Celle-ci n'est donc modifiée que par l'atmosphère et la pression réémise qui ont une action faible, et durant ces périodes, la période varie peu.

Ceci est illustré par le tracé de courbes relatives aux satellites PAGEOS A et 63.30D, représentant les variations du demi grand axe a en fonction du temps. Fig. 1 et 2.

a varie beaucoup lorsque la pression de radiation solaire agit sur lui, c'est à dire lorsque le satellite passe dans la zone d'ombre de la terre.

Les paliers correspondent à un éclairement constant du satellite. Seules l'atmosphère et la pression réémise agissent alors sur a.

Sur la Fig. 4, nous représentons les variations du demi-grand axe en fonction du temps pour 63.30D, courbe (3). Elles correspondent à une époque où le satellite passe dans zone d'ombre de la terre.

La courbe (1) représente les variations théoriques de a dues à la pression de radiation directe, calculées à l'aide d'un programme mis à notre disposition par M. Barlier, de Meudon.

La courbe (2) représente les variations théoriques des a si on tient compte de la pression réémise. Nous en avons calculé l'effet en utilisant les formules de Wyatt [1]. L'accord avec la courbe (3) est meilleur.

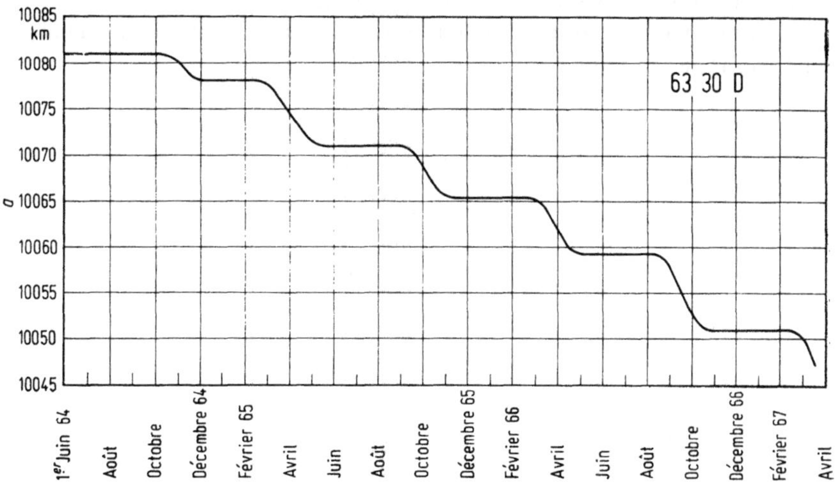

Fig. 1. Variations du demi-grand axe a en fonction du temps.

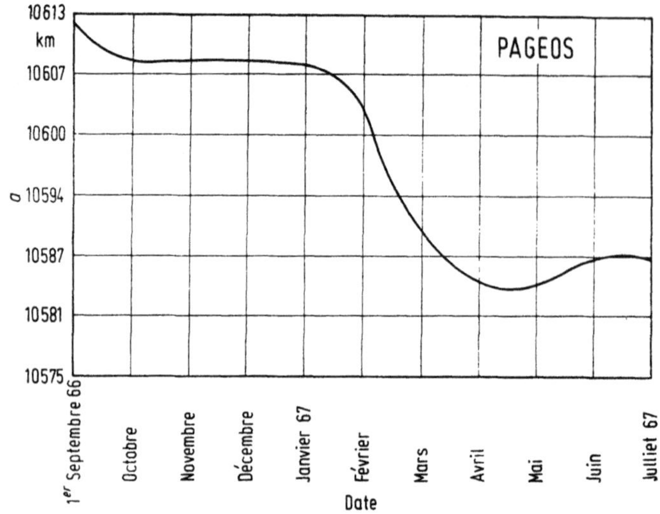

Fig. 2. Variations de *a* en fonction du temps.

Pour donner une idée des ordres de grandeur des 3 effets, nous avons fait le calcul pour 63.30D sur une période de 1 mois, entre le 13 septembre et le 13 octobre 1966. On obtient:

Effet calculé de la pression directe sur a: -3200 m
Effet calculé de la pression réémise sur a: -500 m
Effet calculé de l'atmosphère: -20 m

le perigée étant à une altitude de 2500 km.

Ceci montre bien, que dans de telles situations, l'effet de l'atmosphère est très faible, en valeur relative.

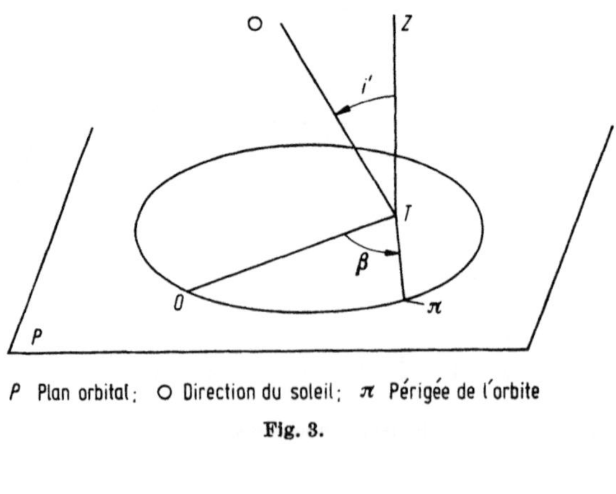

P Plan orbital; O Direction du soleil; π Périgée de l'orbite

Fig. 3.

II. Le satellite reste éclairé

L'accélération séculaire $\Delta P/P$ est alors nulle.

On peut voir facilement que lorsque le satellite reste éclairé la distance zénithale du soleil vue du plan de l'orbite du satellite est assez faible.

Rappelons que selon les formules de WYATT [1] l'accélération séculaire due à la pression réémise est traduite par des formules qui contiennent $\sin i' \sin \beta$ en facteur, β étant l'angle de la projection de la ligne

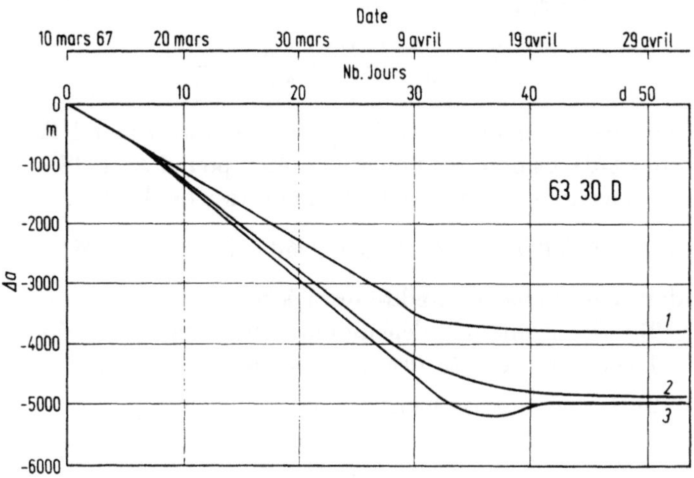

Fig. 4.

terre – soleil, sur le plan orbital et de la direction terre – périgée π de l'orbite du satellite (Fig. 3) i' n'étant déjà pas très grand, l'effet de la pression réémise sera faible. Ces époques se retrouvent bien dans les paliers des courbes des Fig. 1 et 2.

A titre d'exemple, pour un satellite d'éléments $a = 1{,}57$ rayons terrestres, $e = 0{,}1$ l'éclairement est continu tant que $i' < 45°$. Si $e = 0{,}2$ il faut $i' < 37°$, i' pouvant varier de façon assez importante, on conçoit que l'on trouve des époques assez longues d'éclairement continu du satellite.

III. Méthodes de calcul pour la détermination des effets de la pression réémise et de l'atmosphère

Ces deux effets vont provoquer une accélération séculaire $\Delta P/P$ du satellite.

Un moyen d'étude est de considérer l'expression de l'anomalie moyenne sous la forme suivante $M = M_0 + M_1 t + M_2 t^2$.

La période anomalistique est alors $P = 1/(M_1 + 2M_2 t)$.

Il est cependant nécessaire de tenir compte du mouvement du périgée π; l'argument du périgée, ω, étant écrit

$$\omega = \omega_0 + \omega_1 t + \omega_2 t^2 .$$

la période anomalistique corrigée est

$$P_c = \frac{1}{M_2 + \omega_1 + 2(M_2 + \omega_2 t)}$$

ce qui permet d'écrire

$$\frac{d P_c}{dt} = \frac{-2(M_2 + \omega_2)}{[M_1 + \omega_1 + 2(M_2 + \omega_2)t]^2} .$$

Les valeurs de M et ω ont été prises dans les Spécial Reports du S.A.O. pour 30D, et dans les bulletins S.A.O. pour PAGEOS.

Nous allons donner deux résultats d'analyses de M et ω.

1. Satellite 63.30D entre le 20 juin 1964 et le 28 août 1964

Pendant trois mois, le satellite reste éclairé.

Fea [2] a déterminé une densité moyenne durant cette époque.

Le Tableau 1 représente des valeurs discrètes de $\sin i' \sin\beta$ et de la quantité $2(M_2 + \omega_2)$, sensiblement proportionnelle à $\frac{d}{dt} P_c$.

Tableau 1. (63.30D)

Date	MJD	$\sin i' \sin\beta$	$2(M_2 + \omega_2)$ en (10^{-6})
20 juin	38566	+0,258	
24 juin	38570	+0,180	4,6
	38574	+0,108	
	38578	+0,057	
	38582	−0,096	
	38586	−0,128	
	38590	−0,185	
	38594	−0,241	
	38598	−0,295	2,06
	38602	−0,326	
	38606	−0,351	
	38610	−0,367	
	38614	−0,366	
	38618	−0,354	
	38622	−0,316	1,32
	38626	−0,276	
	38630	−0,220	
27 août	38634	−0,230	

Rappelons que l'accélération séculaire $\Delta P/P$ causée par la pression réémise,

$$\left(\frac{\Delta P}{P}\right)_{PR} = - \lambda^2 \sin i' \sin \beta .$$

Entre MJD 38566 et 38578, l'atmosphère et la pression réémise tendent toutes deux à faire diminuer la période.

Entre MJD 38582 et 38606, la pression réémise fait augmenter P. Les deux effets sont opposés, et on lit sur le Tableau 1 une accélération moindre.

Entre MJD 38610 et 38638, la pression réémise fait augmenter P plus fortement. On trouve une accélération séculaire encore moindre.

Conclusion: Sur le plan qualitatif, nous trouvons un accord satisfaisant entre les données et les formules.

Calcul de l'effet de l'atmosphère

Une circonstance particulière se trouve dans l'exemple traité; l'angle β passe par la valeur 0 en changeant de signe.

La pression réémise a alors des effets de sens contraire, et on peut délimiter des époques durant lesquelles son effet sera très faible. C'est ce qui arrive ici dans l'intervalle, (MJD 38566, MJD 38594). La valeur moyenne de $\sin i' \sin \beta$ est $-$ 0,006, ce qui nous permet de négliger la pression réémise sans faire d'erreur appréciable.

On peut alors, considérant le seul effet de l'atmosphère, calculer une densité moyenne.

Dans le calcul que nous avons fait, on a pris $S/M = 40$ cm²/g, $C_D = 4$; l'orbite est supposée circulaire ($e = 0{,}04$), hypothèses analogues à celles faite par Fea [3] dans un calcul où il ne tient pas compte de la pressions réémise.

Nous trouvons $\varrho = 2{,}4 \times 10^{-20}$ g/cm³ pour une altitude de 3000 km.

Effet de la pression réémise

Ayant déterminé une valeur moyenne de densité, nous pouvons par différence calculer l'effet de la pression réémise et faire une comparaison avec les formules données par Wyatt. Nous sommes conduits au tableau suivant.

Date		Date [en MJD]		Valeurs que nous déterminons	Valeurs calculées par les formules de Wyatt
20 juin	2 juil.	38566	38578	$-2{,}6 \times 10^{-8}$	$-2{,}1 \times 10^{-8}$
6 juil.	26 juil.	38582	38602	0	$-2{,}94 \times 10^{-8}$
3 août	27 août	38610	38634	$1{,}8 \times 10^{-8}$	$-4{,}2 \times 10^{-8}$

On ne peut pas tirer de conclusion sur les formules de Wyatt d'une comparaison des chiffres.

2e analyse:

La courbe des variations de a pour PAGEOS (Fig. 2) présente un palier très net entre le 1er octobre et le 23 déc. 1966. Le satellite est constamment éclairé, et la pression de radiation directe n'agit pas sur la période.

Nous suivons le même méthode que précédemment, avec ici, une accélération du périgée beaucoup plus forte que dans l'exemple précédent. Les bulletins S.A.O. sont utilisés ici.

Dans le tableau ci-dessous, nous faisons figurer $n_1 = 1/P_c$ écrit en révolutions par jour.

MJD	Date	$\sin i' \sin \beta$	$\left(\frac{\Delta P}{P}\right)$ P.R.	$n_1 = 1/P_c$	Δn_1 (en 10^{-6})
39399	30 sept. 66	+0,365	−	7,944291	
					+116
39416	17 oct.	+0,074	−	7,944407	
					+69
39427	28 oct.	−0,077	+	7,944476	
					+7
39441	11 nov.	−0,290	+	7,944483	
					+7
39455	25 nov.	−0,420	+	7,944490	
					−3
39469	9 déc.	−0,380	+	7,944487	
					+27
39483	22 déc.	−0,060	+	7,944514	

La pression de radiation solaire directe agit sur P au début de l'époque étudiée. Son effet est traduit par la valeur $\Delta n_1 = +116 \times 10^{-6}$ qui est beaucoup plus élevée que les suivantes qui correspondent à des cas où le satellite reste éclairé.

Entre 39416 et 39427 (17 octobre, 29 octobre) se produit la circonstance décrite plus haut, à savoir que l'angle β va passer par 0 et changer de signe. Nous allons pouvoir dans cet intervalle déterminer une densité moyenne.

Ensuite l'effet de la pression réémise va contrarier celui de l'atmosphère, et nous constatons que n_1 prend des valeurs plus faibles. Qualitativement, nous pouvons conclure que les chiffres obtenus sont en bon accord avec ce que permettent de prévoir le sens des effets physiques.

Etude quantitative:

Nous avons calculé une densité moyenne, dans l'intervalle (39416, 39427), puis étendant cette valeur à toute l'époque étudiée, nous en tirons, par différence, l'effet de la pression réémise. Nous la comparons avec celui que permettent de déterminer les formules de Wyatt. Nous obtenons les résultats suivants:

P moyen = 2,05 × 10^{-20} g/cm³ à une altitude de 3,500 km

Valeurs calculées de $\Delta P_e/P_e$ pour la pression réémise	Valeurs données par les formules de Wyatt	Rapport	Epoque
9,2 × 10^{-8}	21 × 10^{-8}	2,3	28 oct. à 25 nov.
10,2 × 10^{-8}	32 × 10^{-8}	3,1	25 nov. à 10 dec.
8,3 × 10^{-8}	25 × 10^{-8}	3	10 déc. à 23 déc.

Les $\Delta P_c/P_c$ sont obtenus à partir des n_1.

Nous trouvons des chiffres moins forts que ceux donnés par les formules de Wyatt.

Cependant, là aussi, nous avons adopté la densité moyenne obtenue durant un intervalle de temps pour toute l'époque étudiée.

3e analyse:

Nous avons utilisé les bulletins Goddard relatifs à 63.30.D, dans l'intervalle (15 avril, 10 mai 1967).

La calcul conduit au tableau suivant:

Date	Bulletin	i'	β	$\sin i' \sin\beta$	Intervalle entre les bulletins
15 avril 67	187	30°30′	33°	0,26	31 j
	188	28°	25°	0,20	2 j
	190	25°30′	14°	0,10	
	191	24°30′	−17°30′	−0,12	5 j
	193	21°	−39°30′	−0,22	10,9 j
10 mai 67	194	21°	−53°	−0,28	4,6 j

Le satellite reste constamment éclairé. Nous remarquons aussi que β passe par la valeur 0. Le demi grand axe varie peu dans l'époque étudiée, $a = -80$ m selon les bulletins.

L'altitude du périgée varie entre 2365 et 2320 km.

L'excentricité est de l'ordre de 0,13.

Nous avons calculé l'effet de la pression réémise en utilisant les formules de Wyatt.

Nous obtenons $(\Delta a)_{P.R} = +55$ m, ce qui nous donne pour l'atmosphère un effet de − 135 m sur a.

On ne peut plus négliger l'excentricité, ce qui conduit dans les formules relatives à l'effet de l'atmosphère à des expression liées aux fonctions de Bessel.

Nous avons pris pour échelle de hauteur $H = 700$ km.

Le calcul nous conduit à une valeur moyenne de la densité au périgée.

Nous obtenons $\varrho = 1{,}5 \times 10^{-19}$ g/cm^3 à une altitude de 2350 km environ, valeur dont la précision est liée à celle des bulletins et à l'analyse de la pression réémise.

Nous pouvons remarquer que les deux effets sont ici du même ordre de grandeur.

Conclusion:

Les deux premières analyses qualitatives donnent des résultats en accord avec ce que permettent de prévoir les formules des phénomènes.

Références

1. Wyatt: IUTAM Symposium Paris 1962, Dynamics of Satellites, Springer 1963.
2. Fea: Planetary and Space Science, vol. 14, n° 3 (1966).
3. Fea: Planetary and Space Science, vol. 13 (1965).

The Earth's Shadowing Effects in the Short-Periodic Perturbations of Satellite Orbits

By

P. Lála and L. Sehnal

Astronomical Institute, Czechoslovak Academy of Sciences, Prague, ČSSR

Abstract. The authors investigate the short-periodic perturbations of the orbit of an artificial satellite, during one revolution of the satellite around the Earth, caused by the direct solar radiation pressure. Special interest is paid to the study of the influence of the Earth's shadow, which is considered as a special mathematical function in the equations for the perturbations. The position of the satellite orbit according to the Sun is considered to be constant during one orbital period of the satellite, so the quasi-secular perturbations are obtained under those conditions too. Analytical formulae were found for computation of both quasi-secular and short-periodic terms in the perturbations of an arbitrary orbital element. A corresponding algorithm is used in the FORTRAN program for the IBM 360/65 computer. Practical application of this semi-analytical theory is shown for perturbations of semi-major axis and eccentricity of Echo 1, Echo 2 and Explorer 19 satellites. A comparison is made with the previous work neglecting the shadow, and quasi-secular changes of eccentricity are compared directly with changes in the SAO Mean Elements. Satisfactory agreement with observations is obtained.

This paper has been published in full in the Bulletin of the Astronomical Institutes of Czechoslovakia, 20 (1969) 327.

Earth-Reflected Radiation Pressure

By

D. E. Smith*

Wolf Research and Development Corporation, Space and Planetary Sciences Department. Applied Sciences Division, Riverdale, MD, USA

Abstract. A large proportion of the sunlight that falls on the Earth is reflected or re-radiated back into space and any of this radiation that shines on a satellite will cause a small force to be exerted on the satellite thus perturbing its orbit. The development of theories that can be used to calculate this perturbing force is discussed, the assumptions that need to be made and the difficulties in finding a solution are considered. Several theoretical studies are reviewed and the importance of this perturbing force on balloon type satellites is investigated.

1. Introduction

The perturbing effect of direct solar radiation on the orbit of an earth satellite is well known and understood but comparatively little attention has been given to the indirect effect of solar radiation scattered or reflected from the Earth. This lack of attention is probably due, in part, to the fact that all estimates of the latter effect suggest its magnitude is considerably smaller than that of direct radiation, and therefore of less importance, and also due to the problem being much more difficult and defying a simple analytical solution to a realistic problem.

The Earth radiates energy over a wide range of wavelengths and particularly in the infrared and optical regions of the spectrum. The infrared is largely a re-emission of radiation received at other wavelengths while the radiation in the optical wavelengths is a reflection, both diffuse and specular, of the incident sunlight. The reflected radiation, measured as a fraction of the incident radiation, is called the albedo and is composed of diffuse and specular components. If we assume equilibrium, the Earth returns to space as much radiation as it receives; thus we have

$$\gamma_D + \gamma_S + \gamma_{IR} = 1$$

where γ is the albedo and the suffixes D, S and IR indicate diffuse, specular and infrared, respectively.

* Now at NASA goddard space flight center, greenbelt, MD, USA.

The optical albedo ($\gamma_D + \gamma_S$) is usually quoted [1] as 0.4 implying that $\gamma_{IR} = 0.6$ and that equal consideration should be given to both the optical and infrared effects on the orbit. It is very difficult to say how much of the optical albedo is due to diffuse reflection and how much is due to specular reflection. However, since the latter can really only be expected from very calm seas or lakes and since clouds, snow fields and the continents all tend to produce diffuse reflections it seems reasonable to estimate the specular albedo (γ_S) to be only about 0.04, that is 10% of the total of the diffuse and specular components.

The albedo of the Earth is, of course, a variable in both time and position but in order to be able to derive analytical expressions for the forces on a satellite we need to make assumptions about the albedo that depart considerably from the real world and could possible lead to some false conclusions. This, in essence, is the difficulty with Earth-radiation pressure calculations: how do we simplify the disturbing function such that the problem becomes tractable without the model becoming unrealistic.

In the following sections the development of theories to account for Earth-radiation pressure perturbations of a satellite orbit are discussed, the assumptions that need be made, the difficulties in finding a solution are considered, and the importance of this force on satellites of large area to mass are considered.

2. Earth-Radiation

Let us briefly consider the assumptions that are normally made in the computation of direct solar radiation pressure perturbations. First we assume the sun's output of energy is known and is constant, second, the cross-sectional area to mass of the spacecraft can be represented by some mean value, third, the properties of the reflection occurring at the satellite are fully known or can be represented by some known average constant, fourth, the sun can be represented as a point source and, fifth, the entry and exit of the satellite into the shadow cast by the earth can be represented by some simple model. These are just a few of the more important assumptions that need to be made about the perturbing function and all carry over directly into the Earth-radiation problem. In addition, if we wish to express analytically the perturbations of the mean orbit elements an integration must be performed and the assumption that is usually made at this stage is that all the orbital elements (except mean anomaly) remain unchanged during one revolution of the spacecraft. This assumption also needs to be carried over into the Earth-radiation problem.

The additional assumption in the Earth-radiation problem concerns the variation of the albedo both temporally and spatially. Somehow the variation of the albedo must be modeled in such a way as to make (or keep) the problem tractable. For the infrared albedo the simplest model that can be introduced is a uniform flux acting radially from the centre of the Earth and varying according to the inverse square law. Such a force, however, is indistinguishable from the central force term in the gravitational field and produces no perturbations of the orbit. WYATT [2] has considered the next simplest model in which every point on the Earth's surface emits infrared radiation according to Lambert's law. WYATT shows that by symmetry the transverse component of the perturbing acceleration is zero and upon integrating the radial component over the visible cap of Earth obtains a total radial acceleration of

$$\frac{1}{4}\frac{A\Phi}{M}\frac{I}{c}\gamma_{IR}\left(\frac{a}{r}\right)^2$$

where A is the area and M is the mass of the satellite, Φ is the reflectivity of the satellite, I is the solar constant, c the velocity of light, a the radius of the earth and r the radial distance of the satellite.

This expression is identical to the perturbing acceleration due to a uniform flux from the centre of the Earth and consequently produces no perturbations of the orbit, just a small reduction in the gravitational force.

The next stage of development of an infrared model by WYATT permits latitude variations in the infrared radiation of the form (1.) $\cos\delta$, and (2.) $C_1 + C_2\cos\delta$, where δ is the latitude and C_1 and C_2 constants. In order to keep the integration tractable WYATT assumes the Earth's obliquity is zero and that the photons move radially from the Earth's centre. For these models WYATT finds no secular or long period perturbations in either the anomalistic period or eccentricity.

The next logical step in the infrared model would seem to be the introduction of some simple longitudinal variation but, as far as I know, this has not been attempted. However, the experience of WYATT does tend to suggest that simple models which only permit smooth, gentle variations in the radiation with respect to position all lead to the conclusion that infrared radiation has little or no perturbing effect on the orbit. It may therefore be that any long term infrared perturbation of a satellite will be due to the asymmetries and irregularities of the radiation and that any kind of simple modelling will automatically indicate little or no effect.

One of the earliest treatments of the contribution of diffuse reflection was by CUNNINGHAM [3] who calculated the quantity of diffusely reflected radiation incident upon a spherical satellite based on a uniformly diffusing sunlit side of the Earth and no radiation from the dark side.

The disturbing acceleration (δF) of the satellite due to radiation received from an elementary area is

$$\delta F = \frac{1}{\pi} \frac{A\Phi}{M} \frac{I}{c} \gamma_D \left(\frac{a}{\varrho}\right)^2 \sin\alpha \cos\beta \cos\Phi \, d\alpha \, d\lambda$$

where ϱ is the distance of the satellite from the elementary area (see Fig. 1), α is the angle subtended at the Earth's centre between the elemen-

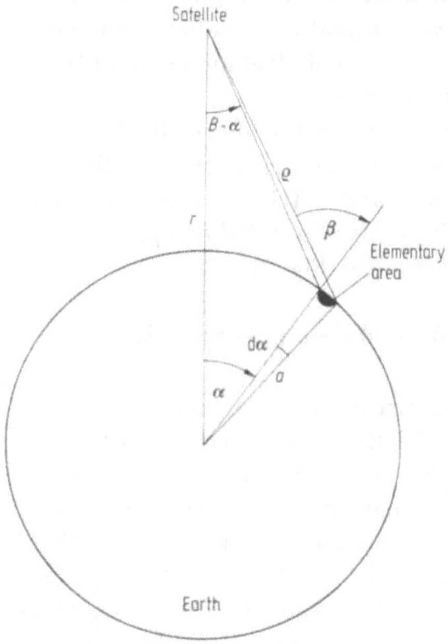

Fig. 1. Geometry of earth radiation.

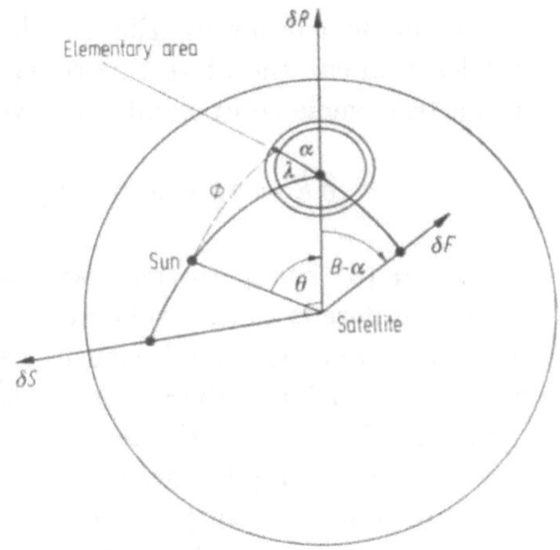

Fig. 2. Components of the disturbing acceleration.

tary area and the satellite (see Fig. 2), β is the zenith angle of the satellite from the elementary area (see Fig. 1), φ is the angle between the sun and the elementary area, and λ is the azimuthal angle of the elementary area from the plane containing the radius vectors to the sun and to the satellite (see Fig. 2).

Cunningham's model, uniform diffuse reflection on the sunlit side of the earth, is perhaps the simplest model that can be constructed and yet an analytical solution is extremely complex. In nearly all cases, additional assumptions are made to simplify the algebra.

Cunningham has integrated δF over the visible and illuminated portions of the Earth to obtain the *total* radiation falling on the satellite. This, as Baker [4] has pointed out, is quite different from the radial component for near earth satellites and is not directly related to the perturbing forces. In a subsequent paper Cunningham [5] has tried to assess the average reflected solar radiation incident on the satellite during one revolution. The position of the satellite appearing in the δF equation is written in terms of the orbital elements and the incident radiation integrated numerically round the orbit. The results of the integration are given in tabular form by Cunningham for a whole range of orbits from a few hundred kilometers altitude to several thousand. Since Cunningham computes the *total* reflected radiation falling on the satellite his figures are applicable to thermal but not dynamical studies. At low altitudes Cunningham's figures over estimate the radial component by some 20 to 30%. However, at the altitudes of 1000 km or more there is probably very little difference between them.

In all models in which there is angular symmetry of the albedo about the sun we know that the instantaneous force acting on the satellite lies in the plane containing the Earth, sun and satellite. The disturbing acceleration can therefore be written in terms of a radial component, δR and a transverse component, δS given by (see Fig. 2).

$$\delta R = \delta F \cos(\beta - \alpha); \quad \delta S = \delta F \sin(\beta - \alpha) \cos\lambda.$$

Levin [6] has performed an integration of these equations for a slightly different diffuse reflection model than that used by Cunningham [3, 5] and described earlier. Levin does not make any allowance for the diminishing apparent cross-sectional area of any element in the direction of the sun as the terminator is approached. Levin argues that this as a gross simplification of an extremely complex reflection mechanism and that in reality we do not have reflection from a geometrical surface but rather in depth through an atmosphere. The effect on the equations of the neglect of any diminishing as we approach the terminator is to remove the $\cos\varphi$ term in the disturbing acceleration which, in addition, simplifies the subsequent integrations. One further consequence of Levin's assumption is that the transverse component of the reflected radiation is zero everywhere except where the satellite can see the terminator. Levin's results for the radial and transverse components are shown in Fig. 3 where the ratios of the radial component, I_r and the transverse component I_θ, of the reflected radiation to the solar constant, I, is plotted against zenith angle of the sun for three values of σ $(= a/r)$.

From Fig. 3 we see that the diffusely reflected radiation is about 10 to 20% of the direct solar radiation for near earth satellites and that the transverse component of the radiation is only about 15% of the radial component for Levin's model.

Levin does not attempt to compute the perturbations of the orbit and a glance at his expression for the radial component of the radiation, given below, suggests that this would be a formidable task analytically. LEVIN obtains

$$I_r = \gamma_D I \sigma^2 \left\{1 - \frac{1}{\pi} \cos^{-1}\left[\frac{\sigma \cot\theta}{(1-\sigma^2)^{1/2}}\right] + \frac{1}{\pi} I^1\right\}$$

where $\sigma = a/r$ and

$$I^1 = \frac{\cos\theta}{[(1+\sigma^2)^2 - 4\sigma^2 \sin\theta]^{1/2}} \left\{\frac{\pi}{2} + \sin^{-1}\left[\frac{\sin^2\theta - (1+\sigma^2)/2}{\sin\theta\,(1-\sigma^2)/2\,\sigma}\right]\right\}.$$

The first study that derived expressions for the perturbations of the orbital elements by diffuse reflection was by SEHNAL [7] who obtained

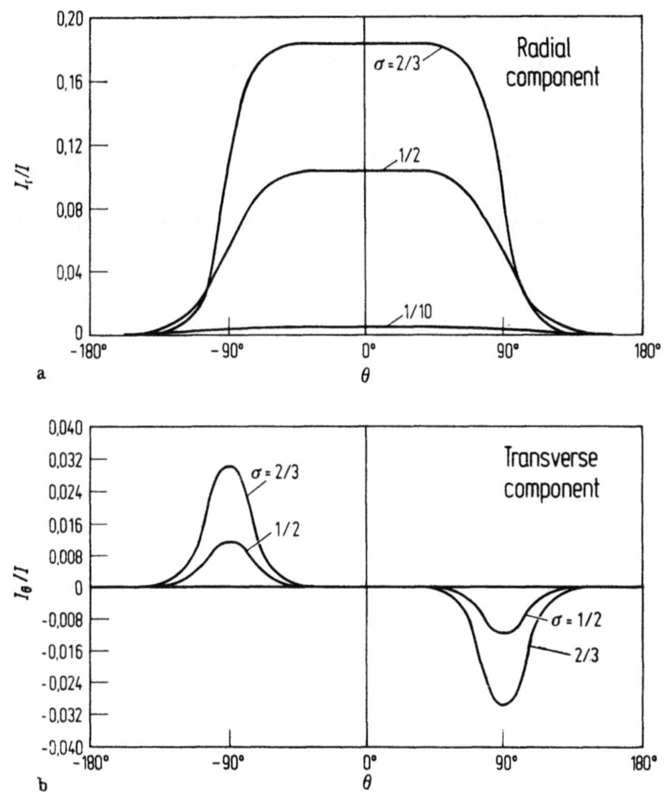

Fig. 3. Radial and transverse components of reflected radiation, LEVIN [6]. a) Radial component; b) Transverse component.

the changes in period and eccentricity for a uniform albedo with the additional assumption that the force on the satellite is radial, the sun lies in the orbit plane, and that the force is zero when the satellite is above the dark side of the Earth. For a satellite of the Echo 1 type in a near circular orbit at about 1600 km altitude SEHNAL finds a change of 0.005 sec in the period of revolution per revolution; a figure that it should be possible to check by observation.

WYATT [2] has also considered the effects of diffuse reflection on the eccentricity and period of a satellite for three simple albedo models, in all of which the mean albedo is constant. The first model assumes the force is radial and constant when the satellite is over the sunlit side of the Earth and zero for the rest of the time (same as SEHNAL [7]), in the second model the force varies as the cosine of the zenith angle of the sun and in the third case WYATT tries to account for the Earth's phase function being more peaked than is predicted by diffuse Lambert reflection. In all three cases WYATT finds a secular perturbation to both the eccentricity and the period and if, in the first model, the sun is made to lie in the orbit plane we obtain the results of SEHNAL [7]. The most significant results of Wyatt's study is, however, that all three models predict about the same magnitude for the perturbations, that is, about, 8 to 10 percent of those of direct sunlight, together with the same basic variation of the perturbations with varying geometry of the satellite, Earth and sun positions. Furthermore, it is worth noting that the perturbations of the period in all three cases are proportional to the eccentricity, supporting the general idea that "smooth" model albedo's and "symmetric" orbits do not produce perturbations.

The problem that nearly all researchers have endeavored to solve has been essentially the computation of the perturbations arising from diffuse reflection by Lambert's law but, as we have seen, the calculations are formidable and further assumptions have usually been made. BAKER [4], however, although not calculating the orbital perturbations, has studied in considerable depth the computation of the flux of radiation received by the satellite. BAKER has considered the infrared, the diffuse and the specular components of the flux and after some extremely heavy algebra has obtained expressions for both radial and transverse components. These results, so far as the author is aware, are the most complete treatment of the problem of Earth radiation forces to date and are derived in a form suitable for application in numerical forms. The expressions for the transverse and radial components of diffuse reflection obtained by BAKER [4] are

transverse:

$$F_T = \frac{2}{\pi} I \gamma_D \left(\frac{a}{r}\right)^2 \{U(\theta_T - \theta) I_4 + U(\theta - \theta_T) U(\theta'_T - \theta) I_.\}$$

radial:

$$F_R = \frac{2}{\pi} I \gamma_D \left(\frac{a}{r}\right)^2 \{U(\theta'_T - \theta)[I_6 - U(\theta - \theta_T) I_7]\}$$

where

θ is the angle between the sun and the satellite,
θ_T is the value of θ when the terminator first enters the spacecraft's horizon,
θ'_T is the value of θ when the terminator leaves the spacecraft's horizon,

$$U(\theta_T - \theta) = 1 \quad \text{for} \quad \theta < \theta_T,$$
$$U(\theta_T - \theta) = 0 \quad \text{for} \quad \theta \geq \theta_T,$$
$$U(\theta - \theta_T) = 0 \quad \text{for} \quad \theta < \theta_T,$$
$$U(\theta - \theta_T) = 1 \quad \text{for} \quad \theta \geq \theta_T,$$
$$U(\theta'_T - \theta) = 1 \quad \text{for} \quad \theta \leq \theta'_T,$$
$$U(\theta'_T - \theta) = 0 \quad \text{for} \quad \theta \geq \theta'_T,$$

and

$$I_4 = \frac{\pi}{8}\left(\frac{a}{r}\right) - \frac{\pi}{15}\left(\frac{a}{r}\right)^2 + \left(\frac{4}{3} - \frac{7\pi}{16}\right)\left(\frac{a}{r}\right)^3 - \frac{2\pi}{105}\left(\frac{a}{r}\right)^4$$

$$I_5 = \frac{\pi}{16}(1 + \cos\theta)\left(\frac{a}{r}\right) + \frac{1}{15}(4\cos\theta\sin\theta - (\pi - \theta))\left(\frac{a}{r}\right)^2$$
$$+ \left(\frac{2}{3} - \frac{7\pi}{32} - \frac{\pi}{8}\cos^3\theta\right)\left(\frac{a}{r}\right)^3 + \left[\frac{-2}{105}(\pi - \theta)\sin\theta\right.$$
$$\left. + \left(\frac{9}{35} - \frac{\pi}{16}\right)\cos\theta - \frac{8}{15}\cos^3\theta + \frac{64}{105}\cos^5\theta\right]\left(\frac{a}{r}\right)^4 / \sin\theta \ldots$$

$$I_6 = \pi\cos\theta\left[\frac{1}{3} + \frac{1}{4}\left(\frac{a}{r}\right) - \frac{1}{15}\left(\frac{a}{r}\right)^2 + \left(\frac{1}{12} - \frac{5}{18\pi}\right)\left(\frac{a}{r}\right)^3 - \frac{1}{105}\left(\frac{a}{r}\right)^4 \ldots\right]$$

$$I_7 = \frac{1}{3}(\theta\cos\theta - \sin\theta) + \frac{\pi}{16}(1 + 2\cos\theta - 3\cos^2\theta)\left(\frac{a}{r}\right) +$$
$$+ \left[\left(\frac{107}{120} - \frac{\pi}{4}\right)\sin\theta - \frac{\theta}{15}\cos\theta - \frac{8}{15}\sin\theta\cos^2\theta\right]\left(\frac{a}{r}\right)^2 +$$
$$+ \pi\cos\theta\left[\left(\frac{1}{24} - \frac{5}{36}\pi\right) - \frac{1}{8}\cos\theta + \frac{5}{24}\cos^3\theta\right]\left(\frac{a}{r}\right)^3 +$$
$$+ \left[\left(\frac{11\pi}{16} - \frac{361}{168}\right)\sin\theta + \frac{\theta\cos\theta}{105} - \frac{8}{21}\sin\theta\cos^2\theta + \right.$$
$$\left. + \frac{32}{35}\sin\theta\cos^4\theta - \frac{1}{24}\cos^2\theta/\sin\theta\right]\left(\frac{a}{r}\right)^4 \ldots$$

Perhaps the largest source of error in any of the albedo models is the lack of latitudinal variations. Much of the albedo variation is due to changing cloud cover which has well known variations with latitude and, in addition, there are the polar ice caps which can add to other seasonal variations in the albedo. Some variations, such as the asymmetry produced by the polar ice fields, could possible be to modeled adequately and included in the analysis but the complexcity of Baker's results suggest that in most cases any further refinement of the albedo model ought to be conducted numerically. Such a technique is currently being attempted by FEA [8] who is dividing the surface of the Earth into areas of a few degrees square and permitting different values of the albedo in each square. The acceleration of the satellite due to the radiation from each of these "visible" areas can then be computed and subsequently integrated round the orbit to obtain the perturbations. The strength of such a method lies in the large number of different local albedo distributions that can be investigated.

Finally, the work of SEHNAL [9] needs to be mentioned. SEHNAL adopts a model of the diffuse albedo such that the radiation pressure is not zero inside the shadow of the Earth but only very small and zero only at the point on the orbit with the greatest angular distance from the sun. For this model SEHNAL computes the perturbations to all spatial elements of the orbit. Sehnal's results are very complex but represent the most detailed study of the orbital perturbations of Earth-radiation pressure at present available.

The importance of specular reflection from the Earth has been discussed extensively by WYATT [2]. WYATT states that the size of the solar image on the surface of the Earth for near Earth satellites in only a few kilometers in radius and argues that waves and ripples on the surface of the sea will tend to smear out the radiation. If this is the case it seems that the specular component of the reflected radiation could easily be lost in the irregularities of the diffuse reflection and therefore hardly worthy of separate treatment. Assuming that $\gamma_S = 0.02$ WYATT finds negligible perturbation or either the period or eccentricity by specular reflection.

3. Orbital Perturbations

For many purposes the perturbing effects of *direct* radiation pressure can be considered negligible on all except satellites of large area to mass ratio. By large area to mass we normally mean larger than about $10\,cm^2/g$ but this is a rather arbitrary figure and depends on the height of the satellite and the accuracy of the orbit calculations. However, since Earth reflected radiation pressure is considerable less than direct pressure we may feel sure that significant albedo perturbations are only likely to be observed on satellites with very large area to mass ratios, say, $50\ cm^2/g$ or greater.

We have seen in the previous sections that the estimates of the effect of infrared radiation indicate little or no perturbations, largely because the force is predominantly radial and varies as the inverse square, thus being indistinguishable from the central force term. The radial force of infrared radiation on a balloon type satellite is of order 10^{-7} that of gravity and is therefore sufficiently small to be neglected even in gravitational field studies.

The forces caused by diffuse radiation do, however, cause perturbations of the eccentricity and period of revolution and the latter could be important in air drag studies. Between 700 km and 1000 km the forces due to direct solar radiation pressure become comparable to the forces of air drag and presumably at slightly greater heights, say 1500 km, the forces due to albedo radiation must become comparable to or even greater than air drag. At these heights direct radiation pressure will be the dominant force but direct radiation pressure does not change the energy of the satellite if the orbit is in complete sunlight thus enabling the air drag to be computed. From the studies on diffuse reflection it appears that the perturbation of the period, representing the energy, is always non-zero for non-circular orbits and will therefore always contaminate the drag results for balloon satellites at altitudes above about 1000 km. At still greater heights, where the albedo perturbation is probably much larger than air drag and for orbits in complete sunlight, the albedo could be the major perturbing force of the semi-major axis.

4. Conclusions

The analytical development of theories to account for Earth radiation perturbations of a satellite orbit are very complex for any reasonable model of Earth radiation and can only partially represent the forces on the satellite. In the theories so far developed it has always been necessary to smooth out the Earth-radiation to such an extent that it can be considered uniform and mean over the whole of the sunlit side of the Earth. The result of this smoothing appears to be a reduction or a cancelling of the perturbative forces leading to a conclusion that Earth-radiation effects on the orbits are small. However, the Earth's albedo is far from uniform and if any perturbations exist they are likely to be a consequence of the irregularities and asymmetries of the Earth's albedo rather than its uniformity.

A full and complete treatment of Earth radiation perturbations can probably only be carried out numerically and when the results of meteorological information from satellites can be incorporated.

References

1. Allen, C. W.: Astrophysical Quantities, 2nd Edition, Athlone Press, 1962.
2. Wyatt, S. P.: IUTAM Symposium, Dynamics of Satellites, Paris, May 1962, Berlin/Göttingen/Heidelberg: Springer 1963.

3. Cunningham, F. G.: ARS Journal, July 1962.
4. Baker, R. M. L.: Trajectories of Artificial Celestial Bodies as Determined from Observations, Proc. of Sympos. Paris, April 1965; Berlin/Heidelberg/New York: Springer 1966.
5. Cunningham, F. G.: NASA TN D-1472, February 1963.
6. Levin, E.: ARS Journal, September 1962.
7. Sehnal, L.: The Use of Artificial Satellites for Geodesy, Edit. G. Veis, Amsterdam: North-Holland Publish. Comp. 1963.
8. Fea, K.: Private Communication.
9. Sehnal, L.: Trajectories of Artificial Celestial Bodies as Determined from Observations, Proc. of Sympos. Paris, April 1965; Berlin/Heidelberg/New York: Springer 1966.

The Orbital Accelerations of High Balloon Satellites

By

K. Fea

University of London Observatory, London, England

Abstract. The orbital accelerations of two balloon satellites have been studied in detail. The satellite 1963-30D is investigated during the year 1967, and 1966-56A (PAGEOS) during the year 1968. During the "no-shadow" phases some indication of air drag is seen. During the shadow phases large accelerations are seen which are only partially explained through solar radiation pressure. It is suggested that two additional sources of acceleration are present, earth reflected radiation pressure and some effect related to the spin of the satellites.

Earlier studies of the orbital accelerations of the balloon satellite 1963-30D, Fea [1, 2, 3], led to statements on likely exospheric densities at a height of 3500 km for the year 1964. Subsequent more detailed analysis of the orbit data indicated that the accelerations seen were not completely explained; Fea [4]. More recent examination of long periods of good data primarily provided by the Smithsonian Astrophysical Observatory (SAO), has provided insight into the various sources of acceleration. A detailed discussion of the behaviour of 1963-30D in the period November 1963, through June 1965, has been given by Fea and Smith, [5]. It is concluded there that the high quality of the acceleration data, giving $\dot{T}$ to about $\pm 1 \times 10^{-8}$ days/day, requires evaluation of the following perturbations:

1. Solar radiation pressure (SRP), during those times when the orbit includes shadow.

2. A second-order solar radiation pressure effect (SRP 2) due to the orbit not being strictly closed; it is significant during the periods when the orbit is in full sunlight, and periodic with a period of about one year.

3. Lunar gravity. It is shown in [5] that an acceleration due to lunar gravity does exist and is significant when the overall acceleration is small. The period is approximately 14 days.

4. Solar gravity. If long periods of data are examined it is possible to demonstrate the existence of this source of acceleration, of period about 180 days, analogous to that due to lunar gravity.

5. Air drag. During the "no-shadow" phases, when the massive contribution due to SRP is not present in the acceleration, some general negative trends in $\dot{T}$ appear. It is possible to estimate upper limits, at least, for hydrogen and helium densities at these heights (above 2000 km).

6. Terrestrial radiation pressure, or earth albedo radiation pressure (EARP). It is shown in [5] that the perturbations due to EARP are probably present in all the elements for 1963-30D and are significant for the acceleration.

7. Additional disturbances not identified in [5].

In forming the acceleration correctly, that is in defining $\ddot{M}$ [the second differential of $M(t)$, the mean anomaly] one must allow for the contribution from $\ddot{\omega}$, the acceleration in the argument of perigee, ω. If $\ddot{\omega}$ is not taken as zero in the computation of $\ddot{M}$ from the observations, then it is necessary to replace $\ddot{M}$ with $(\ddot{M} + \ddot{\omega})$. Similarly $(M + \omega)$ and $(\dot{M} + \dot{\omega})$ are used. This simple device for transforming anomalistic values to "corrected" or "sidereal" values is discussed, and justified for the orbits studied here, in [5]. The procedure closely follows that used by the SAO.

To act as a "control", the orbit of the satellite 1961 Alpha delta 1 is also discussed in [5]. It is shown that this heavy object, of small area/mass ratio, display only the lunar and solar gravity accelerations stated in 3. and 4. above. The complex behaviour of 1963-30D, also exhibited by 1966-56A, (PAGEOS) is due almost certainly therefore to the action of radiation pressures and, to a minor extent, of atmosphere drag. The orbit of 1961 Alpha delta 1 is very similar to that of the two balloon satellites; the inclination is near 90°, the mean height about 3000 km and the eccentricity near zero, as initially were the eccentricities of both 1963-30D and 1966-56A.

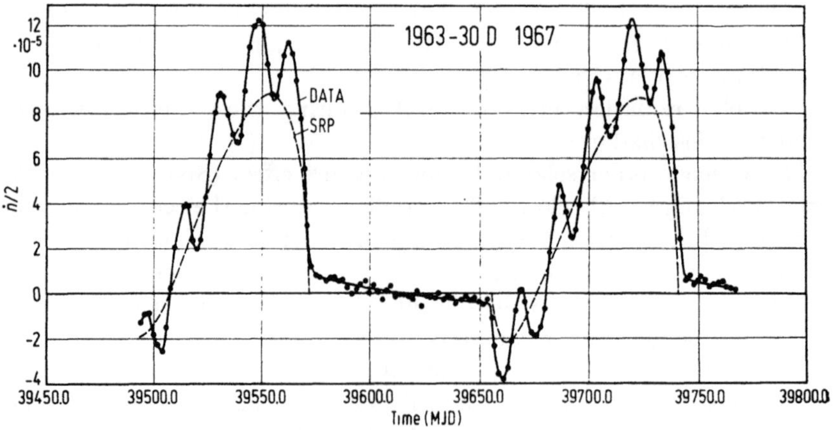

Fig. 1.

In Fig. 1 the orbital acceleration of 1963-30D is shown during the year 1967. The data terminates in October as SAO tracking on this object ceased at that time. The dates shown are Modified Julian Dates. The orbit of 1963-30D has increased steadily in eccentricity since launch. The orbit is in the resonant category; a discussion of resonance, with respect to solar radiation pressure action, has been given by Shapiro [6]. Similarly the eccentricity of 1966-56A has increased during the period 1966 through 1968. The perigee heights of both these balloon satellites had therefore been depressed by SRP to below 2500 km by 1967/68. Solar activity also increased to a maximum in the same period and exosphere temperatures had risen well above the 1964 low values. Consequently, in 1967/68, both 1963-30D and 1966-56A experienced atmospheric drag due to helium, at least an order of magnitude above the drag experienced in 1964 by 1963-30D. The hydrogen region is now far above the orbits of these satellites. Since the 1963-30D satellite will not survive into the next solar minimum period, the 1964 data for 1963-30D remain of particular value for studies of the debated hydrogen content in the exosphere. Improved estimates are given in [5] for the hydrogen density near 3500 km in 1964.

It has been shown in [5] that a fairly acceptable fit can be made to the observed values of the elements through one year by allowing the area/mass ratio, A/m, to vary. For 1963-30D a reasonable value for A/m is 40.0 g/cm². However the best value of A/m differs according to the element selected. It is believed that EARP causes these discrepancies.

A discussion of this last point has also been given by Prior [7]. He concludes, "... values of $K(A/m)$ inferred from best fits to eccentricity and argument of perigee variations are lower than the best $K(A/m)$ fitting variations in the semi-major axis".

In Fig. 1 two curves are given. The data points and continuous line give the acceleration obtained from the orbits specially prepared by the SAO. The standard deviations quoted in the SAO data almost all lie in the range 1 to 3×10^{-6} for $\dot{n} (= \ddot{M})$. Therefore the details on the curve are probably mostly real. The dashed curve gives the acceleration computed for the SRP component with A/m taken as 40.0. The program was run at the Radio and Space Research Station (Slough) and was originally prepared by D. E. Smith.

It is seen that the agreement of the two curves is only approximate. However a point worth emphasizing is the agreement in the times of onset and cessation of the large disturbances. Clearly the additional sources of acceleration are related to the shadow phase in the orbit geometry cycle. Note that the shadow (orbit includes shadow) and no-shadow phases are long, for the satellite is high and the precession of the orbit slow.

The most remarkable features of the acceleration curve in Fig. 1 are the pronounced fluctuations of a nearly periodic nature. These were first noticed in the 1963/4/5 data discussed in [5]. It seems very unlikely that EARP can account for such oscillations and it would be difficult to devise a model for terrestrial radiation able to reproduce these rapid (2 to 3 week period) variations.

The behaviour of 1966-56A in 1968 is shown in Fig. 2. It resembles to some extent that in Fig.1. Note though that the amplitude is much larger, since the A/m value for PAGEOS is about 136 g/cm², more than three times that for 1963-30D. An interesting observation is the increasing amplitude of the periodic fluctuations during the shadow phases. The first shadow phase shows only minor anomalies while the third, incomplete, shadow phase shows striking fluctuations.

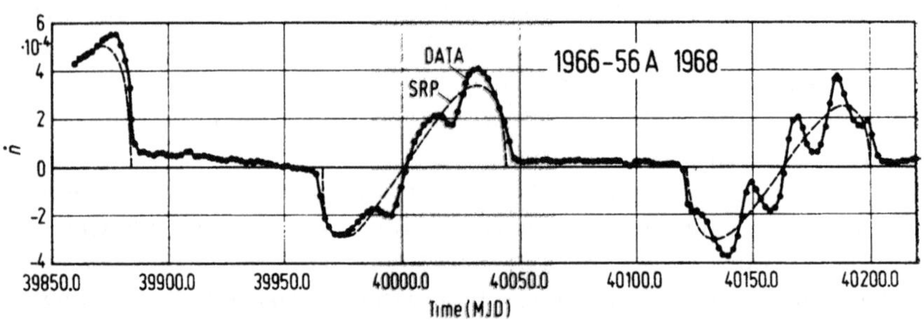

Fig. 2.

It is noted that the general trend for 1963-30D in the observed accelerations is always to exceed the magnitude of the SRP perturbation. It is worth noting that the tendency for EARP will be in this direction, to reinforce the effect due to SRP in $\dot{T}$ (or $\dot{n}$). Wyatt has previously given a general discussion of these perturbations for special cases [9]. Especially relevant is the discussion of EARP perturbations given by Prior [7], and the study and review of the problem by Smith [8]. However, in the case of 1963-30D here it is clear that an oscillation is superimposed on any EARP component which may be present.

However, in the case of 1966-56A, the SRP curve seems to fit rather well the mean curve through the observed acceleration data. Is it possible then that the discrepancy in the case of 1963-30D is due to the use of an erroneous A/m value? While this seems at first sight an obvious explanation three facts suggest otherwise. First the value of 40.0 is close to the maximum possible value for the fully inflated balloon with total loss of the original inflating gas (or subliming powder). Second, the value of 40.0 is a good mean value suggested by the studies, in [5], of the perturbations in all the elements. Third, there is reason to believe that the A/m value usually assumed for PAGEOS (1966-56A) is *too large* – the balloon being significantly prolate in form since late 1966. To fit the observed points in Fig. 1 one would require an increase in A/m for 1963-30D from 40.0 to 46.0 i.e. about 15%. This does not seem acceptable.

To test the interpretation more quantitatively, study of the radar signature of both 1963-30D and 1966-56A is in progress.

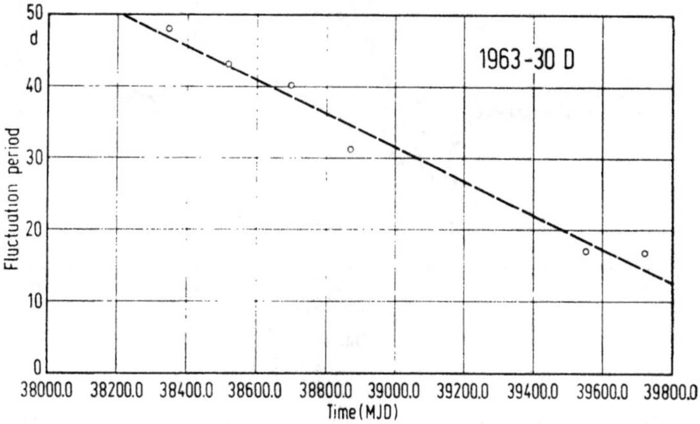

Fig. 3.

In Fig. 3 an indication is given of the decrease in the period of the fluctuations in $\dot{n}$, within each shadow phase, for 1963-30D from November 1963 to October 1967. Data is not presently available between July 1965 and December 1966. Nevertheless a definite and fairly steady decrease in the ocsillation period is seen here. It is believed that this represents a change in the rotation of the satellite itself and may imply a spin-up, though some very recent re-analysis suggests that the precession period for the spin axis is the relevant parameter.

Fig. 4 gives a possible analysis of the observed behaviour of $\dot{n}$ during the second shadow phase for 1963-30D in 1967, taken from Fig. 1. The SRP primary perturbation is shown again by a dashed curve. It is noted that the perturbations described under 2., 3., 4. and 5. above are very small in comparison with the large SRP effect. One may state approximate amplitude and periods for these small components.

The 14-day lunar gravity perturbation seen in $\dot{n}$ has an amplitude of $\pm 1.0 \times 10^{-6}$ rev/day for the degree of smoothing in the 1963-30D data in 1967; each determination of the elements is obtained from 6 days observations. The one year SRP 2 perturbation has maxima approximately in the middle of the no-shadow phases and goes to zero approximately

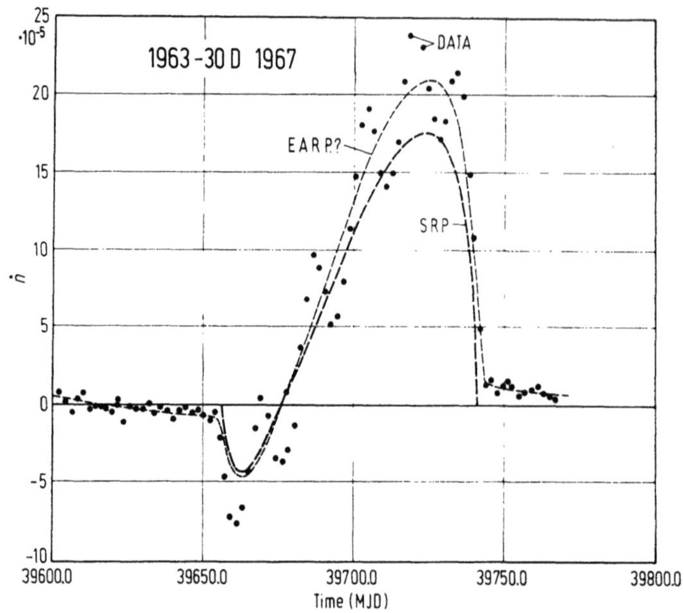

Fig. 4.

in the middle of shadow phases. The amplitude for 1963-30D in this case is $\pm 2.2. \times 10^{-6}$ in $\dot{n}$ (i.e. in $\dot{n} + \ddot{\omega}$). The solar gravity perturbation in $\dot{n}$ is roughly one order of magnitude smaller than the analogous lunar gravity term, and the period is six months.

The air drag (neutral drag) is probably best estimated through comparison with the no-shadow phase accelerations in Fig. 2, for the much larger balloon 1966-56A. If the assumption is made that drag forces other than neutral atmospheric drag are absent then one estimates an average drag $\dot{n}$ for 1966-56A of approximately $+3 \times 10^{-5}$, implying an exospheric density at a height of about 2500 km of 6×10^{-20} g/cm³, in 1968). Scaling down in the ratio of the A/m values, 1963-30D should show a drag $\dot{n}$, in 1967, of about 1×10^{-5}. Examination of the no-shadow phase trend in $\dot{n}$ for 1963-30D in Fig. 1 suggests that the air drag effect is smaller than this; however, superimposed on any quasi-constant drag acceleration, is a steady decrease in $\dot{n}$ from positive into negative values during the no-shadow phase. A similar behaviour is seen for 1966-56A in Fig. 2, though not during the second no-shadow phase in that case. It is very possible that this variation in $\dot{n}$ is due to EARP which acts during no-shadow phases to some extent. Simple models suggest that the EARP perturbation in $\dot{n}$ is smallest when the orbit of 1963-30D lies over the terrestrial terminator, i.e. near the middle of a no-shadow phase, for example (in Fig. 1) at about

MJD 39620.* In this case $\dot{n}$ is very near zero, suggesting an undetected drag acceleration. Three factors may be involved:

a) The perigee height of 1963-30D in 1967 was above that for 1966-56A in 1968. But the relevant heights are 2350 km and 2250 km respectively, a difference much less than the helium scale height.

b) The exospheric temperature was lower in 1967 than in 1968. But the difference would not appear to produce a variation in the helium density as much as a factor of two.

c) Magnetic induction drag, of the type discussed by DRELL et al. [10], is effective for 1966-56A, which has a diameter of about 30 m, and less effective for 1963-30D, which has a diameter of only 2 m.**

In Fig. 4a second dashed curve has been added tentatively, smoothing through the fluctuations with the assumption that these represent roughly a sinusoidal term. The elevation of the second dashed curve over the computed SRP curve is possibly explained by the action of EARP. A model is currently being constructed to investigate the perturbations by terrestrial radiation; the model explores numerical simulations for the earth's albedo varying with latitude, which may be significant. However PRIOR [7] has demonstrated good accord with the observed orbital variations using a simple model for the albedo radiation.

Finally, it is necessary to comment on the quite rapid oscillations apparent in the $\dot{n}$ data for both 1963-30D and 1966-56A especially during the shadow phases.

The short period fluctuations, with about two week periods, seen in Figs. 1 and 2 are tentatively thought to arise through some mode of *transverse* SRP action. Models are being investigated, with SMITH, to determine if accelerations of this magnitude can be produced by a slowly precessing and non-spherical balloon satellite, with the changing precession rate of the satellite modulating the magnitude. Currently we believe such a mechanism to be promising.

Acknowledgments

This work was supported in part by a grant from the Science Research Council. The author wishes to thank Professor C. W. ALLEN for his continued encouragement and to thank Dr. D. E. SMITH for valuable discussion outside those areas of direct collaboration. Thanks are also due to Dr. L. G. JACCHIA of the Smithsonian Astrophysical Observatory and to Professor IRWIN SHAPIRO of the Massachusetts Institute of Technology for helpful suggestions.

* PRIOR [7] has discussed this point and his findings are in agreement with the conclusions here.

** Very recent analysis of additional data for both satellites supports the evidence for a contribution by magnetic induction drag.

References

1. FEA, K.: Determination of the Density of Air at an Altitude of 3500 km. Nature, Lond., 205 (1965) 379.
2. FEA, K.: A Photographic Technique for the Precise Observation of Artificial Satellites and some Results from Data on High Satellites. Planetary and Space Science 13 (1965) 1289.
3. FEA, K.: Exospheric Conditions to a Height of 3500 km, derived from Satellite Accelerations in 1964. Planetary and Space Science 14 (1966) 291.
4. FEA, K.: Satellite Accelerations and Air Densities at extreme Altitudes. Philosophical Transactions, Roy. Soc. Lond., A, 262 (1967) 200.
5. FEA, K., SMITH, D. E.: Further studies of satellite accelerations at extreme altitudes. Submitted to Planetary and Space Science.
6. SHAPIRO, I. I.: Chapter on Prediction of Satellite Orbits. In: Dynamics of Satellites (ed. M. Roy), Berlin/Göttingen/Heidelberg: Springer-Verlag 1963.
7. PRIOR, E. J.: Earth Albedo Effects on the Orbital Variation of Echo 1 and PAGEOS 1. Paper presented at XIIth Plenary Meeting of COSPAR, Prague, May 1969. (Following paper in this Volume.)
8. SMITH, D. E.: Earth reflected Radiation Pressure. Paper presented at XIIth Plenary Meeting COSPAR, Prague, May 1969.
9. WYATT, S. P.: Chapter on Terrestrial Radiation Pressure. In: Dynamics of Satellites (ed. M. Roy), Berlin/Göttingen/Heidelberg: Springer-Verlag 1963.
10. DRELL, S. D., FOLEY, H. M., RUDERMAN, M. A.: Drag and Propulsion of large Satellites in the Ionosphere; an Alfven Propulsion Engine in Space. J. Geophys. Res. 70 (1965) 3131.

Earth Albedo Effects on the Orbital Variations of Echo I and PAGEOS I

By

E. J. Prior

NASA Langley Research Center, Langley Station, Hampton, VA, USA

Abstract. A study is performed to determine the effects on satellite orbits of the force of solar radiation diffusely reflected from the Earth. In particular, the orbits of Echo I and PAGEOS I are studied to determine whether unexplained variations in their orbital elements are due to Earth albedo. Ground-based photometric measurements have confirmed the essential stability of the average radii of curvature of these two balloon satellites during the time interval studied. Expressions are developed which give the effects of albedo on the orbital elements. It is found that this perturbing effect generally acts to decrease the effect of direct solar radiation on the eccentricity and the argument of perigee, but to increase the effect of direct solar radiation on the semi-major axis. Unexplained orbital variations are observed for Echo I from January 1963 to December 1964 and for PAGEOS I from April 1967 to October 1968. Corrections are made for earth oblateness, lunar and solar gravitational fields, atmospheric drag, and direct solar radiation. Using mean albedo measurements from TIROS VII radiometer data, these unexplained variations are identified as earth albedo radiation force perturbations. These albedo perturbations are significant and must be accounted for in order to infer accurate density variations at various global positions at over 1000 km from the Echo I orbital decay, and at over 2000 km from the PAGEOS I orbital decay.

1. Introduction

A study is made of the pertúrbing effects of albedo radiation pressure on the orbital behavior of Echo I and PAGEOS, two balloon satellites highly sensitive to radiation pressure effects. Diffuse Lambert reflection of sunlight from the Earth is assumed. Previous studies of albedo perturbations have not attempted to actually discern the effect in orbital data.

2. Orbital Predictions

Mean orbital elements for Echo I and PAGEOS were obtained from the Smithsonian Astrophysical Observatory. These elements were based on Baker-Nunn camera position measurements [1].

To obtain theoretical orbital predictions, a digital computer program was obtained from the Rand Corporation [2]. This program, originally developed by SHAPIRO and JONES [3], was designed to study long-period and secular orbital motions of balloon satellites such as Echo I. The following perturbations are included:

1. The second through the fifth harmonics of the earth's gravitational field.
2. Lunar and solar gravitational fields.
3. Direct solar radiation pressure.
4. Atmospheric drag.

The direct solar radiation pressure perturbation on an orbital element over one orbit is proportional to the acceleration of the satellite along the sun-satellite line, A_{sr}

$$A_{sr} = \left(\frac{A}{M}\right)\left(\frac{I}{C}\right)K$$

and (A/M) is the area-to-mass ratio of the satellite, I the solar flux near the earth, C the speed of light, and K a scattering constant which is dependent on the reflective properties of the satellite surface [3]. In the case of both Echo I and PAGEOS, the (A/M) was uncertain during the initial satellite lifetime due to the slow escape of sublimating vapors through small holes in the surface.

By the beginning of the intervals of this study, the vapors should have entirely escaped. The nominal (A/M) of Echo I for this condition is 134.08 cm^2/g [4, 5], and for PAGEOS 134.71 cm^2/g. Both satellites, with surfaces of 0.5 mil aluminum-coated Mylar, have been considered specular reflectors with $K = 1$; however, observations of both in orbit with the NASA Mobile Satellite Photometric Observatory indicate that $K(A/M)$ is slightly less than the nominal value [6, 7]. The average radii of curvature of both satellites were found to remain essentially stable during the intervals of this study.

The most recent determination of I give a value of about 1.35×10^6 $ergs/cm^2 sec$ [8], although it appears to be subject to slight irregular changes [9].

3. Albedo Radiation Pressure Perturbations

Theoretical studies have been made of earth albedo radiation pressure [10, 11, 12]. In this paper, the radial and transverse components of earth albedo radiation flux on a satellite surface are found in the following way. The albedo flux reflected from a unit surface area in the direction of the satellite is determined using the Lambert cosine law, as shown in Fig. 1. Then the components of this flux are found and nu-

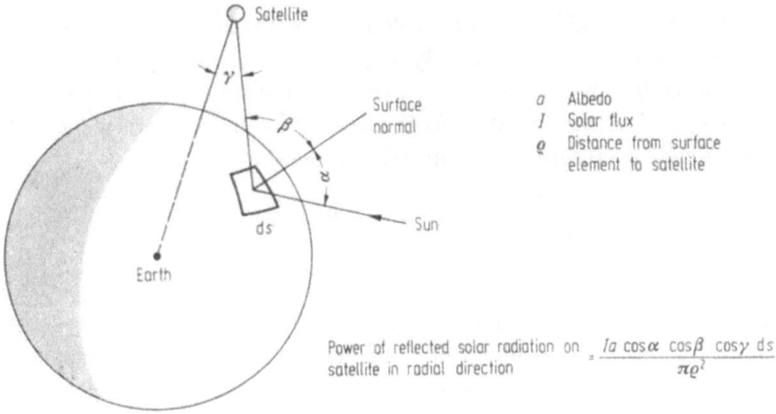

Fig. 1. Diffuse albedo model.

merical integration is performed to determine the components of albedo flux on the satellite from the entire Earth. The basic equation used is given in [13]. A mean albedo over the earth of 0.32 is assumed, based on TIROS VII measurements [14] from July 1963 to June 1964. The albedo varies as a function of season and latitude, so that the values of albedo radiation pressure calculated for this work must be considered approximate. The specular component, which WYATT [12] finds to be small, has not been included in this analysis.

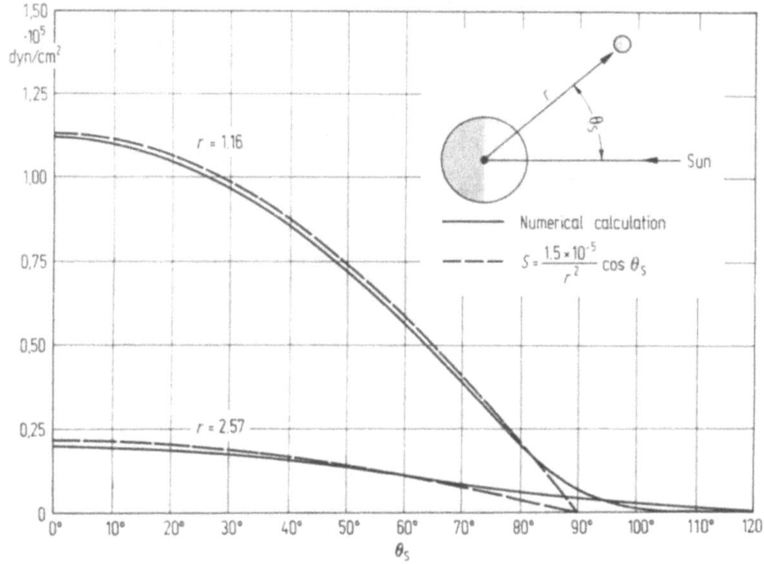

Fig. 2. Radial component of albedo radiation force.

In Fig. 2, the solid curves represent the computed radial component of albedo radiation force S (in dynes) acting on a unit area (1 cm²) as a function of θ_s, the angle between the earth-sun line and the earth-satellite line, and r, the distance from the earth center in earth radii. The dashed curves represent an empirical expression

$$\left.\begin{aligned} S &= \frac{1.5 \times 10^{-5}}{r^2} \cos\theta_s \qquad && 0 \leq \theta_s \leq 90^\circ \\ S &= 0 && 90_0 < \theta_s \end{aligned}\right\} \tag{1}$$

which agrees quite well with the integrated results. This expression is similar to those developed by WYATT [12] and SEHNAL [11]. In the numerical evaluation of the albedo perturbations, all the components of albedo radiation force were used. Generally, only the radial component S causes a significant change in the eccentricity e or the argument of perigee ω. The transverse component, however, can become important for the semi-major axis a.

To determine the approximate effects of albedo on an orbit, we can assume that only S is acting on the satellite. From the perturbation Eqs. (10), (11), and (14) of COOK [16], and considering only the radial component

$$\frac{d\omega}{d\theta} \cong -\frac{r^2 R^2}{n^2 a^3 e}\left(\frac{A}{M}\right) S \cos\theta \tag{2}$$

$$\frac{de}{d\theta} \cong \frac{r^2 R^2}{n^2 a^3}\left(\frac{A}{M}\right) S \sin\theta \tag{3}$$

$$\frac{da}{d\theta} \cong \frac{2 r^2 R^2 e}{n^2 a^2 (1-e^2)}\left(\frac{A}{M}\right) S \sin\theta \tag{4}$$

where θ is the true anomaly, R the earth radius in cm, (A/M) the frontal area-to-mass ratio of the satellite in cm²/g, and mean motion n.

After substituting Eq. (1) for S, the change in each element over one orbit is given by

$$\Delta\omega \cong -\frac{R^2}{n^2 a^3 e} 1.5 \times 10^{-5}\left(\frac{A}{M}\right)\int_0^{2\pi} \cos\theta_s \cos\theta \, d\theta \tag{5}$$

$$\Delta e \cong \frac{R^2}{n^2 a^3} 1.5 \times 10^{-5}\left(\frac{A}{M}\right)\int_0^{2\pi} \cos\theta_s \sin\theta \, d\theta \tag{6}$$

$$\Delta a \cong \frac{2 R^2 e}{n^2 a^2 (1-e^2)} 1.5 \times 10^{-5}\left(\frac{A}{M}\right)\int_0^{2\pi} \cos\theta_s \sin\theta \, d\theta \,. \tag{7}$$

Notice that the dependence on r cancels out. At this point, the expressions are general, requiring only that θ_s be known as a function of θ. Let us consider only those orientations producing the maximum perturbations. For $\Delta\omega$, the maximum value of the integral in Eq. (5) occurs for the solar orientation shown in Fig. 3, with the perigee of the orbit lying on the earth-sun line. For Δe and Δa, the maximum perturbations occur for the solar orientation in Fig. 4. Solving the integrals for these cases and with $n^2 = \mu/a^3$,

$$\Delta\omega_{\max} \cong -90\left[\left(\frac{R^2}{\mu}\right) 1.5 \times 10^{-5}\right]\left(\frac{A}{M}\right)\frac{1}{e} \text{ degrees per orbit} \qquad (8)$$

$$\Delta e_{\max} \cong -\frac{\pi}{2}\left[\left(\frac{R^2}{\mu}\right) 1.5 \times 10^{-5}\right]\left(\frac{A}{M}\right) \text{ per orbit} \qquad (9)$$

$$\Delta a_{\max} \cong -\frac{\pi}{2}\left[\left(\frac{R^2}{\mu}\right) 1.5 \times 10^{-5}\right]\left(\frac{A}{M}\right)\frac{2ae}{1-e^2} \text{ per orbit} \qquad (10)$$

with μ = gravitational constant times mass of earth.

Also given in Fig. 3 is the numerical value of the albedo perturbation $\Delta\omega_{\text{albedo}}$ expected for Echo I, compared to the perturbation of direct solar radiation pressure, $\Delta\omega_{\text{direct solar}}$ for this orientation. Note that $\Delta\omega_a$ and $\Delta\omega_{ds}$ have opposite signs. The predicted albedo effect on ω for this orientation is of the same magnitude as perturbations due to low-order zonal harmonics of the Earth's gravitational field.

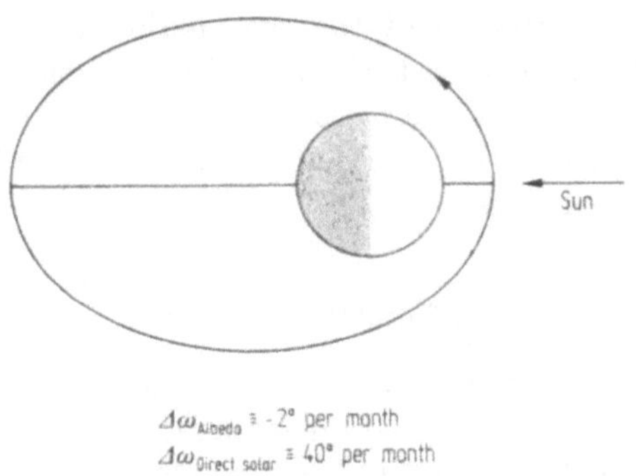

Fig. 3. Maximum perturbation of ω for Echo I orbit.

The Δe_a is quite small and also has the opposite sign of Δe_{ds}. However, Δa_a has the same sign as Δa_{ds}. These qualitative relationships between the maximum perturbations of albedo and direct solar radiation are also valid in general. That is, albedo perturbations generally act to decrease the effect of direct solar radiation pressure on ω and e, but act to increase the direct solar effect on a.

For Echo I and PAGEOS, the albedo effect on a can be as large as 10 percent of the maximum direct solar effect. This effect of albedo should manifest itself in attempts to infer an A/M which best fits the observed e, ω, and a variations. The best A/M to fit a variations should be larger than the best A/M for e or ω. It is interesting that COOK [17] found, in a study of Echo I orbital variations through 1961, that the best inferred A/M for variations in a was approximately 10 percent higher than the best for e or ω variations. As he suggests, difficulties in the drag model could contribute to such a discrepancy.

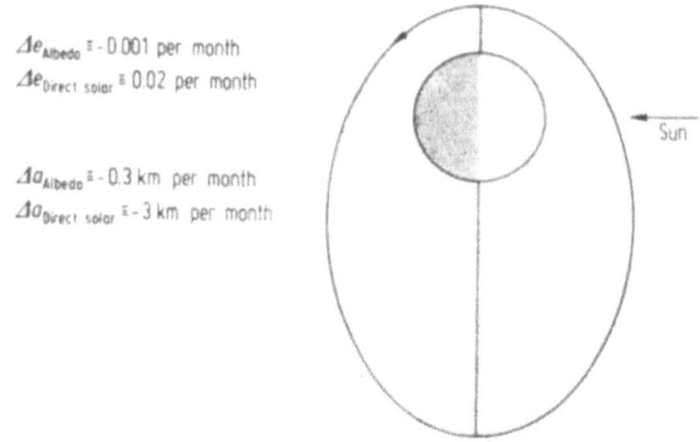

Fig. 4. Maximum perturbation of e and a for Echo I orbit.

The albedo perturbation of the semi-major axis is proportional to ae in Eq. (10). The albedo perturbation of ω is much greater for Echo I than for PAGEOS. This is due to the smaller eccentricity of Echo I and the $(1/e)$ dependence in Eq. (8). Finally, the effect on e is approximately the same for both satellites.

4. Numerical Results

In the numerical evaluation of Cook's perturbation integrals, all three components of albedo radiation acceleration were used. The approximation was made that the orbital parameters and solar orientation remain constant during the interval between epoch dates. Such an approximation is adequate for this paper. Only the orbital elements least affected by atmospheric drag were included in this study, so that errors in the assumed model atmosphere would not seriously affect the results. The argument of perigee of both Echo I and PAGEOS satisfies this criterion, as well as the eccentricity of PAGEOS. Decreases of approximately 0.3 km per 14 days in the semi-major axis of PAGEOS during periods of continuous sunlight show that drag is indeed affecting a. Mean densities at PAGEOS altitudes may be inferred during these periods, but ignoring the albedo effects could result in interpreting them as apparent density variations. However, near the midpoint of each period of continuous sunlight, the effect of albedo on a for PAGEOS becomes practically negligible, so that observed decreases in a at these times are the best for mean density determinations.

Fig. 5 shows the variation of eccentricity residuals during the last 250 days of the PAGEOS interval. Each data point represents the average of two residuals over one month. A residual or Δe observed is defined as

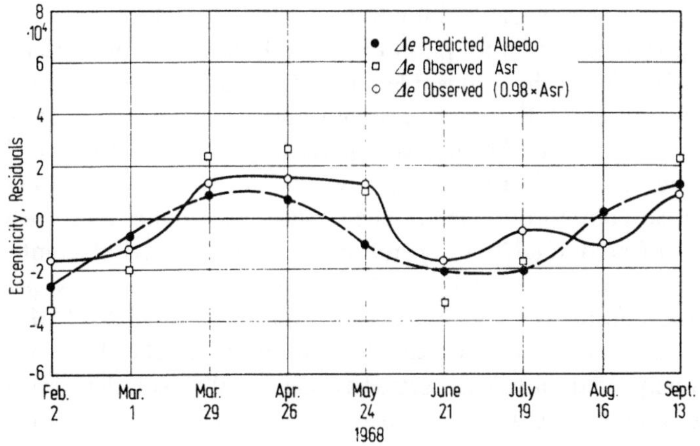

Fig. 5. Pageos eccentricity residuals and predicts.

the observed change in e for each period between epochs minus the predicted change in e due to all the known perturbations listed previously. The square data points correspond to residuals generated assuming the nominal value of A_{sr}. The dark circles connected by a dashed line are the predicted changes due to the albedo perturbation over each period. Let us make the assumption that the observed residuals in e are the result of a combination of earth albedo perturbations and a slightly smaller A_{sr} than the nominal value. Then the best inferred A_{sr} for the PAGEOS eccentricity variation is 0.98 times the nominal value of A_{sr}. The circular data points, connected with solid curves, correspond to residuals generated using this new value of A_{sr}.

Table 1. *Variations in Argument of Perigee for (0.98)* A_{sr}

	PAGEOS	Echo I
Number of Days	532	504
Dates: from	April 14, 1967	Jan. 29, 1963
to	Sept. 27, 1968	Oct. 26, 1964
$\Delta\omega_{\text{Albedo Predict}}$	+1.6°	−14.2°
$\Delta\omega_{\text{Observed}}$	+2.0°	−18.0°

In Table 1, the results of the analysis on ω for both satellites are given, using the A_{sr} inferred from the eccentricity data of PAGEOS. Over a period of 532 days, the predicted effect of albedo is to increase ω by just 1.6°; the observed unexplained increase after this period of time is 2.0°. For Echo I, the predicted effect of albedo over a period of 504 days is to decrease ω by 14.2°; the observed unexplained decrease is 18.0°. Because the observed decrease is greater than expected, A_{sr} appears to be somewhat less for Echo I than PAGEOS.

Zadunaisky, Shapiro, and Jones [3], in their study of Echo I, considered $K(A/M)$ to be a variable parameter and inferred its value when the best agreement between observed and predicted changes in eccentricity was obtained. During this period, (A/M) was indeed changing due to the escape of sublimating powders. An effect of albedo radiation pressure, as pointed out earlier, is to cause discrepancies in the value of the best inferred $K(A/M)$ depending on which orbital element is used. Zadunaisky, et al., points out that such discrepancies appear early in the lifetime of Echo I, and states, "Since we infer $K(A/M)$, as well as an air density model from the data, we may possibly be 'covering up' small unexpected physical perturbations ... We feel that the discrepancy between theory and experiment is probably due to the inadequacy of our atmospheric model and our inadequate treatment of the sunlight scattered by earth."

5. Conclusions

A study was performed to determine albedo radiation pressure perturbations on the orbits of Echo I and PAGEOS. For the simple albedo model developed assuming only diffusely reflected sunlight from the Earth following Lambert's cosine law, it may be concluded that:

1. Albedo radiation pressure perturbations act in such a fashion as to

a) increase the effect of direct solar radiation perturbations on the semi-major axis,

b) decrease the effect of direct solar radiation perturbations on the eccentricity and argument of perigee.

2. The magnitude of these perturbations is great enough so that the effect should be considered in gravitational harmonic studies or drag analyses using Echo I or PAGEOS.

3. Variations are observed in the eccentricity and argument of perigee of PAGEOS over 532 days that can be explained as a combination of the effects of albedo radiation and a mean acceleration due to solar radiation pressure which is 2 percent less than the nominal value.

4. A decrease in the argument of perigee over 504 days for Echo I is observed which is consistent with the albedo prediction and a mean solar radiation pressure acceleration which is somewhat less than that acting on PAGEOS.

5. The most discernible effect predicted for this diffuse albedo model is that values of $K(A/M)$ inferred from best fits to eccentricity and argument of perigee variations are lower than the best $K(A/M)$ fitting variations in the semi-major axis.

6. It appears that it may well be within our capability to determine a realistic albedo model of the Earth through study of balloon satellite orbital perturbations.

Acknowledgement

The help of GERALD KEATING, WALTER BRESSETTE, JAMES MULLINS, ROBERT B. LEE, and WARD SCHOONOVER were much appreciated by the author.

References

1. Satellite Orbital Data: Spec. Rep. No. 225; Smithsonian Inst. Astrophys. Obs., Oct. 1966.
2. BELCHER, S. J., ROWELL, L. N., SMITH, M. C.: Satellite Lifetime Program. RAND Memorandum RM-4007-NASA, April 1964.
3. ZADUNAISKY, P. E., SHAPIRO, I. I., JONES, H. M.: Experimental and Theoretical Results on the Orbit of Echo I. Spec. Rep. No. 61; Smithsonian Inst. Astrophys. Obs., March 1961.
4. CLEMMONS, D. L.: The Echo I Inflation System. NASA TN D-2194, 1964.
5. TEICHMANN, L. A.: The Fabrication and Testing of PAGEOS I. NASA TN D-4596, June 1968.
6. HOSTETLER, R. L., EMMONS, R. H., PRESKI, R. J., ROGERS, C. L., ROMICK, D. C.: Ground-Based Photometric Surveillance of the Passive Geodetic Satellite. NASA CR-791, 1967.
7. ROMICK, D. C., EMMONS, R. H., PRESKI, R. J., KALASKY, E. D.: The Modification and Use of a Ground-Based Photometer for Evaluation of Satellite Materials. Goodyear Aerospace Corporation, GER-13830, June 1968.
8. THEKAEKARA, M. P., KRUGER, R., DUNCAN, C. H., ROGERS, J. F., WEBB, J. J., MCINTOSH, R., WARD, P., MCNUTT, A., LESTER, D. L., STAIR, R.: The Solar Constant and the Solar Spectrum as Measured From a High Altitude Research Aircraft. IES-AIAA-ASTM Space Simulation Conference, Seattle, Wash., Sept. 16–18, 1968.
9. BAKER, R. M. L., Jr.: Radiation on a Satellite in the Presence of Partly Diffuse and Partly Specular Reflecting Body. In: COSPAR-IAU-IUTAM-Symp. 1965 Trajectories of Artificial Celestial Bodies, pp. 85–150. Berlin/Heidelberg/New York: Springer 1966.
10. SEHNAL, L.: The Perturbations of the Orbital Elements Caused by the Pressure of the Radiation Reflected From the Earth. In: COSPAR-IAU-IUTAM-

Sympos. 1965, Trajectories of Artificial Celestial Bodies, pp. 80–84, Berlin/Heidelberg/New York: Springer 1966.
11. SEHNAL, L.: The Effect of Re-Radiation of the Sunlight From the Earth on the Motion of Artificial Satellites. The Use of Artificial Satellites for Geodesy, Amsterdam: North-Holland Publishing Co. 1963.
12. WYATT, S. P.: The Effect of Terrestrial Radiation Pressure on Satellite Orbits. In: IUTAM Symposium, Dynamics of Satellites, Paris, May 28–31, 1962, Berlin/Göttingen/Heidelberg: Springer 1963.
13. CUNNINGHAM, F. G.: Earth-Reflected Solar Radiation Incident Upon Spherical Satellites in General Elliptical Orbits. NASA TN D-1472. Feb. 1963.
14. ARKING, A., and LEVINE, J. S.: Earth Albedo Measurements. July 1963 to June 1964. J. Atmospher. Sci. 24 (Nov. 1967), pp. 721—724.
15. LEVIN, E.: Solar Radiation Pressure Perturbations of Earth Satellite Orbits. AIAA J. 6 (Jan. 1968), pp. 120—126
16. COOK, G. E.: Luni-Solar Perturbations of the Orbit of an Earth Satellite. Geophys. J. Roy. Astronom. Soc. 6 (April 1962), pp. 271—291.
17. COOK, G. E.: The Use of Simplified Orbital Theory for Satellites of Large Area-to-Mass Ratio. Royal Aircraft Establishment, Tech. Note Space 28. March 1963.

MIX
Papier aus verantwortungsvollen Quellen
Paper from responsible sources
FSC® C105338

If you have any concerns about our products,
you can contact us on
ProductSafety@springernature.com

In case Publisher is established outside the EU,
the EU authorized representative is:
Springer Nature Customer Service Center GmbH
Europaplatz 3, 69115 Heidelberg, Germany

Printed by Libri Plureos GmbH
in Hamburg, Germany